Tidal Friction and the Earth's Rotation II

Proceedings of a Workshop Held at the Centre for Interdisciplinary Research (ZiF) of the University of Bielefeld, September 28 – October 3, 1981

Edited by
P. Brosche and J. Sündermann

With 112 Figures

Springer-Verlag Berlin Heidelberg New York 1982

Professor Dr. Peter Brosche
Observatorium Hoher List
Universitätssternwarte Bonn
5568 Daun/Eifel, FRG

Professor Dr. Jürgen Sündermann
Universität Hamburg
Institut für Meereskunde
Heimhuder Straße 71
2000 Hamburg 13, FRG

ISBN 3-540-12011-4 Springer-Verlag Berlin Heidelberg New York
ISBN 0-387-12011-4 Springer-Verlag New York Heidelberg Berlin

Offsetprinting and Bookbinding: Beltz Offsetdruck, Hemsbach/Bergstr.
2131/3130-543210

To the Memory of
H. Gerstenkorn (1923–1981)

Preface

In the four years which elapsed between our first workshop on *Tidal Friction and the Earth's Rotation* and the second, the proceedings of which are presented here, many of the disciplines involved made advances which we felt should be exchanged. We were encouraged by the good reception our first report met with.

Of course, more insight often means more problems. Therefore, this volume contains new results and revisions of matters which previously appeared settled. We are certainly far from "final answers". For this reason, differing opinions on some issues are to be found in this book. Moreover, we have refrained from making mathematical symbols uniform to avoid the risk of errors and non-compatibility with the earlier work of an author.

The two workshops have stimulated collaboration between participants working in various fields. The final versions of the contributions have already profited from these discussions. We are convinced that they will also influence further investigations.

This advancement of our past and future aims is based on the very existence of the Centre for Interdisciplinary Studies and on the support we have received from its leading members.

The authors have agreed to dedicate this book to Dr. H. Gerstenkorn, who died in 1981. He contributed fundamental ideas to the scientific fields treated in the meeting.

Daun, September 1982

P. Brosche
J. Sündermann

Contents*

*The contribution of P. Stumpff on the combined motion of the Earth-Moon-System appeared already in Astronomy & Astrophysics *101*, 52 (1981)

Julius Robert Mayer's Ideas on a Theory of Tidal Friction

H. P. Münzenmayer[1]

The key to the understanding of his general mode of proceeding lies in a more or less casual remark of Mayer's in a letter to his friend Carl Baur from July 24th 1841[1]. Here he speaks of a "System of Physics", which he has brought back with him from a voyage to the East Indies. This might sound somewhat conceited for a young doctor with scanty knowledge of physics, were it not for Mayer's almost proverbial modesty. Yet exactly this non-professionalism is a necessary condition for his further investigations, as he discloses the gaps in established science, without déformation professionelle, and on the other hand is compelled to become acquainted with doctrinal physics well enough to be able, if necessary, to recognize his new ideas as chimeras. His point of departure "from physical and pathological investigations" is thereby not so much his oft-quoted discovery of bright red veinous blood, and his (in fact incorrect) explanation whereafter the reduced loss of heat in the human body in the tropics effects a lowered combustion process in the blood, but the clearly stated question "Can the mechanical effect produced in the animal by the expenditure of combustion develop or not?"[2]. His intention to answer this question for physiology leads Mayer into physics without constraint. But first of all his "system" consists only of the strong conviction that "forces" are not less indestructible than substances and only appear in different forms, namely as "movement, electricity and heat"[3]. His "system" in fact consists therefore of a basic principal — which we might well call metaphysical, if he had not always so strongly objected to this word — according to which all natural phenomena are to be understood as actions and balances of these "forces", whereby we need not attribute to Mayer any particular special knowledge of these "forces".

Mayer's thoughts are finally summarized to a system in the true sense of the word in his first printed treatise, "Bemerkungen über die Kräfte der unbelebten Natur" (Annalen der Chemie und Pharmacie, Vol. 42, 1842, pp. 233-240). With the prerequisites of the Gay-Lussac experiment (1807), which Mayer unhappily does not mention explicitly, and the principle of conservation of vis viva, his presentation is absolutely exact, a genuinely new system of physics.

After having formulated the energy principle in his first work, the following work is characterized by his sailing on the ocean of know-

1 Amselweg, D-7141 Oberstenfeld, Fed. Rep. of Germany

Tidal Friction and the Earth's Rotation II
ed. by Brosche/Sündermann

ledge of nature using this principle or law as a compass, to use one of Mayer's favourite comparisons with navigation. He explains the physiological processes, the effect of the steam engine[4] and finally the origin of the solar heat[5], applying his law of energy. Examining natural phenomena with the instrument of his law of energy, he gives his opinion to the question of Earth's rotation for the first time in 1846, in a mémoire which he presents to the Paris Academy, but which they do not publish[6]. Here he still presumes the absolute constancy of the sidereal day, certainly "un fait très curieux", for the acceleration deriving from the volume contraction of the Earth by cooling and the retardation of the Earth's rotation by the tides compensate each other so well. At this time Mayer gives no further elucidation of his thought process. Only in his work "Beiträge zur Dynamik des Himmels", which appeared two years later, does Mayer state his considerations precisely. He by no means accepts the classical theory of tides or the problem of the secular acceleration of the Moon. He compares the rising and sinking of the water to a pendulum, which, because of the friction, would came to rest "if by a living force the existing resistances were not always overpowered anew". By this living force Mayer understands the "rotation effect", that is more or less the angular kinetic energy of the Earth. As the high tide lags 2 1/2 hours in the East behind the meridian transit of the Moon, there occurs a general flow from East to West. By the total friction and the flow resistance, this current, which runs counter to the Earth's rotation, would necessarily act as a brake to this latter.

To this for him typical summarizing consideration Mayer adds a more classical one. He regards the two tidal bulges and clearly points out that the upper one, because of its shorter distance and lesser inclination against the Earth's tangent, will be more attracted than the lower, and that the retarding East-West current results from this difference. In a rough estimate he concludes that the sideral day would have had to grow longer by 1/16 s within 2500 years "if the volume of the Earth were unalterable". The too high value results first from completely arbitrary assumptions, on the other hand Mayer takes no flow mechanics into account. His only concern is expressiveness.

Essentially Mayer's argumentation resembles that of Kant, apart from the fact that for Mayer the balance of "forces" is always the only point of departure. He indeed did not know Kant's work of 1754[7], and afterwards he only learns of it by hearsay. In any case he did not occupy himself with it, for in 1870 he still finds the excuse that Kant had "not understood the matter fundamentally right" and had come to "no further conclusions"[8]. Mayer's original ignorance is clear from his correspondence of 1848 with Carl Gustav Reuschle who — more an expert than Mayer — still does not call his attention to Kant. Mayer's theory is thus subjectively "only new because occasionally no one had yet focused attention upon it". But he still has doubts, as the correspondence discloses. If Laplace originally had mentioned the braking effect of the tides but estimated the effect as negligibly small, Mayer is ready to drop his whole theory as an error[8]. This doubt simply results from the fact that Mayer had no chance to study the "Mécanique céleste" himself.

Mayer's most important and genuinely new thoughts on the theory of tidal friction, the reaction on the Moon, is not broadly presented[8]. He only explains in a few lines that the upper tidal bulge evokes a tangential acceleration of the Moon, whereby its average distance and its time of revolution increases. This important point is also not touched upon in Mayer's correspondence with Reuschle, whom he often asked for advice. It becomes particularly obvious here that Mayer does not use the professional language, and for this reason is often not understood, and indeed does not penetrate to the experts. In this case Mayer leaves unmentioned that his conclusions are only possible on the basis of the third Kepler law. Similarly he does not indicate that the upper tidal bulge precedes the revolution of the Moon in a co-ordinate system moving with the radius vector of the Earth.

In his modesty, Mayer always claimed for himself only "to show", "to indicate" or "to call attention to". Thus he understood himself as a servant of science, but never really as a scientist, as he clearly felt his inability to master the language. Tragically, Mayer himself did not understand how to insert his indubitably brilliant ideas into the scientific tradition of secular acceleration. Even when John Cough Adams later proved the prolongation of the sidereal day, Mayer accepted this as a confirmation, but without real scientific participation. These facts are partially the reason for the tragedy of the lack of recognition under which he suffered so grievously in the early stages.

Notes

1 Kleinere Schriften und Briefe von Robert Mayer, ed. Jacob Johann Weyrauch, Stuttgart 1893, p. 110.
2 loc.cit., p. 209.
3 loc.cit., p. 111.
4 "Die organische Bewegung in ihrem Zusammenhange mit dem Stoffwechsel". In: J.R.M., Die Mechanik der Wärme, ed. Hans Peter Münzenmayer, Heilbronn 1978, pp. 41-155.
5 "Beiträge zur Dynamik des Himmels", loc.cit., pp. 167-236.
6 "Sur la production de la lumière etc.", loc.cit., pp. 157-166.
7 "Untersuchung der Frage, ob die Erde in ihrer Umdrehung um die Achse ... einige Veränderungen erlitten habe". In: Immanuel Kant's sämtliche Werke, ed. Karl Rosenkranz & Friedrich Wilhelm Schubert, Part 6, Leipzig 1839, pp. 3-12.
8 Kl. Schr., pp. 293ff.
9 Mech. d. Wärme, pp. 49ff.

Tidal Friction Parameters from Satellite Observations

A. Cazenave[1]

1. Introduction

Mutual differential gravitational attraction between two close bodies gives rise to the well-known tidal deformations, which are generally small compared to the dimensions of the interacting bodies. Repeating configurations make these deformations periodic. However, since the deformations are never perfectly elastic, energy is dissipated, causing irreversible evolution processes such as orbits and spin secular changes.

In the Earth-Moon system, tides have various effects. First they appear as periodic fluctuations in the oceans and solid Earth surface height. A direct consequence of this is a periodic change in the surface mean gravity. Changes in the Earth inertia tensor caused by the periodic deformations give rise to fluctuations in the Earth rotation rate, well observed at the 28-day and 14-day frequencies. The tidal deformations create also periodic forces which perturb the motion of close earth satellites. Finally tides are responsible for the secular increase of the length of day and of the lunar orbital deceleration. Table 1 summarizes observable tidal effects in the Earth-Moon system.

Basically, observation of these various tidal phenomena tell us about the elastic parameters of the Earth and Moon (Love numbers) and about the amount of energy dissipated in the solid Earth, oceans and in the Moon.

It is mainly the latter aspect that has motivated recent tidal studies. To determine the dissipation parameters three complementary approaches are necessary:

a) analyses of Earth satellites orbital perturbations,
b) development of numerical models for ocean tides,
c) analyses of lunar-laser ranging to the Moon.

---

1 CNES/GRGS, 18, Avenue Edouard Belin, F-31055 Toulouse Cedex, France

Tidal Friction and the Earth's Rotation II
ed. by Brosche/Sündermann

Table 1. Main observable phenomena of tidal origin in the Earth-Moon System

Phenomena	Amplitude	Period
Radial deformation of the ocean surface and of the solid Earth surface	1 m - 3 m 30 cm - 40 cm	12 h - 24 h , 14 days - 28 days
Variations of gravity	0.1 mgal	12 h - 24 h
Variations of the length of day	0.2 - 0.3 ms	14 days - 28 days
Orbital perturbations of close Earth satellites	1 m - 50 m	10 days - 300 days
Secular deceleration in Moon's longitude	$-25''/(cy)^2$	Secular
Secular increase in the length of day	2 ms/century	Secular

Method (a) informs on the total dissipation in the Earth (solid Earth plus oceans) whereas method (b) informs only on the amount of energy dissipated in the oceans; total dissipation in the Earth-Moon system can be deduced from method (c).

By comparing (a) and (b), it is possible in principle to deduce the dissipation in the solid Earth while comparison between (c) and (a) gives the dissipation inside the Moon. Energy dissipated in solid bodies at a global scale is often expressed through the dimensionless parameter Q defined by $Q^{-1} = \Delta E/2\pi E$ where ΔE is the amount of energy dissipated per cycle and E is the peak energy stored.

The Q parameters are of interest in that thet may be used as constraints in thermodynamical models of Earth and Moon mantles.

In the last 3 years, a number of geodesy satellite observations have been analysed for tidal purposes. At the same time, improved numerical models have been developed for the principal ocean tides. In addition 10 years of lunar laser ranging observations now available have produced significant results on a wide range of astronomical and geophysical problems, including tidal effects. It is these recent results that we review in this paper.

2. Tidal Perturbations on Satellites Orbits

Tidal deformations of the solid Earth and oceans give rise to variable gravity forces which perturb the trajectories of artificial satellites. Semi-diurnal and diurnal tides produce orbital perturbations on a wide frequency spectrum but the largest are essentially of long periods (for example the semi-diurnal M_2 tide gives orbital perturbations around a 14-day period).

2.1 Solid Tide

At the satellite orbit the solid tide perturbation is expressed as a function of the Love number k_2 and of the phase lag ε between the Moon passage and maximum deformation. In the present Earth-Moon system this phase lag is less than 1° and is related to the energy dissipated inside the Earth mantle through the relationship

$$\mathrm{tg}\varepsilon \simeq Q^{-1} = \frac{\Delta E}{2\pi E} \quad .$$

The main perturbation in the orbital inclination caused by the solid M_2 tide is, for example,

$$\Delta i \simeq -\frac{9}{2}\frac{M_{☾}}{M_{\oplus}} \frac{\sin i(1+\cos i_{☾})^2}{(2\dot{\ell}_{☾} - 2\dot{\Omega})} \frac{R^4}{a \cdot a_{☾}^3} k_2 \frac{\left(1 - \frac{5}{2} e_{☾}^2\right)}{(1-e^2)^2}$$

$$\times \cos[2\ell_{☾} - 2\Omega + \varepsilon] \tag{1}$$

with

$M_{☾}$: Moon's mass,

$M_{\oplus}$: Earth's mass,

R : mean Earth radius,

$a_{☾}$, $e_{☾}$, $i_{☾}$: lunar semi major axis, eccentricity, and inclination on the Earth' equator,

a, e, i : satellite semi major axis, eccentricity and inclination,

k_2 : solid Earth Love number ($\simeq 0.30$),

$\ell_{☾}$: mean longitude of the Moon ($2\pi/\ell_{☾} = 28$ days),

Ω : node longitude of the satellite,

ε : solid phase lag.

Expression (1) is simple to derive from the tidal potential, which is itself proportional to the direct attraction potential through the Love number k (see for example Lambeck et al. 1974). In (1) above, it is assumed at that each tidal frequency is associated a different set of (k,ε) values. However, up to now observations have not been accurate enough to show up any variation with frequency of these parameters.

2.2 Ocean Tide

Unlike the solid tide, the ocean tide cannot be simply expressed from the astronomical potential because of the complex response of the oceans to the Moon's attraction. The global tide $\xi(\phi,\lambda)$ can be obtained by solving the Laplace tidal equations over the whole ocean surface.

At the satellite orbit, the ocean tide potential is expressed by

$$U = GR^2 \iint_S \rho_W \frac{\xi_\sigma(\phi,\lambda)}{r} \, dS \tag{2}$$

G : gravitational constant,

r : distance from the satellite to a surface element dS, located at latitude ϕ and longitude λ,

$\xi_\sigma(\phi,\lambda)$ is the ocean tidal elevation at (ϕ,λ), of frequency σ($\xi = 0$ over the continents),

ρ_W is the sea water density.

Since $1/r$ expands in series of Legendre polynomials, it is convenient to express $\xi_\sigma(\phi,\lambda)$ in spherical surface harmonics (because of orthogonality properties of Legendre polynomials when integrated over the sphere, this leads to a simple expression for U):

$$\xi_\sigma(\phi,\lambda) = \sum_{n=0}^{\infty} \sum_{m=0}^{n} P_n^m (\sin\phi) \Big[C_{nm}^+ \sin(\sigma t + m\lambda + \varepsilon_{nm}^+) + C_{nm}^- \sin(\sigma t - m\lambda + \varepsilon_{nm}^-) \Big] \tag{3}$$

$C^\pm$ = harmonic expansion amplitudes,

$\varepsilon^\pm$ = harmonic expansion phases,

$P_n^m(\sin\phi)$ = associated Legendre polynomials,

t = time.

Including (3) into (2) and integrating over the Earth's surface gives a simple expression for the ocean tide potential at the satellite orbit from which orbital perturbations are easily deduced. The main perturbation in inclination due to the M_2 ocean tide is

$$\Delta i \approx \frac{12\pi}{5M_{\oplus}}\left(\frac{R}{a}\right)^3 \text{Ra} \sin i \, \rho_W \frac{n}{(2\dot{\ell}_{\mathbb{C}} - 2\dot{\Omega})} C^+_{22} \sin(2\ell_{\mathbb{C}} - 2\Omega + \varepsilon^+_{22})$$

$$+ \text{ smaller contributions due to } (C^+_{42}, \varepsilon^+_{42}) \text{ and } (C^+_{62}, \varepsilon^+_{62}) \quad . \quad (4)$$

In (4) above, the contributions of higher degree terms of the ocean tide expansion ($\ell = 4,6$), decrease because each contribution has a $(R/a)^{\ell+1}$ term in factor. In addition, amplitude coefficients $C^{\pm}_{\ell m}$ decrease as ℓ increases. Note that the main orbital perturbation depends only on the (C^+, ε^+) coefficients. It should also be added that only those coefficients with m = 2 for semi diurnal tides and m = 1 for diurnal tides give significant effects at the satellite orbit. Thus as far as satellite orbits are concerned the important parameters are $(C^+_{22}, \varepsilon^+_{22})$ for the M_2 ocean tide whereas for the O_1 tide, $(C^+_{21}, \varepsilon^+_{21})$ are the important parameters. The quantity $C^+_{22} \cos \varepsilon^+_{22}$ is called the dissipation parameter and is a measure of the dissipation at the M_2 frequency (and similarly $C^+_{21} \sin \varepsilon^+_{21}$ at the O_1 frequency).

On close satellite orbits, ocean tide perturbations are on the average ten times smaller than solid tide perturbations and have an amplitude of a few meters on the satellite position in space. Although small, such effects have become detectable in the past few years owing to the increasing accuracy in orbit computations.

Solid tide perturbations are on the other hand easily detectable. However, it is still impossible to determine the solid phase lag ε from orbital analyses, the corresponding perturbation being much too small. Solid and ocean tides have the same frequency spectrum and at the satellite orbit both perturbations are inseparable. Recent satellite orbit analyses have been devoted to show up the dissipation parameters. As it is not yet possible to derive the solid phase lag, recent investigations have removed solid tides perturbations (using the solid Love number k_2 value 0.301, assuming $\varepsilon_{\text{solid tide}} = 0$) and then attributing perturbations at the tidal frequencies to ocean tides.

Among the lunar tides, the M_2 ocean tide is the most easily detectable. Several satellite investigations have proposed solutions for the M_2 ocean tide parameters C^+_{22}, ε^+_{22} (among the most recent, Goad and Douglas 1978, Felsentregger et al. 1978, Cazenave and Daillet 1981). Several geodesy satellites observations have been analysed: Doppler satellites of the US Transit navigation system and Laser observations of the US altimeter satellite Geos 3. The laser satellite STARLETTE (launched by France in 1975) especially designed for tidal studies is also currently used for ocean tides analyses. Figure 1 shows the observed residuals in inclination of Starlette due to ocean tides. Table 2 summarizes the most recent results for the M_2 ocean tide parameters. For comparison the $(C^+_{22}, \varepsilon^+_{22})$ values of the equilibrium tide are also given.

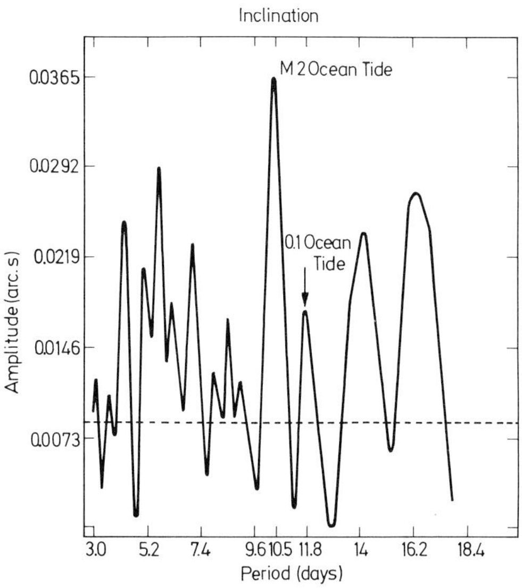

Fig. 1. Harmonic analysis of the residual perturbations in the inclination of the Starlette satellite. Dashed line represents the noise level

Table 2. M_2 tide harmonic coefficients determined from recent satellite solutions

M_2 Tide / Source	C^+_{22} (cm)	ε^+_{22} (deg)	$C^+_{22} \cos \varepsilon^+_{22}$ (cm)	$\dot{n}''_{☾}/(cy)^2$
Goad and Douglas (1978)	3.2 ± 0.25	331 ± 6	2.8 ± 0.37	-21.9 ± 2.9
Felsentregger et al. (1978)	3.4 ± 0.24	325 ± 4	2.8 ± 0.32	-21.9 ± 2.5
Cazenave and Daillet (1981)	3.1 ± 0.30	333 ± 5	2.8 ± 0.38	-21.9 ± 2.9
Weighted mean value				-21.9 ± 1.6
Equilibrium tide	4	270°	0	

3. Global Ocean Tide Models

Many new numerical models of global ocean tides (M_2, O_1, N_2, etc.) have been produced since 1977 (Accad and Pekeris 1978, Schwiderski 1979, Parke and Hendershott 1980). They differ from previous models in that they have incorporated terms for ocean loading and gravitational self-attraction into the Laplace tidal equations. Indeed it had been gradually recognized the importance of these terms

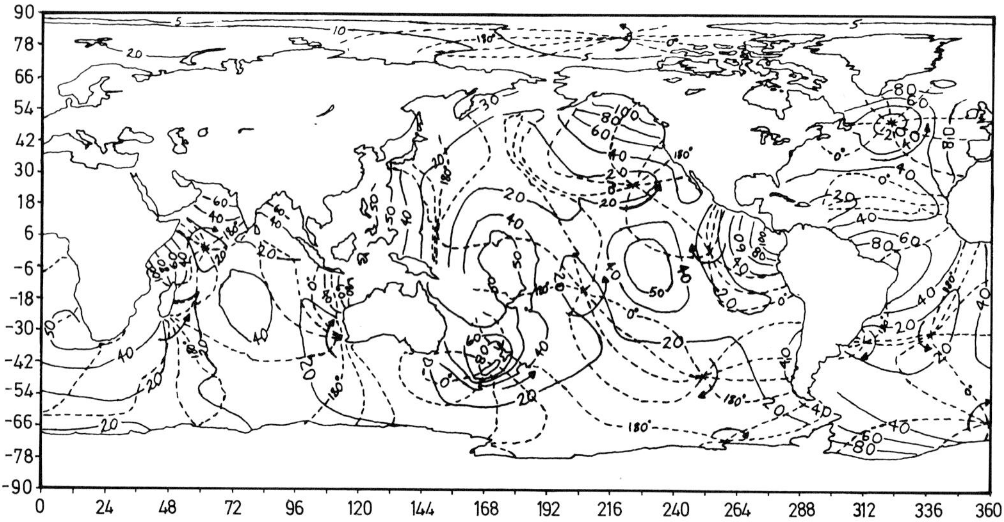

Fig. 2. The Schwiderski (1979) M_2 tidal model. Many shallow seas and shelves such as the Patagonian shelf show spatial structure much too fine to be resolved at these scales and so these regions have been left blank (amplitudes in cm, phases in degrees)

Table 3. Ocean tides harmonic coefficients determined by global numerical models

M_2 ocean tide	C^+_{22} (cm)	ε^+_{22} (deg)	$C^+_{22}\cos\varepsilon^+_{22}$	$\dot{n}_{☾}$ ("/cy)^2
Accad and Pekeris (1978)			2.55	-19.9
Schwiderski (1979)	2.95	311	1.92	-15.0
Parke and Hendershott (1980)	3.38	313	2.3	-18
O_1 Ocean tide	C^+_{21}	ε^+_{21}	$C^+_{21}\sin\varepsilon^+_{21}$	$\dot{n}_{☾}$
Parke (1979)			1.12	-1.82
Schwiderski (1979)	2.3	225	1.63	-2.65
N_2 Ocean tide	C^+_{22}	ε^+_{22}	$C^+_{22}\cos\varepsilon^+_{22}$	$\dot{n}_{☾}$
Schwiderski (1979)	0.65	322	0.51	-1.15

especially for the M_2 tide. Although the new models show still some differences, they present now very similar characteristics in many of the world's oceans. Figure 2 presents the map of the M_2 ocean tide by Schwiderski (1979). The characteristics of the recent M_2 solutions are summed up in Parke (1978). The tidal dissipation at the M_2 frequency deduced from the new models is significantly lower than estimates from previous models. Recent values are much more in agreement with satellite results and with recent estimates based on Lunar Laser ranging data. Estimates for C^+_{22}, ε^+_{22} deduced from the recent ocean tide models are given in Table 3. Also shown in Table 3 the coefficients for the O_1 and N_2 ocean tide models by Parke (1979) and Schwiderski (1979).

4. Lunar Tidal Deceleration and Tidal Earth' Q

It has been shown by Lambeck (1975, 1977) that the tidal parameters $C^+_{\ell m}$, $\varepsilon^+_{\ell m}$ determined from satellites orbit analyses or from ocean tide modelling appear directly in the expressions for the secular changes of the Moon' orbit: the formalism developed for artificial satellites perturbations is also applicable to the Moon, But instead of giving periodic perturbations, lunar tides produce secular perturbations in the Moon's orbit (this is so because the perturbed body is also the tide-raising body). The secular change in the lunar semi-major axis due to the M_2 solid tide is:

$$\dot{a}_☾ = \frac{GM_☾}{\sqrt{G(M_\oplus + M_☾)a_☾}} k_2\left(\frac{R}{a}\right)^5 \frac{3}{16} [1 + \cos i_☾]^4 \sin \varepsilon(M_2) \tag{5}$$

Similarly, the secular change due to the M_2 ocean tide is

$$\dot{a}_☾ \approx \frac{9}{5}\sqrt{\frac{G(M_☾ + M_\oplus)}{R^2 a_☾}} (1 + \cos i_☾)^2 \frac{\rho_W}{\bar{\rho}} \left(\frac{R}{a_☾}\right)^2 \cdot C^+_{22} \cos \varepsilon^+_{22} \tag{6}$$

where $\bar{\rho}$ is the mean Earth density (all other parameters have been defined above). The corresponding secular change in the mean orbital motion of the Moon is simply given by

$$\dot{n}_☾ = -\frac{3}{2}\left(\frac{n_☾}{a_☾}\right) \dot{a}_☾ \quad . \tag{7}$$

Ocean tide components contributing mainly to $\dot{n}_☾$ are M_2, O_1 and N_2 through the dissipation factors $C^+_{22} \cos \varepsilon^+_{22}$ for semi-diurnal and $C^+_{21} \sin \varepsilon^+_{21}$ for diurnal tides.

Ignoring for the moment internal dissipation in the Moon the theoretical secular change in $\dot{n}_{☾}$ is easily deduced from (5) and (6):

$$
\begin{aligned}
\dot{n}_{☾} = & -843\ (k_2 \sin \varepsilon)\ M_2 \\
& -151\ (k_2 \sin \varepsilon)\ O_1 \\
& -\ 47\ (k_2 \sin \varepsilon)\ N_2
\end{aligned}
\quad \Bigg\} \quad \text{solid Earth tides}
$$

$$
\begin{aligned}
& -\ 7{,}81\ (C^+_{22} \cos \varepsilon^+_{22})\ M_2 \\
& -\ 1{,}63\ (C^+_{21} \sin \varepsilon^+_{21})\ O_1 \\
& -\ 2{,}25\ (C^+_{22} \cos \varepsilon^+_{22})\ N_2
\end{aligned}
\quad \Bigg\} \quad \text{Ocean tides} \qquad (8)
$$

$\dot{n}_{☾}$ is in arc s/(century)^2 if C^+_{2m} is in centimeters. Contributions to $\dot{n}_{☾}$ due to the M_2, O_1 and N_2 ocean tides are given in Tables 2 and 3. The weighted mean value of the recent satellite solution for M_2 gives (see Table 3):

$$\dot{n}_{☾} = -21.9 \pm 1.6 \text{ arc s/(century)}^2 \quad . \qquad (9)$$

This is the contribution of solid plus oceanic dissipation. The value of $\dot{n}_{☾}$ derived from the M_2 ocean tide models is (see Table 3)

$$
\dot{n}_{☾} = \begin{cases} -\ 15''\ /(\text{cy})^2 & \text{(a)} \\ -\ 19.9 & \text{(b)} \\ -\ 18 & \text{(c)} \end{cases} \qquad (10)
$$

with (a), (b), (c) corresponding to Schwiderski, Accad and Pekeris and Parke and Hendershott M_2 ocean tide models.

Comparison of the above values with the satellite derived estimate confirms clearly that about 90% of the Moon's secular deceleration is caused by dissipation in the oceans.

As mentioned above, since no dissipation in the solid Earth is taken into account when modelling the effect of solid Earth tides in orbital perturbation analyses, satellite solutions in fact give the total dissipation in the Earth. Thus, the difference between $\dot{n}_{☾}$ from satellite solutions and $\dot{n}_{☾}$ from ocean models can be interpreted as dissipation in the solid Earth. At the M_2 frequency, it comes:

$$
k_2 \sin \varepsilon = \begin{cases} 8.18 \times 10^{-3} & \text{(a)} \\ 2.37 \times 10^{-3} & \text{(b)} \\ 4.62 \times 10^{-3} & \text{(c)} \quad . \end{cases} \qquad (11)
$$

Assuming $k_2 = 0.301$ gives for the solid phase lag

$$(\varepsilon)\ M_2 = \begin{cases} 1^\circ 55 & (a) \\ 0^\circ 45 & (b) \\ 0^\circ 88 & (c) \ . \end{cases} \tag{12}$$

The above phase lag can be converted in term of tidal mantle Q using the relation tg $\varepsilon = 0.612/Q$ derived by Zschau (1979). This relation accounts for self gravitation and hydrostatic prestress and is available for constant Q and Gutemberg Earth model:

$$(Q)\ M_2 \sim \begin{cases} 25 & (a) \\ 78 & (b) \\ 40 & (c) \ . \end{cases} \tag{13}$$

5. Lunar Laser Ranging Results

Analyses of nearly 10 years of Lunar Laser ranging (LLR) observations have yielded an impressive quantity of physical parameters of the Earth-Moon system, with astronomical geophysical and cosmological implications (see Mulholland 1980 for a review). Table 4 summarizes the physical parameters that are determined from LLR observations. Among other parameters, the secular deceleration of the Moon $\dot{n}_{☾}$ has been determined recently from LLR (Calame and Mulholland 1978, Williams et al. 1978, Ferrari et al. 1980). Solutions for the solid body elasticity and dissipation parameters $k_{☾}$, $Q_{☾}$ have also been obtained (Ferrari et al. 1980, Cappallo et al. 1981).

The solutions for $\dot{n}_{☾}$ by Calame and Mulholland (1978) and Williams et al. (1978) are only based on LLR data whereas those by Ferrari et al. (1980) has been obtained by a combination of LLR data and Doppler tracking data from Lunar Orbiter 4. Corresponding values for $\dot{n}_{☾}$ are given in Table 5.

In principle, the difference between $\dot{n}_{☾}$ determined by LLR and $\dot{n}_{☾}$ determined by satellite orbital perturbations permits an estimate of the dissipation inside the Moon since the former is due to total dissipation in the Earth-Moon system while the latter represents only dissipation in the solid Earth and oceans. In addition $\dot{n}_{LLR}$ includes all tidal frequencies contributions. Thus to do comparison with satellite results it is necessary to add the effect of O_1 and N_2 (solid plus oceans) tides to $\dot{n}_{M_2,\ sat.}$. No reliable solution for O_1 and N_2 have yet been obtained from satellites analyses. Nevertheless the contribution of the O_1 and N_2 ocean tides can be deduced from the Schwiderski numerical models (see Table 3). Solid Earth dissipation at the O_1 and N_2 frequencies is added, using the theoretical expression (8) for $\dot{n}$, using $k_2 = 0.301$ and $\varepsilon(O_1) = \varepsilon(N_2) = \varepsilon(M_2)$ (with $\varepsilon(M_2) = 0^\circ 35$. This value is based on a Q value of 100).

Table 4. Parameters determinable from Lunar Laser Ranging

Observatory coordinates and baselines	Lunar dissipation factor Q
Reflector coordinates and baselines	Lunar Love number k
Mass of the Earth-Moon system	Long period terms in Earth rotation
Lunar orbit model and initial conditions	Apparent universal time UTO
Lunar libration model	Variation of latitude
Lunar moment of inertia ratio	Universal time UTI and polar motion
Lunar gravity field coefficient	Dynamical equinox and obliquity of ecliptic
Free libration amplitudes and phases	Tectonic motion and continental drift
Lunar tidal deceleration $\dot{n}$	Time variation of gravitation

Table 5. Lunar Laser Ranging results for the secular deceleration $\dot{n}$ of the moon

Source	$\dot{n}_{☾}$ "/(century)^2
Calame and Mulholland (1978)	-24.6 ± 1.6
Williams et al. (1978)	-23.8 ± 4
Ferrari et al. (1980)	-23.8 ± 3.1

Then, the $\dot{n}_{☾}$ value due to total dissipation in the Earth is

$$
\begin{aligned}
\dot{n}_{M_2+O_1+N_2} = & - 21.9 \ \text{"/(cy)}^2 && \text{(satellite solutions for } M_2) \\
& - 2.65 && (O_1 \text{ Ocean tide}) \\
& - 1.15 && (N_2 \text{ Ocean tide}) \\
& - 0.28 && (O_1 \text{ solid Earth tide}) \\
& - 0.08 && (N_2 \text{ solid Earth tide})
\end{aligned}
$$

or

$$\dot{n}_{M_2+O_1+N_2} = -26.06 \ \text{"/(cy)}^2 \quad . \tag{14}$$

This value may be compared with $\dot{n}_{LLR} = -23.8$ "(cy)^2 (if we consider only Williams et al. and Ferrari et al. latest results based on longer data span). This gives in principle the contribution to $\dot{n}_{☾}$ due to dissipation inside the moon:

$$(\dot{n}_{☾})_{\text{lunar dissipation}} = +2.26 \ ''/(\text{cy})^2 \ (\text{observed}) \quad . \tag{15}$$

The theoretical $\dot{n}_{☾}$ due to dissipation inside the Moon is (Yoder 1979):

$$(\dot{n}_{☾})_{\text{lunar dissipation}} \simeq \frac{9}{2}\left(\frac{k}{Q}\right)_{☾} \frac{M_{\oplus}}{M_{☾}} \left(\frac{R}{a}\right)^5_{☾} n^2_{☾} \left[7e^2_{☾} + \sin^2 i_{☾}\right]$$

$$\simeq + 370 \ (k/Q)_{☾} \quad . \tag{16}$$

The comparison of observed and theoretical quantities gives an estimate of $(k/Q)_{☾}$:

$$(k/Q)_{☾} = 6.10 \times 10^{-3} \quad . \tag{17}$$

LLR solutions by Ferrari et al. (1980), Cappallo et al. (1981) have provided values for $k_{☾}$, $Q_{☾}$ and $(k/Q)_{☾}$. The values are gathered in Table 6. Cappallo et al. (1981) estimate the lunar Love number $k_{☾}$ equal to 0.024 ± 0.006 for a constant time lag model and 0.027 ± 0.006 for a constant Q. Ferrari et al. (1980) find $k_{☾} = 0.022 \pm 0.013$. Although the latter determination has a 60% error bar, the two solutions are in the same range and are especially consistent will the theoretical values derived by Cheng and Toksöz (1978) for two lunar interior models (the difference between the two models depending on fitting the lunar seismic data). Values of Cheng and Toksöz (1978) are 0.029 and 0.0335, with the difference being a rough measure of the error. Solutions for the lunar $Q_{☾}$ are 14 ± 10 by Ferrari et al. (1980) and 22 ± 6, 24 ± 6 by Cappallo et al. (1981). If we tried to deduce $Q_{☾}$ from results given in (17) assuming $k_{☾}$ in the range 0.022 - 0.033 we would obtain $Q_{☾}$ in the range 4 - 6. These values are lower than the LLR determinations but are also indicative of low lunar Q. A surprisingly small solid body Q for the Moon had been first reported by Yoder (1979).

Table 6. Lunar Love number $k_{☾}$ and solid body $Q_{☾}$

Source	Lunar Love number $k_{☾}$	Lunar $Q_{☾}$
Cheng and Toksöz (1978) (models of lunar interior based on seismic data)	0.0293 0.0335	
Ferrari et al. (1980) (LLR and Doppler tracking data of Lunar orbiter 4)	0.022 ± 0.013	14 ± 10
Cappallo et al. (1981) (LLR)	0.024 ± 0.006 0.027 ± 0.006	22 ± 6 24 ± 6

All these estimates of $Q_☾$ are much lower than the values obtained from studies of lunar seismicity: a seismic $Q_☾$ in the range 200 - 700 for regions 800 - 1.100 km deep has been reported by Toksöz et al. (1974) and according to Ferrari et al. (1980) and Cappallo et al. (1981), there is no proof yet that the $Q_☾$ factor determined by LLR represents actually solid-body dissipation. However, a frequency-dependent $Q_☾$ has been proposed by Cappallo et al. (1981) invoking that dissipative mechanisms may be quite different at seismic and tidal frequencies.

An alternative explanation for the low $Q_☾$ investigated by Yoder et al. (1978) and Ferrari et al. (1980) is rotational dissipation at an interface between a liquid core and a mantle: viscous core-mantle coupling in the Moon dissipates energy and provides a rotational signature similar to that of solid body dissipation. However, the low $Q_☾$ value requires either a large core kinematic viscosity ν ($\sim$ 3000 st for a 230 km core radius) or a large core radius (500 km radius for $\nu \simeq 1.5$ st), the latter possibility being inconsistent with lunar seismic results. Thus the origin of the low lunar $Q_☾$ remains still obscure.

At the present time, tidal studies (satellite orbit analyses and ocean tide modelling) are not yet able to provide reliable constraints on the lunar dissipation factor. This is mainly due to the poor results obtained for the N_2 and O_1 tides. Thus it appears more interesting for now to use LLR estimates for the lunar $Q_☾$ to establish constraints on the amount of dissipation inside the solid Earth and oceans at the O_1 and N_2 frequencies. If we use the mean $(k/Q)_☾$ value estimated by LLR, the theoretical $\dot{n}_☾$ due to dissipation in the Moon is

$$\dot{n}_☾ \simeq 0.45 \ ''/(cy)^2 \quad .$$

Substracting this contribution to the total LLR $\dot{n}$ and comparing it to the $\dot{n}$ value derived from the satellite solutions for M_2 gives the contribution of the total dissipation in the Earth (solid Earth plus oceans) at the O_1 and N_2 frequencies:

$$(\dot{n}_☾)_{O_1+N_2} \simeq - 1.45''/(cy)^2 \quad .$$

This is much less than the *computed* contribution of ocean tides alone (see Table 3).

Future efforts should be devoted to improve the determination of the O_1 and N_2 tides, both by satellite perturbations analyses and by numerical modelling since these tides contribute significantly to the lunar secular deceleration.

References

Accad Y, Pekeris CL (1978) Solution of the tidal equations for the M_2 and S_2 tides in the world oceans from a knowledge of the tidal potential alone. Philos Trans R Soc London 290: 235-266

Calame O, Mulholland D (1978) Lunar tidal acceleration determined from laser range measures. Science 199: 977-978

Cappallo RJ, Counselman CC III, King RW, Shapiro II (1981) Tidal dissipation in the Moon. J Geophys Res 86: 7180-7184

Cazenave A, Daillet S (1981) Lunar tidal acceleration from Earth satellite orbit analyses. J Geophys Res 86: 1659-1663

Cheng CH, Toksöz MN (1978) Tidal stresses in the Moon. J Geophys Res 83: 845-853

Felsentregger TL, Marsh J, Williamson RG (1978) M_2 ocean tide parameters and the deceleration of the Moon's mean longitude from satellite orbit data. NASA Tech Mem 79571

Ferrari AJ, Sinclair WS, Sjogren WL, Williams JG, Yoder CF (1980) Geophysical parameters of the Earth-Moon system. J Geophys Res 85: 3939-3951

Goad C, Douglas B (1978) Lunar tidal acceleration obtained from satellite derived ocean tide parameters. J Geophys Res 83: 2306-2310

Lambeck K (1975) Effects of tidal dissipation in the oceans on the Moon's orbit and the Earth's rotation. J Geophys Res 80: 2917-2925

Lambeck K (1977) Tidal dissipation in the oceans: astronomical, geophysical and oceanographic consequences. Philos Trans R Soc London Ser A 287 (1347): 545-594

Lambeck K, Cazenave A, Balmino G (1974) Solid earth and ocean tides estimated from satellite orbit analyses. Rev Geophys Space Phys 12 (3): 421-433

Mulholland D (1980) Scientific achievements from ten years of lunar laser ranging. Rev Geophys Space Phys 18: 549-563

Parke ME (1978) Open ocean tide modelling, in the Proc of the 9th GEOP conference. OHIO State Univ, Columbus, OHIO

Parke ME (1979) Global ocean tide modelling, paper presented at the Tidal Interact Symp. 17th Gen Assembly IUGG, Canberra, Aust

Parke ME, Hendershott MC (1980) M_2, S_2, K_1 models of the global ocean tide on an elastic earth. Mar Geod 3: 379-408

Schwiderski EW (1979) Detailed ocean tide models of M_2, S_2, K_1 and O_1, Paper presented at the Tidal Interact Symp. 17th Gen Assembly IUGG, Canberra, Aust

Toksöz MN, Dainty AM, Solomon SC, Anderson KR (1974) Structure of the Moon. Rev Geophys Space Phys 12: 539-567

Williams JG, Sinclair WS, Yoder CF (1978) Tidal acceleration of the Moon. Geophys Res Lett 5 (11): 943-946

Yoder CF (1979) Effects of the Spin-spin interaction and the inelastic tidal deformation on the lunar physical librations. In: Nacozy PE, Ferraz-Mello SS (eds) Natural and artificial satellite motion. Texas Univ Press

Yoder CF, Sinclair WS, Williams JG (1978) The effect of dissipation in the Moon on the lunar physical librations. Lunar Sci 9: 1291

Zschau J (1979) Tidal friction in the solid earth: loading tides versus body tides. In: Brosche P, Sündermann J (eds) Tidal friction and the Earth's rotation. Springer, Berlin Heidelberg New York

On Some Topical Problems of the Dynamics of the Earth-Moon System

M. Bursa[1]

1. Introduction

There is no satisfactory agreement (Parijskij 1978), between the recently observed value of the secular deceleration of the Earth's rotation, obtained, e.g., by Lambeck (1977, p. 557)

$$\dot{\omega} = -(5.4 \pm 0.5)10^{-22} \text{ rad s}^{-2} \quad , \tag{1}$$

($\dot{\omega}/\omega = -2.4 \times 10^{-8}$/cy, $\sim + 2 \times 10^{-3}$ s/cy in the length of day), and the value of $\dot{\omega}$ computed from the observed phase lag angles $\Delta\varphi$ of tidal waves in observed gravity at many tidal stations on the territory of Europe and Asia (Parijskij 1978, p. 51)

$$\Delta\varphi(M_2, S_2) = -(0\overset{\circ}{.}15 \pm 0\overset{\circ}{.}09) \quad , \tag{2}$$

$$\Delta\varphi(O_1, K_1) = -(0\overset{\circ}{.}08 \pm 0\overset{\circ}{.}07) \quad . \tag{3}$$

The values (2) are too small to be able to explain the deceleration (1). With the value of gravity tidal factor

$$\delta = 1 - \frac{3}{2} k_2 + h_2 = 1.160 \pm 0.002 \quad , \tag{4}$$

($k_2 = 0.290$, h_2 the Love numbers), we obtain for $\dot{\omega}$, on the basis of M_2 lunar tide phase lag

$$\dot{\omega} \doteq 10.9\, k_2 \frac{GM_{☾}}{\bar{\Delta}_{\oplus☾} C_{rel} a_o^2} \frac{GM_{☾}}{GM_{\oplus}} \left(\frac{\bar{R}}{\bar{\Delta}_{\oplus☾}}\right)^5 \Delta\varphi \quad ; \tag{5}$$

$GM_{☾} = 4902.75 \times 10^9 m^3 s^{-2}$, $GM_{\oplus} = 398\,600.47 \times 10^9 m^3 s^{-2}$ the selenocentric and geocentric gravitational constant respectively; $\bar{R} = 6371 \times 10^3$m the mean Earth's radius; $\bar{\Delta}_{\oplus☾} = 384\,402 \times 10^3$m the mean distance of the Earth-Moon mass centres; $C_{rel} = 0.330\,68$ the largest relative principal moment of inertia of the Earth in $M_{\oplus}a_o^2$ units; $a_o = 6378140$ m

1 Astronomical Institute, Czechoslovakian Academy of Science, Prague, Czechoslovakia

Tidal Friction and the Earth's Rotation II
ed. by Brosche/Sündermann

the mean equatorial Earth's radius. With $\Delta\varphi$ of semi-diurnal wave M_2 (2) with the period $12^h25^m14^s$ we get from (5) $\dot{\omega} \doteq - 1.2 \times 10^{-22}$ rad s^{-2}, i.e., in the length of day $\sim + 0.5 \times 10^{-3}$ s cy^{-1} only. For to explain the observed value (1) of deceleration, we need $\Delta\varphi \sim 0\overset{\circ}{.}7$, i.e., the value of $\Delta\varphi \sim$ four times larger than the observed one (2) from M_2. With the use of O_1, K_1 wave phase lags, the discrepancy is more than five times larger.

The relative effect of the Sun in comparison with the Moon's one ($\Delta_{\oplus\odot}$ = A; the heliocentric gravitational constant $GM_\odot$ = 132 712 496.5 $\times 10^{12} m^3 s^{-2}$)

$$\frac{\dot{\omega}_{Sun}}{\dot{\omega}_{Moon}} = \left[\frac{GM_\odot}{GM_☾}\left(\frac{\Delta_{\oplus☾}}{\Delta_{\oplus\odot}}\right)^3\right]^2 \tag{6}$$

varies from 1/3.1 $[(\Delta_{\oplus\odot})_{min} = 147.1 \times 10^9 m$, $(\Delta_{\oplus☾})_{max} = 406\ 730 \times 10^3 m]$, to 1/7.2 $[(\Delta_{\oplus\odot})_{max} = 152.1 \times 10^9 m$, $(\Delta_{\oplus☾})_{min} = 364\ 400 \times 10^3 m]$.

That is why, the solid Earth tides might be considered as being not responsible for the observed value of $\dot{\omega}$ (Parijskij 1955, 1978).

The effect of the ocean tides is essentially greater. According to Parijskij et al. (1972) it is $\sim + 3.2 \times 10^{-3}$s cy^{-1} in the length of day. This value is about 60% greater than the observed one, corresponding to (1) and, a non-tidal acceleration

$$\delta\dot{\omega}_{non-tidal} \doteq + 3 \times 10^{-22} \text{ rad s}^{-2} \tag{7}$$

[Lambeck's, 1977 estimation is $(+ 1.8 \pm 1.0)\ 10^{-22}\ s^{-2}$] should be explained.

The results in question, if realistic, put the problem of deceleration of the Earth's rotation and, in connection with it, the problem of dynamics and evolution of the Earth-Moon system, as regards the secular and long-periodical terms, back to a position from where it can still be considered as an open problem.

For simplicity, the problem can be divided into four parts: purely astronomical (celestial mechanics); tidal if perfectly elastic bodies; tidal friction influence; other phenomena. Here some topics of the first (astronomical) part will be discussed, as well as some considerations given on the last one.

2. The Earth-Moon Force Function

The starting point for describing the orbital and rotation motion of the Earth-Moon system is the force function $V_{\oplus☾}$ (Fig. 1)

$$V_{\oplus☾} = G \int_{M_\oplus} \int_{M_☾} \frac{dm_☾ \, dm_\oplus}{r} \tag{8}$$

or its perturbing part

$$R_{\oplus☾} = V_{\oplus☾} - G\frac{M_\oplus M_☾}{\Delta_{\oplus☾}} = R_{\oplus☾}\left[J_n^{(k)}, S_n^{(k)}; (J_n^{(k)})', (S_n^{(k)})' ; \psi', \vartheta', \varphi'; \delta_{0'}, T_{0'} \quad (\text{or } \delta_0', T_0')\right] \tag{9}$$

to be expressed in terms with the Stokes constants of the Earth $J_n^{(k)}$, $S_n^{(k)}$ and of the Moon $(J_n^{(k)})'$, $(S_n^{(k)})'$ in the usual definition (n-degree, k-order of harmonics), the Euler angles ψ', ϑ', φ' (positive counter-clockwise when viewing from the positive ends of the rotation axes) defining the mutual orientation of the axes x, y, z; x', y', z' of the ellipsoids of inertia of both bodies, the distance $\Delta_{\oplus☾} = \overline{00'}$ between the mass centres of the bodies and the geocentric equatorial coordinates $\delta_{0'}$, $T_{0'}$ of the Moon's mass centre 0' (or selenocentric coordinates δ_0', T_0' of the Earth's mass centre 0).

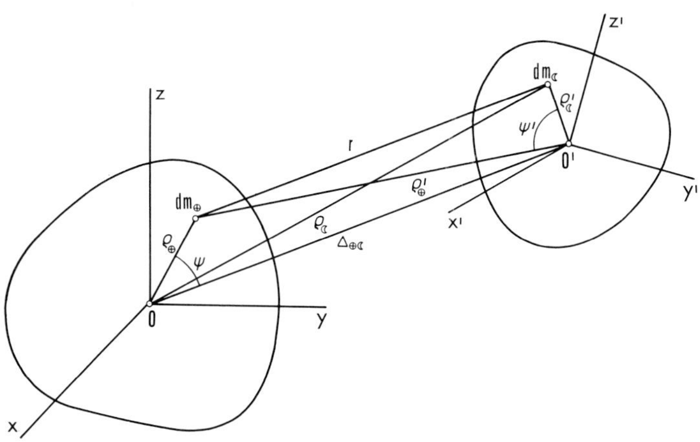

Fig. 1. Force function $V_{\oplus☾}$

We solved the problem keeping the 4th degree terms (i.e., terms containing $(a_0/\Delta_{\oplus☾})^4$, $(a_0'/\Delta_{\oplus☾})^4$ in a concrete form; Šidlichovsky (1978) gave a general solution. The 4th degree solution (Burša 1978) reads (a_0' - the mean equatorial radius of the selenoid)

$$R_{\oplus☾} = G\frac{M_\oplus M_☾}{\Delta_{\oplus☾}}\left\{\sum_{n=2}^{n}\sum_{k=0}^{n}\left(\frac{a_0}{\Delta_{\oplus☾}}\right)^n (J_n^{(k)} \cos k\, T_{0'} - S_n^{(k)} \sin k\, T_{0'})\, P_n^{(k)}(\sin \delta_{0'}) + \sum_{n=2}^{n}\sum_{k=0}^{n}\left(\frac{a_{0'}}{\Delta_{\oplus☾}}\right)^n [(J_n^{(k)})' \cos k\, T_0' - (S_n^{(k)})' \sin k\, T_0']P_n^{(k)}(\sin \delta_0')\right\} + \Delta R_{\oplus☾} \quad , \tag{10}$$

$$R_{\oplus C} = G\frac{M_\oplus M_☾}{\Delta_{\oplus☾}}\left(\frac{a_0}{\Delta_{\oplus☾}}\right)^2\left(\frac{a_0'}{\Delta_{\oplus☾}}\right)^2\Big[F_{22}^{00} J_2^{(0)}(J_2^{(0)})' + F_{22}^{22} J_2^{(2)}(J_2^{(2)})' + G_{22}^{22} S_2^{(2)}(S_2^{(2)})' + F_{22}^{02} J_2^{(0)}(J_2^{(2)})' + F_{22}^{20} J_2^{(2)}(J_2^{(0)})' + G_{22}^{02} J_2^{(0)}(S_2^{(2)})' + G_{22}^{20} S_2^{(2)}(J_2^{(0)})' + H_{22}^{22} J_2^{(2)}(S_2^{(2)})' + K_{22}^{22} S_2^{(2)}(J_2^{(2)})'\Big] \quad , \tag{11}$$

$$F_{22}^{00} = \frac{3}{4}\Big[1 - 5(\sin^2 \delta_0' + \sin^2 \delta_0') + 35 \sin^2 \delta_0' \sin^2 \delta_0' + 2\cos^2 \vartheta' + 20 \sin \delta_{0'} \sin \delta_0' \cos \vartheta'\Big] \quad ,$$

$$F_{22}^{22} = 6\Big\{\cos 2\varphi' \cos 2\psi' (1 + \cos^2 \vartheta') - 2 \sin 2\varphi' \sin 2\psi' \cos \vartheta' + 10 \cos \delta_{0'} \cos \delta_0' [\cos(\psi' - T_0') \cos(\varphi' + T_{0'})] - \cos \vartheta' \sin(\psi' - T_0')\sin(\varphi' + T_{0'})] + \frac{35}{2}\cos^2 \delta_{0'} \cos^2 \delta_0' \cos 2T_{0'} \cos 2T_0'\Big\} \quad ,$$

$$G_{22}^{22} = -\,6\Big\{\sin 2\varphi' \sin 2\psi'(1 + \cos^2\vartheta') - 2\cos 2\varphi' \cos 2\psi' \cos\vartheta'$$
$$- 10\cos\delta_{0'} \cos\delta_0'[\sin(\psi' - T_0') \sin(\varphi' + T_{0'})$$
$$- \cos\vartheta' \cos(\psi' - T_0') \cos(\varphi' + T_{0'})]$$
$$- \frac{35}{2}\cos^2\delta_{0'} \cos^2\delta_0' \sin 2T_{0'} \sin 2T_0'\Big\} \quad ,$$

$$F_{22}^{02} = [10\cos^2\delta_0' \cos 2T_0' + 30\sin\delta_{0'} \cos\delta_0' \sin\vartheta' \sin(\psi' - T_0')$$
$$- 3\sin^2\vartheta' \cos 2\psi' + 35\cos^2\delta_0' \cos 2T_0'\, P_2^{(0)}(\sin\delta_{0'})] \quad ,$$

$$F_{22}^{20} = -\,3[\sin^2\vartheta' \cos 2\varphi' - 10\sin\delta_0' \cos\delta_0' \sin\vartheta' \sin(\varphi' + T_{0'})$$
$$+ \frac{5}{2}\cos^2\delta_{0'} \cos 2T_{0'}(1 - 7\sin^2\delta_0')] \quad ,$$

$$G_{22}^{02} = -\,[10\cos^2\delta_0' \sin 2T_0' + 30\sin\delta_{0'} \cos\delta_0' \sin\vartheta' \cos(\psi' - T_0')$$
$$- 3\sin^2\vartheta' \sin 2\psi' + 35\cos^2\delta_0' \sin 2T_0'\, P_2^{(0)}(\sin\delta_{0'})] \quad ,$$

$$G_{22}^{20} = 3\ \ \sin^2\vartheta' \sin 2\varphi' + 10\sin\delta_0' \cos\delta_{0'} \sin\vartheta' \cos(\varphi' + T_{0'})$$
$$+ \frac{5}{2}\cos^2\delta_{0'} \sin 2T_{0'}(1 - 7\sin^2\delta_0')] \quad ,$$

$$H_{22}^{22} = 6\Big\{\sin 2\varphi' \sin 2\psi'(1 + \cos^2\vartheta') + 2\sin 2\varphi' \cos 2\psi' \cos\vartheta'$$
$$+ 10\cos\delta_{0'} \cos\delta_0'[\sin(\psi' - T_0') \cos(\varphi' + T_{0'})$$
$$+ \cos\vartheta' \cos(\psi' - T_0') \sin(\varphi' + T_{0'})]$$
$$- \frac{35}{2}\cos^2\delta_{0'} \cos^2\delta_0' \cos 2T_{0'} \sin 2T_0'\Big\} \quad ,$$

$$K_{22}^{22} = 6\ \Big\{\sin 2\varphi' \sin 2\psi'(1 + \cos^2\vartheta') - 2\cos 2\varphi' \sin 2\psi' \cos\vartheta$$
$$- 10\cos\delta_{0'} \cos\delta_0'[\cos(\psi' - T_0') \sin(\varphi' + T_{0'})$$
$$+ \cos\vartheta' \sin(\psi' - T_0') \cos(\varphi' + T_{0'})]$$
$$+ \frac{35}{2}\cos^2\delta_{0'} \cos^2\delta_0' \sin 2T_{0'} \cos 2T_0'\Big\} \quad .$$

3. The Effect of $R_{\oplus☾}$ on the Earth's Rotation Dynamics

The problem has not yet been solved, especially as regards the effect of term $\Delta R_{\oplus☾}$. In general, the derivatives $\partial R_{\oplus☾}/\partial\psi$, $\partial R_{\oplus☾}/\partial\vartheta$ according to the precession ψ and nutation ϑ angles could contain secular terms in all three components $\dot{\omega}_1$, $\dot{\omega}_2$, $\dot{\omega}_3$ of ω.

However, in the third Euler's dynamic equation for $\dot{\omega}_3$

$$\dot{\omega}_3 = \frac{1}{c}\frac{\partial R_{\oplus☾}}{\partial\varphi} - \frac{B-A}{C}\omega_1\omega_2 \quad , \tag{12}$$

$\partial R_{\oplus☾}/\partial\varphi$ cannot yield secular terms, because the derivatives $\partial/\partial\varphi$ of all the zonal terms are zero, and the derivatives of the tesseral and the sectorial ones are short-periodic functions; the same may be said about the $\Delta R_{\oplus☾}$ term.

The only source for secular variations in $\dot{\omega}_3$ is the term $\omega_1\omega_2(B-A)/C$, where ω_1 and ω_2 may contain secular terms, through $\partial R_{\oplus☾}/\partial\psi$ and $\partial R_{\oplus☾}/\partial\vartheta$. However, the 2nd term in (12) is relatively small,

$$\frac{B-A}{C}\omega_1\omega_2 < 2.10^{-25}s^{-2} \quad ,$$

and to obtain the needed order $10^{-22}s^{-2}$, secular terms in $\omega_1\omega_2$ ~ three orders greater would be necessary.

However, $R_{\oplus☾}$ generates secular perturbations in orbital elements of the Moon's orbit. The problem has not yet been solved in a final form; as an example, we shall give here the solution for secular terms $\Delta\lambda_☾$ in the mean longitude $\lambda_☾$ of the Moon caused by the fourth zonal harmonic of the Earth with $J_4^{(0)}$ and by the tesseral third degree and the first order harmonic in the selenopotential $(\bar{J}_3^{(1)})'$ (marked with a bar - fully normalized) not covered by the Brown's theory

$$\Delta\dot{\lambda}_☾ \doteq \frac{G(M_\oplus + M_☾)}{(GM_☾)^{1/2}}\frac{a_0^4}{a_☾^{11/2}}J_4^{(0)} + 12\frac{G(M_\oplus + M_☾)}{(GM_☾)^{1/2}}\frac{(a_0')^3}{a_☾^{9/2}}(\bar{J}_3^{(1)})'(1+3e_☾^2) \quad . \tag{13}$$

After integration ($n_\oplus$, $n_☾$ the mean motions, $(\Delta\lambda_☾)_0$ the integration constant)

$$\Delta\lambda_☾ \doteq (\Delta\lambda_☾)_0 + \frac{GM_\oplus[1+(GM_☾/GM_\oplus)]}{GM_\odot}\left(\frac{n_\oplus}{n_☾}\right)^2\left[\frac{A^3 a_0^4}{a_☾^7}J_4^{(0)}\right.$$

$$+ 12 \frac{A^3 (a_0')^3}{a_{\mathbb{C}}^6} (1 + 3e_{\mathbb{C}}^2)(\bar{J}_3^{(1)})' \Bigg] n_{\mathbb{C}} t \doteq 0\overset{''}{.}06 \text{ cy}^{-1} \quad . \tag{14}$$

In general, the secular variation in $\lambda_{\mathbb{C}}$ is not due to the Earth's tesseral and sectorial harmonics; as regards the Stokes constants of the Moon, all the zonal, tesseral and sectorial terms may influence this secular motion; as regards the part $\Delta R_{\oplus\mathbb{C}}$ with products of Stokes constants of both bodies, it also generates secular variations.

4. Theoretical Non-Tidal Acceleration in the Moon's Mean Motion on the Basis of Celestial Mechanics and the Non-Tidal Acceleration in the Earth's Rotation

The expression for theoretical mean longitude $\lambda_{\mathbb{C}}$ of the Moon, computed on the bsis of celestial mechanics only, i.e., not including the large tidal part (Brown 1915, p. 513) contains the terms

$$\partial\lambda_{\mathbb{C}} = + 7\overset{''}{.}14 \ T^2 + 0\overset{''}{.}0068 \ T^3 \quad , \tag{15}$$

(T in Julian centuries), i.e., we have theoretical value for the acceleration $\delta\dot{n}_{\mathbb{C}}$ of the mean motion of the Moon

$$\delta\dot{n}_{\mathbb{C}} = + 14\overset{''}{.}28 \text{ cy}^{-2} = + 0.70 \times 10^{-23} \text{ rad s}^{-2} \quad ; \tag{16}$$

it will be considered here as constant, i.e., the second term in (15) neglected.

Laplace first computed the term in 1787, on the basis of his theory describing the secular evolution of the orbital elements of the Moon caused by the secular (exactly, very long periodic) planetary term in the eccentricity of the Earth's orbit (Laplace 1802, T.III, p.212). The solution by Brown (15) contains also a small term $\sim 1'' T^2$ due to the secular variation of mean equinox.

In 1952, the theoretical Brown's value (15) was corrected for an empirical term (Clemence 1948, p. 172)

$$(\delta\lambda_{\mathbb{C}})_{\text{empirical}} = - 11\overset{''}{.}22 \ T^2 \quad , \tag{17}$$

i.e.,

$$(\delta\dot{n}_{\mathbb{C}})_{\text{empirical}} = - 22\overset{''}{.}44 \text{ cy}^{-2} = - 1.09 \times 10^{-23} \text{ rad s}^{-2} \quad . \tag{18}$$

Later many authors treated the problem; however, large discrepancies exist between individual results (see Lambeck 1977, p. 555) mentioned by Calame and Mulholland (1978, p. 50). The actual accuracy of (18) should be considered still as uncertain and for this reason we do not wish to use this value.

However, for further estimation, we will suppose that the total angular moment $L_{\oplus☾}$ of the Earth-Moon system must remain conserved in time, i.e., that the system is isolated; it holds then with a sufficient accuracy

$$\dot{\omega} \doteq \frac{1}{3} C_{rel}^{-1} \frac{GM_☾}{GM_\oplus} \left(1 + \frac{GM_☾}{GM_\oplus}\right)^{-1} \left(\frac{\rho_☾}{a_0}\right)^2 \dot{n}_☾ \doteq 44.5 \, \dot{n}_☾ \quad ; \tag{19}$$

$$GM_\oplus = 398\,600.47 \times 10^{-9} m^3 s^{-2} \quad , \quad GM_☾ = 4\,902.75 \times 10^9 m^3 s^{-2} \quad ,$$

$$C_{rel} = C/(M_\oplus a_0^2) = 0.330\,677 \quad , \quad \rho_☾ = 384\,400 \text{ km} \quad ,$$

$$a_0 = 6\,378.14 \text{ km} \quad .$$

On the assumption put, the tidal part $(\dot{n}_☾)_{tidal\ only}$ in the deceleration of the mean Moon's motion should correspond to the tidal part $\dot{\omega}_{tidal\ only}$ in the deceleration of the Earth's rotation:

$$\dot{\omega}_{tidal\ only} \doteq 44.5 \, (\dot{n}_☾)_{tidal\ only} \quad . \tag{20}$$

If

$$(\dot{n}_☾)_{tidal\ only} = (\dot{n}_☾)_{total} - \delta\dot{n}_☾ \quad , \tag{21}$$

i.e., if $(\dot{n}_☾)_{total}$ is the result of tidal friction plus celestial mechanics perturbations only, then ($\dot{n}_☾$ expressed in 10^{-23} rad s^{-2})

$$\dot{\omega}_{tidal\ only} = 44.5 \, (\dot{n}_☾)_{total} - 3.1 \times 10^{-22} \text{ rad } s^{-2} \tag{22}$$

$$(0.70 \times 10^{-23} \text{ rad } s^{-2} \times 44.5 = 3.1 \times 10^{-22} \text{ rad } s^{-2}) \quad .$$

The total (observed) value (1), i.e., $\dot{\omega}_{total}$ can be expressed as the sum of the tidal part and the non-tidal one and,

$$\dot{\omega}_{non-tidal} = \dot{\omega}_{total} - \dot{\omega}_{tidal\ only} \quad . \tag{23}$$

From (23, (22) and (1) we have

$$\dot{\omega}_{non-tidal} \doteq - 44.5 \, (\dot{n}_☾)_{total} - 2.3 \times 10^{-22} \text{ rad } s^{-2} \tag{24}$$

(if $\dot{n}_☾$ in 10^{-23} rad s^{-2}). However, because of large discrepances among different determinations of $(\dot{n}_☾)_{total}$, we cannot compute the non-tidal part in $\dot{\omega}$.

Some estimation could be done through the tidal part in $\dot{\omega}$ being

$$\dot{\omega}_{\text{tidal only}} \doteq -(8.5 \pm 1.0) \times 10^{-22} \text{ rad s}^{-2} , \tag{25}$$

[+ 3.2×10^{-3} s cy^{-1} in the length of day, see Parijskij (1978)]; then we get

$$\dot{\omega}_{\text{non-tidal}} \doteq +3.1 \times 10^{-22} \text{ rad s}^{-2} \tag{26}$$

(i.e., in the length of day $\sim -1.2 \times 10^{-3}$s cy^{-1}). If (25) is correct, then it should be concluded that phenomena must exist leading to the acceleration of the Earth's rotation, the absolute value of which is about 60% of the absolute value of the total deceleration in the Earth's rotation observed.

If (26), the $(\dot{n}_{☾})_{\text{total}}$ deceleration can be computed from (24)

$$(\dot{n}_{☾})_{\text{total}} \doteq -1.2 \times 10^{-23} \text{ rad s}^{-2} = -24''\!.9 \text{ cy}^{-2} , \tag{27}$$

as well as the tidal part (21)

$$(\dot{n}_{☾})_{\text{tidal only}} \doteq -(1.9 \pm 0.23) \times 10^{-23} \text{ rad s}^{-2} = -(39''\!.2 \pm 4''\!.6) \text{ cy}^{-2} . \tag{28}$$

5. Other Phenomena

If (25) is correct, there must be processes of a non-tidal and at the same time non-astronomical nature which are responsible for the significant acceleration in the Earth's rotation as (26) is.

They could be connected with the Earth's interior dynamics, as well as, e.g., with the fact that the Earth's atmosphere rotates more quickly than as the Earth's body itself. A supposition could be also made that a differential rotation inside the Earth exists, e.g., that the lithosphere rotates more quickly than the astenosphere, and/or the crust + the mantle more quickly than the core.

In general, the secular variations in the moments of inertia should be connected with the re-distribution of masses in the Earth's interior and/or with the secular variations in figure, which should lead to secular variations in gravity.

6. Conclusion

1. The large acceleration in the Earth's rotation (26) cannot be explained by the secular terms caused by the perturbing function $R_{\oplus\mathbb{C}}$.

2. There must be processes out of celestial mechanics perturbations which are responsible for the acceleration.

Acknowledgement. The author would like to express his thanks to his colleagues, Dr. M. Šidlichovský and Dr. J. Vondrák for inspiring discussions on the given problem, especially as regards to the origin and the actual accuracy of the astronomical part of the acceleration (16).

References

Brown EW (1915) The elements of the Moon's orbit. Mon Not R Astron Soc LXXV: 508-516

Burša M (1978) Theory of the rotation of the Earth retaining higher harmonics. Trav Inst Geophys Acad Tchec Sci XXVI: 9-61

Calame O, Mulholland JD (1978) Effect of tidal friction on the lunar orbit and the rotation of Earth and its determination by laser ranging. In: Brosche P, Sündermann J (eds) Tidal friction and the Earth's rotation. Springer Berlin Heidelberg New York, p 43

Clemence GM (1948) On the system of astronomical constants. Astron J 53: 169-179

Lambeck K (1977) Tidal dissipation in the oceans: astronomical, geophysical and oceanographic consequences. Philos Trans R Soc London 287: 545-594

Laplace PS (1802) Traité de mécanique céleste. T. III. Duprat JBM, Paris

Parijskij NN (1955) Neravnomernost' vrashcheniya Zemli. Trudy geofiz. inst. AN SSSR 26: 131-152

Parijskij NN (1978) K izucheniyu zemnykh prilivov. Fiz Zemli 9:43-54

Parijskij NN, Kuznetsov MV, Kuznetsova LV (1972) O vliyanii okeanicheskikh prilivov na vekovoe zamedlenie vrashcheniya Zemli. Fiz Zemli 2: 3-12

Šidlichovský M (1978) The force function of two general bodies. Bull Astron Inst Czech 29: 90-97

History of the Earth's Rotation Since 700 B.C.

F. R. Stephenson[1] and L. V. Morrison[2]

1. Introduction

The Earth is by no means a reliable clock. Astronomical observations covering roughly 2½ millennia reveal a gradual increase in the length of the day (l.o.d.) of about 2 ms per century (ms/cy). In addition there are irregular fluctuations on time-scales of anything from weeks to centuries with typical amplitudes of the order of 1 ms.

It is well established that the predominant long-term causal mechanism is tidal dissipation - both lunar and solar. Over the period covered by historical records this can reasonably be expected to have produced a rather steady lengthening of the day at roughly the observed rate - see below (Sect. 2). At the other end of the scale, the very short period fluctuations (a year of less) are largely due to changing wind patterns. Perhaps the most interesting fluctuations are the so-called "decade" variations. These have a characteristic period of 10-30 years. The required torques are an order of magnitude greater than the tidal torque. The usual explanation is coupling of the Earth's mantle to fluid motions in the core. This might be electromagnetic or topographic in nature (see Rochester 1974). However, it seems plausible that these fluctuations could at least in part be due to changes in the moment of inertia of the Earth associated with small global sea-level variations (Jeffreys 1982).

Changes in the moment of inertia of the Earth - either of surface or internal origin - may well have a significant effect on its rotation on the time-scale of millennia. However, the evidence as yet has been far from convincing; astronomical analyses have not enabled the contributions of the various individual mechanisms to be separated satisfactorily. It is the object of the present paper to attempt to clarify this issue using both ancient and modern observational data.

---

1 Sub-Department of Geophysics, University of Liverpool, Liverpool L69 3BX, England. Present address: Department of Physics, University of Durham, Durham DH1 3LE, England

2 Royal Greenwich Observatory, Herstmonceux Castle, Hailsham BN27 1RP, England

Tidal Friction and the Earth's Rotation II
ed. by Brosche/Sündermann

The astronomical observations analysed here are of two main types. These are: (a) telescopic timings of occultations and solar eclipses covering the period 1620 to 1955 (extended to 1978 using atomic clock data); (b) unaided eye timings of lunar eclipses recorded in Babylon between about 700 B.C. and 50 B.C. Set (a), which contains many thousands of data, enables the decade fluctuations to be mapped in detail. Set (b) numbers less than 20 observations, but the individual measurements are remarkably accurate. They yield an excellent value for the average rate of lengthening of the day over the past 2½ millennia or so. In addition, they provide the first firm evidence in favour of a secular decrease in the moment of inertia of the Earth.

2. The Lunar Acceleration ($\dot{n}$)

A change in the l.o.d. of, say, 2 ms/cy would produce a cumulative discrepancy in time of 36.5 s over a century. Over 25 centuries, the average change in the l.o.d. would be 0.025 s and the resultant discrepancy would be

$$0.025 \times 36525 \times 25/3600 \text{ h} \quad ,$$

i.e., some 6 h. Individual cumulative discrepancies at different epochs are measured in this paper by comparing the expected and observed times of conjunction of the Moon with the Sun (in the form of eclipses) and with stars (occultations). The expected (i.e., calculated) times are derived from a knowledge of the Moon's motion. The theory of the lunar motion is well established, but it is necessary to make an empirical adjustment for the tidal acceleration ($\dot{n}$). It is important to consider whether $\dot{n}$ has remained constant over the past 2½ millennia.

The lunar acceleration is produced by the reciprocal action of the tidal bulges which the Moon raises on the Earth. This process effectively transfers angular momentum from the Earth (rotational) to the Moon (orbital). The oceans of the Earth play a dominant role; body tidal dissipation is on a much smaller scale (see Lambeck 1980). Regardless of whether the contribution of shallow seas is important or not, studies by Fairbridge (1961) and Mörner (1971) indicate that throughout the historical period global sea-level was never more than about 1 or 2 m above or below the present level. Such small changes would alter the total area of the continental shelves by no more than about 1 percent, which is negligible.

This conclusion is particularly important since there is no reliable method of determining $\dot{n}$ in the pre-telescopic period. Newton (1970), Muller and Stephenson (1975) and Muller (1975) attempted to deduce both $\dot{n}$ and $\langle\dot{\omega}\rangle$ (the average acceleration of the Earth's spin) independently from ancient observations. These are mainly of solar eclipses. We have arrived at the opinion that a joint solution is impractical because of the high degree of correlation between $\dot{n}$ and

Table 1. Recent determinations of the lunar tidal acceleration

Method	Author(s)	$\dot{n}$ (arc s/cy^2)
Transits of Mercury	Morrison and Ward (1975)	-26 ± 2
Lunar laser ranging	Ferrari et al. (1980)	-23.8 ± 2.6
Numerical tidal model	Lambeck (1980, p. 335)	-29.6 ± 3.1
Artificial satellites	Cazenave (1982 this vol.)	-26.1 ± 1.6
Lunar occultations	Van Flandern (1982)	-21.4 ± 2.6

$\langle\dot{\omega}\rangle$. Hence to realise the optimum value of the ancient data, it is necessary to solve for a single parameter only. Since $\dot{n}$ is considered to be constant, we prefer to adopt its value as determined from modern data, and analyse for $\langle\dot{\omega}\rangle$ alone. We thus require $\dot{n}$ to be found using other techniques, e.g., lunar laser ranging, transits of Mercury and tidal studies.

In Table 1 are summarised a number of recent estimates of $\dot{n}$ using a variety of methods. Results from lunar laser ranging and occultations are on Atomic Time (TAI); the others are on Dynamical Time (TD). Assuming that G is constant (Stephenson 1981) there should be no appreciable difference between the two time-scales. This is supported by the general agreement between the individual results although the scatter is rather large. The present situation, although far from ideal, is encouraging. The large disparity between individual determinations of $\dot{n}$ evident 10 years ago is no longer apparent. Looking to the future, the use of lunar laser ranging and artificial satellites seems to promise the best hope for establishing a reliable value for this important parameter.

In our subsequent investigation we have used the lunar tidal acceleration derived by Morrison and Ward (1975), i.e., -26 arc s/cy^2. Their investigation covered by far the longest time-scale (some 250 years) and the result which they deduced lies fairly close to the mean of all the values listed in Table 1. However, we shall discuss the effect on our solutions of altering $\dot{n}$ within reasonable limits.

3. Analyses of Telescopic Observations: 1620-1978

Over the relatively short interval covered by telescopic data, the steady tidal increase in the l.o.d. is masked by the decade fluctuations. As a result it is not possible to derive a reliable value of $\langle\dot{\omega}\rangle$ from modern observations. Instead, there are two main objectives: (1) to map the decade fluctuations in detail and (2) to deduce the torques necessary to give rise to them.

Each observation is recorded either directly or indirectly in Universal Time (UT) which is measured by the diurnal rotation of the Earth. For the period 1620-1955 we have relied on occultations (supplemented by a few solar eclipses in the early years). The TD of each event may be computed using the lunar theory. Hence each measurement leads to a determination of the value of ΔT, the difference TD-UT. During the year 1955, TAI was introduced and this replaced the lunar motion as the long-term standard of time-keeping. Over the short period 1956-1978, ΔT values have been taken from the high precision results published by the Bureau International de l'Heure (BIH).

3.1 Decade Fluctuations: 1620-1860

The values of ΔT in this interval were deduced from an analysis of about 2700 timings, mainly occultations. These observations are published in Morrison et al. (1982). The first occultation to be satisfactorily observed with the aid of a telescope was probably that of α Vir in AD 1623. This was recorded by Boulliau in Paris. Between this date and about 1670 there were only a few regular observers. It was common practice at this early period to determine time indirectly by measuring the altitude of the Moon or a suitable placed bright star. To give an example, Gassendi (1658) records the following observation which he made in 1627:

"June 17. In the evening I observed an occultation of the Heart of the Lion (α Leo) at Digne when the height of the Tail of the Lion (β Leo) in the west was 25° 13', i.e., 10 h 30 min exactly."

Occultations have the important advantage that they are effectively instantaneous events. However, because of the low precision of time measurement in the interval 1620-1670, we have included roughly 100 timings of the first and fourth contacts of solar eclipses. These are again largely based on altitude determination, this time of the Sun. Although eclipse contacts are by no means sharp events, the timing accuracy is quite comparable to that of the occultations. In analysing both sets of data, we have corrected each altitude measurement for refraction, rather than accept the observers' own reductions.

After about 1670, with the rapid dissemination of accurate pendulum clocks, the precision of time determination increased substantially. Consequently, for the later period we have used only occultations. Up to about 1770, almost all observers used local apparent time. After that date, use of local mean time became more and more widespread.

We have analysed all of the occultation and eclipse observations using the lunar ephemeris j = 2 (IAU 1968) but with the following amendment to the mean longitude of the Moon.

$$\Delta L = - 1''.54 + 2''.33T - 1''.78\ T^2 \qquad (1)$$

where T is measured in Julian centuries from the standard epoch 1900.0. The amended quadratic term is equivalent to use of a value for $\dot{n}$ of -26 arc s/cy^2. For further details see Morrison (1979b).

During the interval in question (1620-1860), the individual values of ΔT were smoothed by cubic splines having in all 13 knots. These were spaced at proportionately smaller intervals of time according to the increase in the number of data. This spacing varied from 20 years to 10 years.

3.2 Decade Fluctuations: 1861-1978

After 1860, with the improved quality and quantity of data, it became viable to calculate annual means for ΔT. Up to 1955 occultation observations were still used. Some 40,000 data were reduced (Morrison 1979a) using the amended lunar ephemeris j = 2. Annual means for the last few years (1956-1978) were taken from the BIH figures. TAI values were converted to TD using the relation

$$TD = TAI + 32^s.184 \quad . \qquad (2)$$

Table 2 illustrates the improvement in the accuracy of timing occultations over the entire interval 1620-1955. This gives the standard deviation (σ) for the values of ΔT derived from single observations. The improvement in precision between about 1650 and 1800 is very marked.

Table 2. Standard deviation of ΔT per observation for telescopic data

Period	σ
1620-1669	60 s
1670-1699	15 s
1700-1759	5 s
1760-1819	2 s
1820-1819	1.5 s
1861-1942	1.3 s
1943-1955	1.0 s

3.3 ΔT Curve: 1620-1978

Our derived ΔT curve for the telescopic period is shown in Fig. 1. Uncertainties in the position of this curve vary from ± 10 s at 1650 to 0.05 s at 1950, first reaching 1 s around 1700. The well defined portion after 1860 reveals none of the erratic changes on a time-scale of less than 5 years which Brouwer (1952) noted. We suggest that these were merely data artifacts. Figure 1 is based on a larger sample of data than that used by Brouwer. Additionally, we have applied lunar limb profile corrections in reducing all occultation observations.

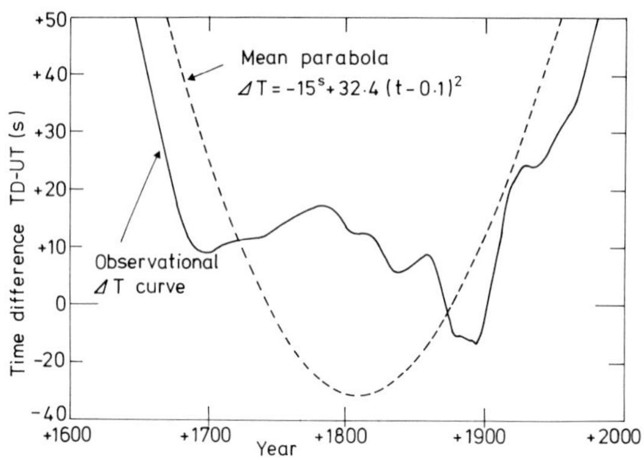

Fig. 1. ΔT curve derived from modern observations (full line), with the parabola from ancient data (broken line) shown for comparison. The constant and linear terms of the parabola were chosen to give the best fit to the modern observations

The above investigation is based on the use of a lunar tidal acceleration of -26 arc s/cy^2, equivalent to a coefficient of T^2 of -13 arc s/cy^2. The mean motion of the Moon is 1.821 arc s per second of time. As a result, if our preferred value of $\dot{n}$ is in error by ± 2 arc s/cy^2, the consequent error in the values of ΔT obtained in this analysis would be about ∓ 1.8 s/cy^2. This would result in substantial corrections of about 2 s at AD 1800, 7 s at 1700 and 14 s at 1620. However, it is not possible to be definite until a better value for $\dot{n}$ becomes available.

The overall shape of the modern ΔT curve is parabolic, which corresponds to a long-term deceleration of the Earth's rotation. However, the decade fluctuations very much distort this trend. For this reason we have adopted the coefficient +32.4 s/cy^2 from ancient observations (see Sect. 5.2) and have determined a constant and linear term to give the best fit to the telescopic data. The resulting expression for ΔT is

$$\Delta T = -15 + 32.4\ (t-0.1)^2\ \text{s} \qquad (3)$$

where t is measured in centuries from 1800.0. This is shown as a dashed curve in Fig. 1. Changing the epoch to 1900.0, the standard reference point, and measuring T in centuries from this date, the equivalent smoothed expression for ΔT over the past 350 years becomes:

$$\Delta T = + 11 + 58.4 \, T + 32.4 \, T^2 \text{ s} \quad . \tag{4}$$

Such formulae are of value in estimating the value of ΔT in the distant past or future, e.g., in the production of eclipse canons.

3.4 Changes in the Length of Day and Associated Torques

Figure 2 shows the first and second derivatives of ΔT in the period 1620-1978. The first derivative with respect to TD (TAI after 1955) is plotted in Fig. 2a. This was obtained by taking the derivatives of the individual spline curves up to 1860. In later years a 5-point convolute was applied to the annual values of ΔT (see Morrison 1979a). The dashed straight line indicates the secular acceleration of the Earth's rotation, i.e., the derivative of the mean parabola. It is clear that over the past 3½ centuries variation in the l.o.d. of several milliseconds have been quite common. Because of the uncertainties in the data before about 1800, Fig. 2a can be regarded as providing only a general indication of the character of the fluctuations in the l.o.d. over this period. However, after 1800 the variations are well established. In particular, the several large and rapid changes which occurred around 1900 are accurately mapped.

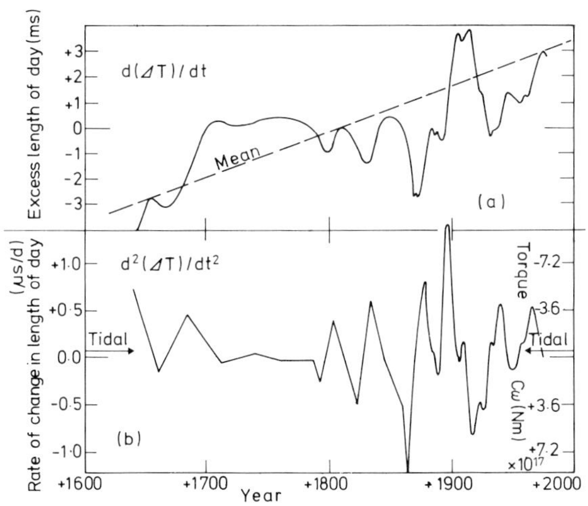

Fig. 2a,b. First derivative (a) and second derivative (b) of the modern ΔT curve. The dashed straight line in (a) indicates the average rate of lengthening of the day determined from ancient data

Figure 2b shows the second derivative of ΔT with respect to TD (latterly TAI). Before 1861, the graph consists of 13 straight lines between the knots of the cubic splines. The smoothed curve after that date was derived by applying an 11-point convolute to the annual values of ΔT (Morrison 1979a). In the diagram, the left ordinate measures the rate of change in the l.o.d. in microseconds per day. The right ordinate measures the estimated torque acting on the mantle in units of Nm. Torques have been computed taking the value of the principal moment of inertia of the mantle (including crust) as 7.2×10^{37} kg m^2. The total tidal torque (5×10^{16} Nm) is shown for comparison. This is obviously heavily outweighed by the non-tidal torques.

Around 1900, the non-tidal torque reached an amplitude of some 20 times the steady tidal component over a period of about 5 years. At present, it is not clear whether electromagnetic coupling is adequate to account for fluctuations on this scale. We await a more thorough assessment of the various mechanisms which have been proposed.

4. Medieval Fluctuations

For the purposes of the present paper we interpret the term "medieval" to mean the interval from AD 500 to 1600. Astronomically this is a disappointing era. There are precious few useful observations of any kind between AD 500 and 800 and again between 1300 and 1600. Data in the latter period would have been particularly useful, coming immediately before the telescopic material.

Between about AD 800 and 1000 Arab astronomers made a series of observations of the first and last contacts for both solar and lunar eclipses (see Newcomb 1875, Newton 1970). In general, these were made in a similar manner to the early 17th century measurements, using altitudes of the Sun, Moon and selected clock stars. However, the precision is quite low, with a standard error about the mean of some 1.0°. Additionally, the presence of appreciable systematic errors cannot be ruled out. We have thus restricted our attention to untimed reports of total and near total solar eclipses in the medieval period. We shall divide these into two groups.

4.1 Observations made by Clavius

In 1560 and 1567, the Jesuit astronomer Christopher Clavius happened to witness two very large eclipses of the Sun. The former he saw at Coimbra in Portugal (he was then a student at the University) and the latter when he had moved to Rome (where he was teaching mathematics at the Collegio Romano). Both accounts are of some value in extending the modern ΔT curve into the past. Clavius' account in his *In Sphaeram Ioannis de Sacrobosco* is worth translating in full (see Stephenson and Clark 1978).

"I will cite two remarkable eclipses of the Sun which happened in my own time and thus not long ago. One of these I observed in the year 1559 about midday at Coimbra in Lusitania in which the Moon was placed directly between my sight and the Sun, with the result that it covered

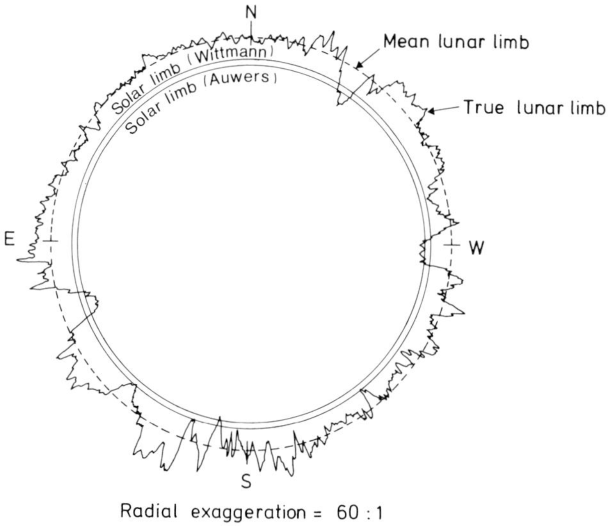

Fig. 3. Obscuration of the Sun by the Moon at the eclipse of AD 1567 April 9, seen at Rome. An observer on the central line is assumed here. Both the lunar limb profile (Duncombe 1973) and the difference between the apparent lunar and solar diameters are exaggerated by 60:1. The size of the inner circle representing the solar limb is based on the mean semidiameter due to Auwers (1891). The outline based on the work of Wittmann (1977) is shown for comparison

the whole Sun for a considerable length of time and there was darkness in some manner greater than that of night. Neither could one see very clearly where one placed one's foot. Stars appeared in the sky, and (miraculous to behold) the birds fell down from the sky to the ground in terror of such horrid darkness. The other I saw in Rome in the year 1567, also about midday, in which although the Moon was again placed between my sight and the Sun it did not obscure the whole Sun as previously, but (a thing which perhaps never happened at any other time) a certain narrow circle was left on the Sun, surrounding the whole of the Moon on all sides (*undique*)."

The precise dates are 1560 August 21 (Clavius has here mistaken the year) and 1567 April 9. In the former instance Clavius is very clearly describing a total eclipse. The latter would seem to allude to an annular obscuration, but this was not in fact the case. Using the standard ephemeris values for the mean solar and lunar semi-diameters, the apparent semi-diameter of the Moon as seen from Rome exceeded that of the Sun by 2.3 arc s. When allowance is made for the lunar limb profile using the charts of Duncombe (1973) - see Fig. 3 - it is evident that at least two beads of photospheric light would always be visible. However, even if the eclipse was partial, the crescent would only have extended less than half-way round the solar limb. Clavius thus seems to be describing a more or less central eclipse. We might note here that recent investigations of Mercury transit and total solar eclipse recorded durations show negligible secular trend in the mean solar semidiameter over the last 250 years (Parkinson et al. 1980, Gilliland 1981). At most, there is some evidence of oscillation of amplitude some 0.2 arc s (0.02%) and period roughly 80 years. Such a small variation is of no significance for the present purpose.

Based on their own observations of total solar eclipses, Muller (1975) and Bennett (1980) independently identified the "narrow circle" which Clavius described as the bright inner corona. This seems to be the most feasible explanation. Both observations by Clavius will be analysed in Sect. 4.3 along with some additional data from the 12th century.

4.2 Other Medieval Observations

In the proceedings of the first Bielefeld symposium, Stephenson (1978) published a list of observations of total and near total eclipses recorded in ancient and medieval history. Further research has not significantly added to this list. However, three of the medieval data are useful in setting limits to the ΔT fluctuations. As in 1560, these relate to instances where the observer chanced to be very close to one edge of the zone of totality. Unfortunately, in most cases the total phase was witnessed from a point well inside a wide shadow zone. Here computation can only confirm observation. Only three records - all from the 12th century AD - fulfil our requirements. All of the other reliable (i.e., place securely known) medieval observations in the list referred to above are satisfied by almost any realistic choice of ΔT.

A useful report of the eclipse of 1133 August 2 is contained in the *Canonici Wissegradensis Continuatio Cosmae* (see Muller and Stephenson 1975). This chronicle was kept by a canon of the cathedral at Vysehrad (Prague) between AD 1126 and 1142. The canon had a remarkable interest in observational astronomy, and among more mundane events he reports aurorae, eclipses, mock Suns, etc. It seems clear that he himself was an eye witness in each case. His description of the solar eclipse of 1133 establishes that as seen from Vysehrad the greatest phase was not quite total, for a "crown like the crescent Moon" remained. Near the opposite edge of the zone of totality of this same eclipse, another chronicler - this time in the monastery of S. Peter at Salzburg - reported that "the Sun suddenly disappeared (*subito disparuit*) and a little afterwards it seemed terribly covered over, like a round bag". This is recorded in the *S. Rudberti Salisburgensis Annales Breves* (see Newton 1972, p. 266). A sudden disappearance of the Sun implies totality. Muller (1975) interprets the subsequent description as a reference to the corona.

Finally we include an observation of the eclipse of AD 1178 September 13 made at Vigeois (France). In *Ex Chronico Gaufredi Vosiensis*, a detailed report, specifically mentioning the *first person plural*, states that "the Sun became like a two- or three-day-old Moon" (Newton 1972, p. 349). Muller (1975) notes that even a very thin crescent would suffice since irradiation would cause it to appear much wider.

4.3 ΔT Variation in Medieval Times

In Table 3 are summarised the five selected observations made in 1133 (two) 1178, 1560 and 1567. We have used large-scale maps to estimate the coordinates of each site accurately. As it happens, positional errors are only significant in 1567 (Rome was, of course, of substantial size, even in the 16th century).

In the above table, for Rome we have given the site of the Collegio Romano where Clavius was based. He could well have been elsewhere in Rome at the time, but, as we shall discuss below, errors of location are much smaller than observational errors. In none of the five records was any useful timing information given. We have to rely entirely on the description of the event itself.

As colleagues who have witnessed the two most recent total solar eclipses (1980 February 16 and 1981 July 31) have confirmed, the unaided eye can define totality with high precision. Hence allowance for the irregular configuration of the lunar limb is desirable, although the effect is small. For this purpose we have made use of the lunar limb profiles drawn by Duncombe (1973) from Watts's (1963) charts of the marginal zone of the Moon. Wittmann (1977) has shown that the adopted standard for the mean solar semi-diameter - due to Auwers (1891) - is slightly too small. Based on measurements made between 1836 and 1975, he deduced a figure of 960".00 ± 0".09 rather than Auwers' value of 959".63. Both of these results are free from irradiation. This increment is confirmed in a recent paper by Wittmann et al. (1982). Use of 960".0 would alter the ΔT limits in Table 3 by only

Table 3. Critical medieval observations of solar eclipses

Date	Place	Lat (N)	Long	Greatest phase	ΔT
1133	Vysehrad	50° 03.7'	14° 25.0' E	Crescent	<1400 s
1133	Salzburg	47 48.2	13 02.8 E	Totality	<1450 s
1178	Vigeois	45 22.9	1 31.2 E	Crescent	>1170 s
1560	Coimbra	40 12.5	8 25.0 W	Totality	< 200 s
1567	Rome	41 53.9	12 28.9 E	"Narrow circle"	95-190 s

about 4 s, lowering upper limits and raising lower limits. Such changes are scarcely significant in the present study.

The Salzburg observation in 1133 gives no hint of the duration of totality. Hence in its reduction we have merely supposed that no part of the solar photosphere was visible at maximum phase. The true limit could thus be appreciably lower than that given in the table. As the reports from Vysehrad and Vigeois both mention the crescent phase, we have assumed that an unbroken arc of at least 20° in extent (i.e., a fairly obvious crescent) was visible. The difference between each result and that needed to render the first bead of sunlight visible is only about 5 s. In both cases there is nothing in the record to prevent the assumption of a larger crescent. This would correspond to a smaller limit for Vysehrad and a larger one for Vigeois. In all three examples (as well as 1560 below) we have only investigated one limit. In each case the zone of totality was wide so that the other ΔT limit is redundant, i.e., impossibly deviant values would be needed to again satisfy observation.

At the eclipse of 1560, Clavius seems to have been situated some distance inside the zone of totality. He noted that the whole Sun was covered "for a considerable length of time" (*non modico temporis intervallo*). Although this statement is uncompromisingly vague, it is difficult to envisage a duration of less than 30 s. The result in Table 3 is based on this. However, for negligible duration the limiting value would only be about 5 s more. Equally, a considerably smaller value of ΔT is possible, corresponding to a longer duration of totality.

The observation in 1567 is more difficult to assess, although it has the important advantage of providing two useful limits for ΔT. The main difficulty lies in the virtual uniqueness of Clavius' description. For comparison, Mathers (1966) saw an almost central eclipse in which the semi-diameter of the Moon was only about $0''.1$ less than that of the Sun, while the lunar limb profile was much the same as in 1567. Although the Sun was reduced to a broken ring of light, no corona was visible in 1966, even with the aid of a telescope. Searching through earlier records, we have noted a useful parallel in an account of the eclipse of 1842 cited by Arago (1846). At Alais (France), which lay very close to the northern limit of totality (computed magnitude 0.9990), "everyone remarked on the circle of pale light which encompassed the Moon when it was almost entirely covered". Several observers of the same eclipse who were nearer the central line stated that the corona first became visible anything from 3 to 6 s before the onset of totality. Judging from the reports of modern observers, including several colleagues, 5 s seems quite typical. The greatest phase at Alais was rather similar to that which would be apparent some 6 s before totality near the centre line. Hence it would seem that this place was situated quite close to the margin of the region in which the corona was visible as a full circle.

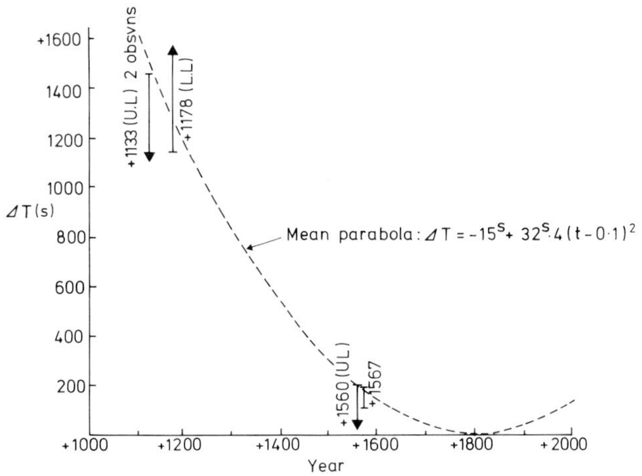

Fig. 4. Range of ΔT values inferred from medieval observations of total and near-total solar eclipses. The mean parabolic representation of ΔT, based on ancient and modern data, is shown for comparison

Applying these observations to the 1567 eclipse and supposing that Clavius was indeed at the Collegio Romano, we arrive at a range of values of ΔT between about 105 and 180 s to satisfy Clavius' description. Assuming that he was merely somewhere in Rome, it is necessary to widen this range by about 20 s (10 s at each limit). This leads to the figures in Table 3. However, it must be conceded that because of the difficulty in interpreting just what Clavius saw, these values are less reliable than those deduced from the data between 1133 and 1560.

Figure 4 shows the deviations from the mean parabola in medieval times deduced from the observations listed in Table 3. There is no evidence to suggest significant variations from this mean parabola throughout the whole of the period considered. Build up to large amplitudes thus seems unlikely, implying that the so-called decade fluctuations are only of short-term importance.

5. Analysis of Ancient Observations

The average long-term acceleration of the Earth's spin $\langle\dot{\omega}\rangle$ can only be satisfactorily be determined from ancient data. For observations on this time scale, i.e., some 2000 years, the irregular fluctuations are likely to be negligible. In this section, late Babylonian timings of lunar eclipses - made in the period 700 B.C. to 50 B.C. - are investigated to determine the average acceleration of the Earth's rotation over the past 2500 years. Other data used includes untimed observations of large solar eclipses - mainly from China. Here we have a number of observations which either allege totally at a particular location or else deny it. We shall commence with an objective investigation of the ancient observations discussed by Fotheringham (1920) in his now classic paper entitled "A Solution of Ancient Eclipses of the Sun".

5.1 Discussion of Fotheringham's Ancient Eclipse Data

In his 1920 paper, Fotheringham analysed 11 eclipses of the Sun recorded in a variety of ancient works. The sheer simplicity of this investigation has aroused much admiration and his well known solution diagram has frequently been used in geophysical studies. However, both Newton (1970) and Muller and Stephenson (1975) have questioned the reliability of the material which Fotheringham used. In Table 4 we have summarised each of the 11 observations and appended our comments on the important question of reliability - with regard to date, place of observation and recorded magnitude.

Table 4. Investigation of the reliability of Fotheringham's eclipse data

Fotheringham's inferences			Our interpretation		
Eclipse name	Year	Location	Date	Place	Magnitude
Babylon	-1062	Babylonia	Doubtful	Vague	Atmospheric phen?
Eponym Canon	- 762	Assyria	Reliable	Vague	Not recorded
Archilochus	- 647	Greece	Doubtful	Vague	Probably total
Thales	- 584	Asia Minor	Doubtful	Vague	Large, if real
Pindar	- 462	Thebes	Doubtful	Reliable	Very large
Thucydides	- 430	Athens	Reliable	Reliable	Only partial
Agathocles	- 309	Off Sicily	Reliable	Vague	Total
Hipparchus	- 128	Hellespont	Doubtful	Reliable	Total
Phlegon	+ 29	Nicaea	Doubtful	Doubtful	Large or total
Plutarch	+ 71	Chaeronia	Doubtful	Doubtful	Total, if real
Theon	+ 364	Alexandria	Reliable	Reliable	Rough times only

It will be apparent from our comments in Table 4 that we regard all of the material which Fotheringham used in his 1920 paper as of doubtful validity. In no case are all three important factors - date, place and eclipse magnitude - securely known. In several instances, two or three distinct dates are equally acceptable. Newton (1970) pointed out that like several of his contemporaries, Fotheringham played a sort of "identification game" to date observations for which only a very approximate date was known. This amounted to starting with a particular set of corrections to the appropriate astronomical parameters and using these to select the most likely eclipse candidate - often over a period of several decades. The newly "dated" material is then used to "improve" the accuracy of the original corrections. The flaws in this kind of reasoning are patently obvious today, but in Fotheringham's time this was a standard technique. Its use probably arose on account of the limited data then available.

5.2 Babylonian Lunar Eclipse Timings

From the 8th century B.C., the astronomers of Babylon made systematic observations of celestial phenomena. These were recorded on clay tablets, many of which are still extant. Virtually all of these texts are in the British Museum. Excellent drawings of many of these have been published by Sachs (1955), but as yet very few observations have ever been published. We are indebted to P.J. Huber of Harvard University for making available to us his manuscript translation of the lunar and solar eclipse records which he has been able to identify on the British Museum texts.

As it happens, only about 10 Babylonian solar eclipse observations have survived. (It is worth emphasising here that only about 5% of the original tablets have been recovered and all of these are damaged, sometimes extensively). Fortunately, lunar eclipse reports are still quite numerous, numbering more than 50. Hence we have concentrated on these. It was the practice of the astronomers to time eclipse contacts in relation to sunrise or sunset - two well-defined moments. In addition, moonrise or moonset was occasionally used as a reference point. The unit of time was invariably the uš equal to the interval for the celestial sphere to turn through 1° (i.e., 4 min). Almost certainly, some kind of water clock was used for time measurement. However, we know little about how the instrument operated.

From Huber's compilation we have abstracted observations of lunar eclipses for which the measured time interval is fairly short - less than about 20 uš (80 min). On a visit to the British Museum during August 1981, one of us (FRS) made a careful study of the tablets bearing the record of these observations. As a result, we have rejected any entries for which critical readings - especially numbers - are in doubt. The 17 selected data are listed in Table 5. This gives the equivalent Julian calendar date, computed eclipse magnitude (for reference only) and a summary of each observation. All dates are accurately known. Most are specified in the texts themselves either in terms of the years of the reigning monarch or later (after -310) from the Seleucid Era. Other dates can be readily identified from occasional references to historical events. The place of observation

Table 5. Timed Late Babylonian lunar eclipses for which the measured interval is shorter than about 20° (80 min)

Julian date	Mag	Summary of observation	Remarks
-684 Oct 3	1.84	Began 20° after sunset	
-562 Sep 5	0.35	Ended 6° after moonrise	Reject
-536 Oct 27	1.50	Began 14° before sunrise	
-482 Nov 19	1.47	Began 10° before sunrise	
-464 Nov 29	1.46	Ended 20° (or 21°) after sunset	
-420 Feb 2	1.64	Began 19° after sunset	
-407 Oct 31	0.18	Began 15° after sunset	Reject
-406 Oct 21	1.38	Moon set 15° after end of totality	
-405 Oct 10	0.97	Began 14° before sunrise	
-352 Nov 22	1.35	Moon set 6° after end of totality	
-239 Nov 3	1.41	Began 3° before sunrise	
-211 Apr 30	0.62	Began 20° before sunrise	
-153 Mar 21	0.85	Began 4° after sunset	
-142 Feb 17	0.88	Began 7° after sunset	
- 79 Apr 11	0.60	Moon set $16\frac{1}{2}^{\circ}$ after mid-eclipse	
- 66 Jan 19	0.80	Mid-eclipse 4° after moonrise	Reject
- 65 Dec 28	0.34	Mid-eclipse 13° after moonrise	Reject

is taken as Babylon. Several of the texts are known to have been excavated there. The provenance of the other texts in the same British Museum collection was shown to be Babylon by Sachs (1948) in an exhaustive investigation.

Four of the eclipses listed in Table 5 are rejected on fundamental observational grounds. In three instances (-562, -407 and -65) the magnitude was small - less than about one third. At small magnitudes the contacts are likely to be less sharply defined since the Moon enters the Earth's shadow quite obliquely. In -79, mid-eclipse (the phase observed) occurred so soon after moonrise that it would be impossible to judge the moment with any precision.

Each selected observation in Table 5 effectively gives a determination of the UT of the contact since local times of sunrise, etc. are readily calculated. In each case we have computed the TD of the event on the assumption that $\dot{n} = -26$ arc s/cy^2 (the figure adopted throughout this paper). Hence each observation leads to a direct estimate

Table 6. Analysis of timed Babylonian lunar eclipses

Year	ΔT (h)	t^2 (cy^2)	c (s/cy^2)
-684	+5.21 ± 0.26	617	+30.4 ± 1.4
-536	5.14 ± 0.13	546	33.9 ± 0.8
-482	4.71 ± 0.09	521	32.6 ± 0.6
-464	4.44 ± 0.19	513	31.2 ± 1.3
-420	4.22 ± 0.16	493	30.8 ± 1.1
-406	4.38 ± 0.13	487	32.4 ± 0.9
-405	4.49 ± 0.13	486	33.3 ± 0.9
-352	4.11 ± 0.07	463	32.0 ± 0.5
-239	3.87 ± 0.06	416	33.3 ± 0.5
-211	3.18 ± 0.17	404	28.3 ± 1.4
-153	3.43 ± 0.07	381	32.4 ± 0.6
-142	3.39 ± 0.08	377	32.4 ± 0.7
- 79	2.96 ± 0.17	353	30.2 ± 1.6

of ΔT at the time. In computing the TD of the various external contacts, we have applied an increment of 7 ± 2% to the radius of the Earth's shadow - in addition to the customary 2% to allow for effect of the Earth's atmosphere. This increment is based on a separate investigation of 17th century timings of lunar eclipses made with the unaided eye (see Stephenson and Morrison 1982). Because of confusion between the deep penumbral shadow and that of the umbra near first and fourth contacts, the 17th century observers systematically overestimated the duration of lunar eclipses. However, as might be expected, durations of *totality* are fairly accurately determined.

From the results for ΔT, we have computed the coefficient c of t^2 (see Sect. 3.3). Here t is measured in centuries from 1800, an epoch conveniently close to the minimum of the ΔT parabola. The units of c are s/cy^2, such that

$$\Delta T = c\, t^2 \quad . \tag{6}$$

Here we are neglecting the small constant term (-15 s) in (3). The results are shown in Table 6.

The standard errors associated with the individual ΔT values in Table 6 each comprise (1) an estimated clock drift of 13% of the elapsed time, (2) a ± 2 min (0.5 <u>uš</u>) rounding error, (3) a ± 1 min error in the observed time of sunset, sunrise, etc. and (4) the previously discussed 2% error in the enlarged umbral radius used in our computations. The uncertainties (1) to (3) were estimated in a separate investigation of Babylonian timed observations, viz., durations of total lunar eclipses and intervals between moonrise/moonset and sunset/sunrise around full Moon. The weighted mean value of c and its standard error are

$$c = +\ 32.4 \pm 0.3\ \mathrm{s/cy}^2\quad . \tag{7}$$

If no penumbral correction is applied, this result must be increased to 33.0. However, 17th century evidence in favour of use of an enlarged umbral radius when analysing unaided eye observations of lunar eclipse contacts seems to be fairly convincing.

The above result is derived assuming that $\dot{n} = -\ 26$ arc s/cy^2. By far the largest source of error in the determination of c lies in the uncertainty in the chosen value for $\dot{n}$. The estimated standard error of $\pm\ 2''/\mathrm{cy}^2$ in this parameter leads to an uncertainty in c of $\pm\ 2$ s/cy^2. With this revised estimate of uncertainty, our finally adopted result for c becomes

$$c = +\ 32.4 \pm 2.0\ \mathrm{s/cy}^2\quad . \tag{8}$$

An improved determination of $\dot{n}$ is obviously necessary before the full potential of the Babylonian observations can be realised.

5.3 Comparison with Ancient Observations of Large Solar Eclipses

Use of observations of large solar eclipses in the study of the Earth's rotation in ancient times has become almost standard practice. It is important to compare the above result for c (7) with individual determinations from total and near-total eclipses around the same period.

The observations listed in Table 7 are taken directly from the paper by Stephenson (1978) in the proceedings of the first Bielefeld symposium. They cover roughly the first 7 centuries before the Christian era, and with one exception (-135) are from China. Negative dates (rather than B.C. dates) are used here, the difference between the two systems being one year (thus -180 = 181 B.C.).

For all Chinese observations except -600, we have assumed that the place of observation is the appropriate capital. In the case of observations made after about -200, this assumption seems particularly reasonable since the records are taken from the astronomical treatise of the offical dynastic history (the *Han-shu*). This treatise appears to be largely compiled from the observations of the imperial astronomers. For the three earliest observations the source is the annals of the small state of Lu (the *Ch'un-ch'iu*). This state attached parti-

Table 7. Analysis of ancient observations of large solar eclipses

Julian date	Place	Observation	Range of c (s/cy^2)
-708 Jul 17	Chü-fu	Total	32.2 - 33.4
-600 Sep 12	Ying ?	Total	32.4 - 33.9
-548 Jun 12	Chü-fu	Total	> 29.3
-197 Aug 7	Ch'ang-an	Annular	< 32.8
-187 Jul 17	Ch'ang-an	Almost complete	< 35.1
-180 Mar 4	Ch'ang-an	Total, dark in daytime	20.0 - 32.5
-146 Nov 10	Ch'ang-an	Almost complete	Any reasonable value
-135 Apr 15	Babylon	Total, 4 planets seen	30.0 - 32.5
- 88 Sep 29	Ch'ang-an	Not complete, like a hook	Any reasonable value
- 79 Sep 20	Ch'ang-an	Almost complete	Any reasonable value
- 34 Nov 1	Ch'ang-an	Not complete, like a hook	Any reasonable value
- 27 Jun 19	Ch'ang-an	Not complete, like a hook	> 28.6
- 1 Feb 5	Ch'ang-an	Not complete, like a hook	Any reasonable value

cular importance to solar eclipses and practised eclipse ceremonies at its capital. In -600, the eclipse cannot have been total within about 700 km of the Lu capital (Chü-fu). The record could have come from Ying, the capital of the important state of Ch'u, far to the south, but this is no more than a presumption. It may be proper to reject this observation.

Several of the partial eclipses in Table 7 fell far short of totality. Hence for "any reasonable value" of c (between 20 and 40 s/cy^2) the observation is confirmed.

Figure 5 shows a plot of c as a function of time for both the Babylonian timed lunar eclipse observation and the data in Table 7. The result derived from the timed data (32.4 ± 0.3) is shown for reference. As the analysis of all observations makes use of a fixed value for $\dot{n}$ (-26"/cy^2) it is appropriate to quote only the observational uncertainty here. In the figure, timed observations, with their associated error bars, are shown by broken lines. Untimed observations of solar eclipses are represented by full lines. Provided the place

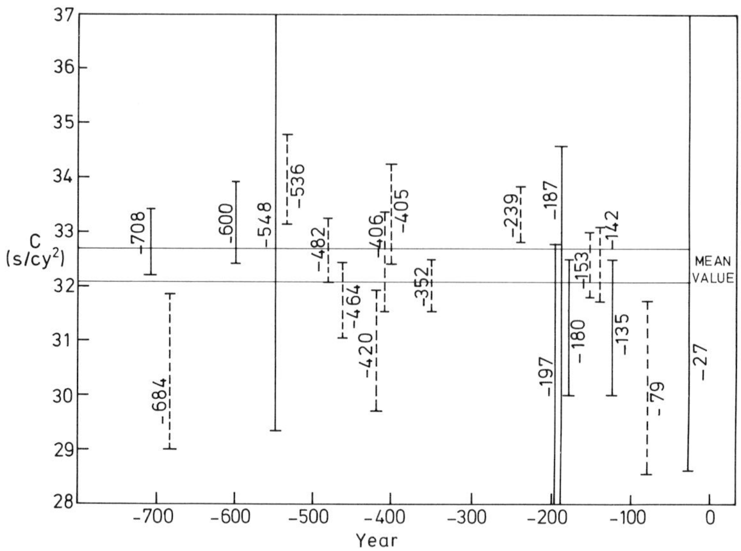

Fig. 5. Coefficient of t^2 (t in centuries from 1800.0) in the mean parabolic expression for ΔT as determined from ancient observations of solar and lunar eclipses. Each vertical line represents a single observation. Full lines represent solar eclipses (hard limits) and broken lines lunar eclipses (with uncertainties arising from various causes). The mean value of c - including standard error - deduced from the lunar eclipse data is represented by the two horizontal lines

of observation is confidently known, these latter limits are very sharply defined. The agreement between the results for c obtained from two quite independent data sets is excellent. In particular, there is no obvious indication of any variation of c during the period in question.

In contrast to the impressive observations just analysed, the ancient material of later date, i.e., up to about AD 500, is rather sparse and in general of dubious quality. Hence we have not included it in the present investigation. This completes our study of the observational data.

6. Geophysical Implications

The result for c (32.4 ± 2.0) derived in Sect.5.2 corresponds to an average rate of lengthening the day over the past 2½ millennia of 1.8 ± 0.1 ms/cy^2. The equivalent value of $\langle\dot{\omega}\rangle/\omega$, where ω is the present rate of spin of the Earth, is

$$\langle\dot{\omega}\rangle/\omega = - 20.6 \pm 1.3 \times 10^{-11}/\mathrm{yr} \quad . \tag{9}$$

From a consideration of conservation of angular momentum in the Earth-Moon system, the expected tidal acceleration of the Earth's rotation, $\dot{\omega}_T$ is given by (see, e.g., Morrison 1978)

$$\dot{\omega}_T/\omega = 1.16\ \dot{n} + 1.1 \times 10^{-11}/\mathrm{yr} \quad . \tag{10}$$

The constant term here represents the contribution of the atmospheric tidal couple. Hence with $\dot{n} = -26''/\mathrm{cy}^2$, we have

$$\dot{\omega}_T/\omega = - 29.0 \pm 2.3 \times 10^{-11}/\mathrm{yr} \quad . \tag{11}$$

Differencing (9) and (11), we obtain the residual non-tidal acceleration of the Earth's spin

$$\langle\dot{\omega}\rangle_{NT}/\omega = + 8.4 \pm 1.0 \times 10/\mathrm{yr} \quad . \tag{12}$$

It is important to remark here that the errors in $\langle\dot{\omega}\rangle/\omega$ and $\dot{\omega}_T/\omega$ both depend largely on the error in $\dot{n}$. These uncertainties tend to cancel since they enter with the same sign. Hence the final error is remarkably small.

The reality of the non-tidal component in the acceleration of the Earth's spin now seems to be well established. This probably arises from a fractional change in the moment of inertia of the Earth of the same magnitude but of opposite sign, i.e.,

$$\dot{I}/I = - 8.4 \pm 1.0 \times 10^{-11}/\mathrm{yr} \quad . \tag{13}$$

This figure is, of course, an average over the past 2½ millennia. Although there have been several determinations of $\dot{I}/I$ in recent years, only now has it proved possible to deduce this quantity with any precision. We welcome discussion of the geophysical consequences of our result.

References

Arago A (1843) Annuaire 1846
Auwers A (1891) Astron Nachr 128: 361
Bennett CA (1980) Sky Telesc 60: 475
Brouwer D (1952) Astron J 57: 125
Duncombe (1973) US Naval Obs Circ 141
Explanatory Supplement to the Astronomical Ephemeris and the American Ephemeris and Nautical Almanac (1974) HMSO, London
Fairbridge RW (1961) Phys Chem Earth 4: 99
Ferrari AJ, Sinclair WS, Sjogren WL, Williams JG, Yoder CF (1980) J Geophys Res 85: 3939

Flandern Van TC (1982) (Pre-print)
Fotheringham JK (1920) Mon Not R Astron Soc 81: 104
Gassendi P (1658) Opera Omnia, IV, 100. Lugduni
Gilliland RL (1981) Astrophys J 248: 1144
IAU (1968) Trans Int Astron Union Ser B 13: 48
Jeffreys H (1982) (Pre-print)
Lambeck K (1980) The Earth's variable rotation. Cambridge Univ Press Cambridge
Mathers JH (1966) New Sci 30: 736
Mörner N (1971) Palaeogeogr Palaeoclimatol Palaeocol 9: 153
Morrison LV (1978) In: Brosche P, Sündermann J (eds) Tidal friction and the Earth's rotation. Springer Berlin Heidelberg New York
Morrison LV (1979a) Geophys J R Astron Soc 58: 349
Morrison LV (1979b) Mon Not R Astron Soc 187: 41
Morrison LV, Ward CG (1975) Mon Not R Astron Soc 173: 183
Morrison LV, Lukac MR, Stephenson FR (1982) Greenwich Obs Bull 156
Muller PM (1975) An analysis of the ancient astronomical observations with the implications for geophysics and cosmology. Ph D Thesis, Univ Newcastle upon Tyne
Muller PM, Stephenson FR (1975) In: Rosenberg GD, Runcorn SK (eds) Growth rhythms and the history of the Earth's rotation. Wiley New York
Newcomb S (1875) Astron Meteorol Obs US Naval Obs 22: App II
Newton RR (1970) Ancient astronomical observations and the accelerations of the Earth and Moon. Johns Hopkins Univ Press, Baltimore London
Newton RR (1972) Medieval chronicles and the rotation of the Earth. Johns Hopkins Univ Press, Baltimore London
Parkinson JH, Morrison LV, Stephenson FR (1980) Nature (London) 288: 548
Rochester MG (1974) Veröff Zentralinst Phys Erde Akad Wiss DDR 30: 77
Sachs AJ (1948) J Cuneiform Stud 2: 271
Sachs AJ (1955) Late Baylonian astronomical and related texts. Brown Univ Press, Providence, RI
Stephenson FR (1978) In: Brosche P, Sündermann J (eds) Tidal friction and the Earth's rotation. Springer Berlin Heidelberg New York
Stephenson FR (1981) J Br Astron Assoc 91: 136
Stephenson FR, Clark DH (1978) Applications of early astronomical records. Hilger, Bristol
Stephenson FR, Morrison LV (1982) Mon Not R Astron Soc (in press)
Watts CB (1963) Astron Pap Washington 17
Wittmann A (1977) Astron. Astrophys 61: 225
Wittmann A, Bonet Navarro JA, Wöhl H (1982) (Pre-print)

Deceleration of the Earth's Rotation from Old Solar Observations

A. D. Wittmann[1]

1. Introduction

1.1 ET and UT

The rotation of the Earth, when suitably averaged over its short-period fluctuations, provides us with a time scale known as Universal Time (UT). The orbital motion of the Earth around the fictitious mean Sun provides us - in principle at least - with a time scale known as Ephemeris Time (ET). Since the semi-major axis of the Earth's orbit, and hence its period of revolution, are constant to a very high degree of approximation, the ET scale is uniform and represents an almost ideal clock.

The UT scale is non-uniform mainly for two reasons: the (presently) decreasing excentricity of the Earth's orbit - caused by planetary perturbations - gives rise to an acceleration of the retrograde precessional motion of the (vernal) equinox, and hence of the fictitious mean Sun from which is reckoned the rotation of the Greenwich meridian. Secondly, the rotation rate (angular velocity ω) of the Earth is slowly decreasing due to tidal interaction in the Earth-Moon system, and it is the magnitude of the latter effect that we shall try to determine.

For the sake of brevity we shall also write t instead of UT, and T instead of ET (note that T will also be used to denote Julian centuries reckoned from 1900.0). The epoch 1900.0 (i.e., 31 December 1899 at 12:00:00 ET) has been generally adopted as the initial epoch for reckoning time in ephemerides calculations. At this epoch the difference $\Delta T = T - t$ is small (theoretically zero, practically about -4 s), and one has:

mean sidereal day: $D'_o = 23^h56^m04\overset{s}{.}090540$

mean solar day: $D_o = 24^h00^m00\overset{s}{.}000000$

period of rotation: $P_o = 23^h56^m04\overset{s}{.}098904$

1 Universitäts-Sternwarte, Geismarlandstr. 11, D-3400 Göttingen, Fed. Rep. of Germany

Tidal Friction and the Earth's Rotation II
ed. by Brosche/Sündermann

sidereal ratio: $D_0/P_0 = 1.0027378119060$

tropical year: $Y_0 = 8765^h48^m45^s.9747$

1.2 Relation Between Time Scales

In what follows we shall consider epochs large enough that $\Delta T \ll |T|$. Then the angular velocity of the Earth, which is given by

$$\omega(T) = \omega_0 + \dot{\omega}T \quad , \tag{1}$$

where $\dot{\omega} = -\omega_0\dot{P}/P_0 < 0$, may also be written as

$$\frac{\omega(T)}{\omega_0} = 1 + \frac{\dot{\omega}}{\omega_0}\, t \simeq \frac{1}{1 - \dot{\omega}/\omega_0 \cdot t} \quad . \tag{2}$$

Then the relation between ET and UT becomes

$$T(t) = T_0 + \int_0^t \frac{2\pi D_0/P_0}{\omega(T)}\, dt \quad , \tag{3}$$

where D_0/P_0 is practically independent of time. Since $2\pi D_0/(\omega_0 P_0) = 1$ day, and since T_0 is sufficiently small, we finally have:

$$T(t) = t - \frac{1}{2}\frac{\dot{\omega}}{\omega_0}\, t^2 \qquad (\dot{\omega} < 0) \quad . \tag{4}$$

Instead of expressing T (or t) in days, it is customary to express $\Delta T = T - t$ in seconds, $\dot{\omega}/\omega_0$ in centuries^{-1}, and T in Julian centuries of 36525 days. By adopting these units we get

$$\Delta T = -\ 1.57788 \times 10^9 \cdot (\dot{\omega}/\omega_0)T^2 \quad . \tag{5}$$

From solar eclipse observations Fotheringham (1920), Curott (1966), Newton (1970), Stephenson (1978) and others have concluded that $\dot{\omega}/\omega_0 \simeq -\ 2 \times 10^{-8}\ cy^{-1}$. By inserting, say, T = 20 cy into (5) we get ΔT = 12623 s or 3.5 h. From this we can draw the important conclusion that an ancient observational epoch on UT must be known with an accuracy of a few hours or better if a meaningful value of $\dot{\omega}$ is to be derived from the observation in question. As far as the Sun is concerned, the day (or date) of a positional observation is not a question at all. To quote a famous example, it was recognized as early as in 1753 by Tobias Mayer (cf. Forbes 1971) that Ptolemy's alledged equinox observations were actually calculated and not observed (they differ by more than a day from the correct epoch, for details cf. Newton 1977). In fact Ptolemy calculated his epochs very accurately (which is why they can be proved to be fabricated), although (un)-fortunately he adopted some incorrect input data.

1.3 Longitude and Motion of the Sun

On ET the motion of the Sun is uniform, and the mean longitude L may be expressed as

$$L(T) = L_0 + nT \quad (\text{modulo } 2\pi) \quad , \tag{6}$$

where $n \simeq 0.985647$ deg/day is the mean motion. Hence according to (4) we have on UT:

$$L(t) = L_0 + nt - \frac{1}{2} n \frac{\dot{\omega}}{\omega_0} t^2 \quad . \tag{7}$$

The (positive) coefficient of t^2 is usually called $\dot{n}$, it is seen to be proportional to the mean motion of the celestial body under consideration (in our case the Sun), and it is conveniently expressed in seconds of arc per century^2:

$$\dot{n} = -\frac{1}{2} n \frac{\dot{\omega}}{\omega_0} = \frac{1}{2} n \frac{\dot{P}}{P_0} \quad , \tag{8a}$$

or, with $n \simeq 1.2960276 \times 10^8$ "/cy

$$\dot{n} = -6.480138 \times 10^7 \cdot \frac{\dot{\omega}}{\omega_0} \quad (''/\text{cy}^2) \quad . \tag{8b}$$

Now the longitude L is measured from the vernal equinox, which itself exhibits an accelerated motion with a speed called general precession in longitude:

$$\dot{L}_\gamma = -5025''\!.6 - 2''\!.2\ T(\text{cy}^{-1}) \quad . \tag{9}$$

Hence the longitude of the (mean) Sun is subject to a net dynamical acceleration of $+1''\!.1$ per cy^2, and (6) becomes

$$L(T) = L_0 + nT + 1''\!.1\ T^2 \quad . \tag{10}$$

1.4 Length of the Tropical Year

A closely related parameter is the length of the tropical year, which is defined as

$$1 \text{ trop. yr} = Y = t(L + 2\pi) - t(L) \quad , \tag{11}$$

where L is reckoned from the moving mean equinox of date. Y is a decreasing function of time because the motion of the Sun is accelerated:

$$Y = 360^\circ/n = 365^{\rm d}\!.25 - \Delta \quad , \tag{12a}$$

$$n = 0.98562628336755646 8/(1 - \Delta/365.25) \quad , \tag{12b}$$

where $\Delta \simeq 0.008$ days is increasing with time. Thus the acceleration of the equinox gives rise to a decrease of the length of the tropical year of $\dot{Y} \simeq -0.54$ s/cy (or $\dot{\Delta} \simeq + 6.3 \times 10^{-6}$ d/cy), and any observed excess decrease of Y (or excess acceleration of L) may be attributed to tidal braking effects:

$$\dot{Y} = -0.487\ \dot{n}_{total}(''/cy^2)(s/cy) \quad , \tag{13a}$$

and

$$\frac{\dot{\omega}}{\omega_0} = -\ 1.54318 \times 10^{-8}\ \dot{n}_{tidal}(''/cy^2)$$

$$= 3.16874 \times 10^{-8}\ (\dot{Y} + 0.54)(cy^{-1}) \quad . \tag{13b}$$

2. Analysis of Solar Tables

2.1 Description of Tables

The solar tables used in the present paper are: (1) the theory of Ptolemy (ca. 132), which is known to be defective and is used here for the purpose of illustration only, cf. Manitius (1963); (2) the Hakemite Tables of Ibn Junis (1008), cf. Caussin de Perceval (1804); (3) the Alfonsine Tables of Rey Alfonso X. (1250), cf. Santritter (1492); (4) the Rudolphine Tables of Johannes Kapler (1627), cf. Hammer (1969); and the solar tables or theories of (5) Philippe Gabriel de Lahire (1687); (6) Jacques Cassini (1740); (7) Nicolas Louis de Lacaille (1754); (8) Johann Tobias Mayer (1767); (9) Joseph Jerome le Français de Lalande (1771); (10) Franz Xaver von Zach (1804); and (11) Simon Newcomb (1895).

With the exception of Newcomb, the mean longitudes given by the tables represent results derived from a linear formula like (6). The mean longtiude of the Sun at any epoch is simply obtained by summing up the net motions corresponding to the number of (Julian) centuries, years, days, hours, and minutes that have elapsed between a tabulated epoch and the epoch in question, and by adding the sum to the tabulated longitude. Next, from the tabulated position of the apogee (L_A) one obtains the mean anomaly, and from the equation of the centre one gets the true anomaly and thus the true longitude of the Sun.

Apart from obvious typographical errors or numerical inaccuracies the various solar tables can be re-condensed into their underlying formulae. This has been done in Table 1, which so to speak represents a "table of tables".

Most of the columns of Table 1 are self-explanatory, for instance L_A denotes the longitude of apogee of the Sun (which of course is the same as the longitude of perigee of the Earth), and ΔL_A is its annual increment. Concerning the numerical excentricity e it is to be borne in mind that excentric circles were in use before the Keplerian ellipse, and that in case of the Sun (Earth) to a very good approximation the circle offset corresponds to twice the numerical excentricity.

2.2 Longitude Comparison

The epoch adopted for Newcomb's theory is 0 Jan 1900 (JD 2415020), at which $L = 279^\circ 41' 27''.6$, $\Delta L = 59' 08''.33040739$ per day ($27''.6813$ per year), $L_A = 101^\circ 13' 15''.0$, $\Delta L_A = 61''.890300$ per year, and $e = 0.0167510$. The Newcomb longitudes $L_{Newc.}$ (column 9) have been calculated by disregarding the quadratic term in the expression

$$L = 279^\circ 41' 27''.6 + 36000^\circ.768925\ T + 1''.089\ T^2 \quad , \tag{14}$$

and the coefficient $\dot{n}$ given in the last column of Table 1 was obtained from

$$\dot{n}_{total} = (L - L_{Newc.})/(\Delta JD/36525)^2\ ''/cy^2 \tag{15}$$

where ΔJD denotes the difference between the mean epoch of the observations (column 2 of Table 1) and Newcomb's epoch.

By disregarding the $\dot{n}$ implied by Ptolemy's theory, and by including $\dot{n}_{Newc.} = 1''.09$ we finally get from longitude comparison:

$$\dot{n}_{LC} = (2.59 \pm 1.11)''/cy^2 \tag{16a}$$

or

$$\dot{n}_{tidal} = 1.50''/cy^2 \quad . \tag{16b}$$

3. The Length of the Year (Motion Comparison)

3.1 Reduction of Equinox Observations

An equinox observation consists of one or more measurements of the declination δ of the Sun in the vicinity of the vernal (March) or autumnal (September) equinox, from which the epoch of equator crossing ($\delta = 0$) can be inferred. To a good approximation the proper motion of the Sun in declination near the equinoxes is given by

Table 1. Table of tables (for a description see text)

Author	Epoch Julian Day	L (1900.0)	ΔL per day year cy	L_A (Epoch)
Ptolemy	25 Sep 132 1769539	271°00'42".6	59'08".28700239 11.82762334 19 42.76233413	65°30'00".0
Ibn Junis	0 Jan 1000 2086307	279 37 03.8	59 08.32893322 27.14285714 45 14.28571429	86 07 25.7
Alphonsus	0 Jan 1250 2177620	279 29 27.7	59 08.32703348 26.44897948 44 04.89794800	88 40 00.0
Kepler	0 Jan 1590 2301805	279 39 51.8	59 08.32908966 27.2 45 20.0	95 34 03.7
Lahire	0 Jan 1680 2334667	279 41 24.1	59 08.32991102 27.5 45 50.0	97 45 58.5
Cassini	0 Jan 1700 2341972	279 40 47.9	59 08.32968026 27.41571429 45 41.57142857	97 35 55.1
Lacaille	0 Jan 1750 2360234	279 42 16.1	59 08.33107921 27.92667980 46 32.66798000	98 39 33.3
Mayer	0 Jan 1760 2363886	279 41 49.6	59 08.33081451 27.83 46 23.0	98 48 33.9
Lalande	0 Jan 1770 2367539	279 41 17.1	59 08.33006458 27.55608695 45 55.60869500	98 59 53.0
Zach	0 Jan 1800 2378496	279 41 02.1	59 08.32985626 27.48 45 48.0	99 28 20.0

$L_A(yr^{-1})$	e(Epoch)	L(Epoch)	$L_{Newc.}$	$\dot{n}(cy^{-2})$
36″.0	0.0208333	182°10′00″.0	63′47″.8	-10″.34
51.428571	0.0175224	284 39 35.2	35 54.4	2.73
36.0 ?	0.0189032	287 02 10.0	00 48.8	1.92
62.0	0.0180000	289 38 33.9	37 40.5	5.56
61.5	0.0168280	279 58 50.8	58 14.4	7.52
61.75	0.0168499	280 07 41.4	07 28.0	3.35
63.0	0.0168241	280 01 09.6	00 57.9	5.20
66.0	0.0168024	279 36 01.8	36 00.6	0.61
65.5	0.0168024	280 10 17.3	10 11.6	3.37
62.0	0.0167935	279 54 22.4	54 27.8	-5.40

$$\dot{\delta}(\text{vernal}) = 59.245 + 0.029\ T\ (''/\text{h}) \quad , \tag{17a}$$

$$\dot{\delta}(\text{autum.}) = 58.430 - 0.091\ T\ (''/\text{h}) \quad , \tag{17b}$$

where T is in centuries from 1900.

The observations are usually made by measuring the meridian zenith distance z of the Sun, which has to be corrected for refraction ($R \simeq 59.''9 \tan z$), geocentric parallax ($P \simeq 8.''8 \sin z$), and nutation in declination (N_δ). The very small fluctuations of ecliptic latitude may in general be neglected. The effect of refraction is to render the observed vernal equinox too early, and the observed autumnal equinox too late (northern hemisphere).

Finally the conversion from apparent to mean equinox has to be made by adding (vernal) or subtracting (autumnal) $(|v - M|)/n$ days, where v and M denote the true and the mean anomaly, respectively. At the vernal equinox one has $M = 360^\circ - L_A$, and at the autumnal equinox one has $M = 180^\circ - L_A$, while v is found from the excentric anomaly by solving Kepler's equation.

3.2 Hipparchus' Equinox Observations

I have analyzed only those nine equinoxes that were explicitly quoted by Ptolemy (Almagest, Lib. 3), and I have rejected the additional ones that might be inferred from Ptolemy's remarks about accuracy (cf. Fotheringham 1918). The results are shown in Table 2. The first column gives the date of observation (all times have been reduced from local time to UT), the next column gives the nutation of date (a somewhat academic correction to ancient naked-eye observations), and the next column gives the Julian day number corresponding to the observed equinox (note that Hipparchus interpolated his original observations). The last column gives the (truncated) Julian day number of the vernal equinox -142 to which all the observations were reduced (the length of the year was found beforehand from separate numerical fits to the vernal and autumnal observations).

Reduced to the vernal equinox of the year -142 one finally has from the observations of Hipparchus:

$$\text{vernal eqx. } -142\text{: } JD_{\text{mean}} = 1669276.645 \quad . \tag{18a}$$

3.3 Arab Equinox Observations

These observations were mainly taken from the work of Newton (1970) and Newton (1972), they are given in Table 3. Reduced to the vernal equinox of the year 870 one finally has from the Arab observations:

$$\text{vernal eqx. } 870\text{: } JD_{\text{mean}} = 2038901.926 \quad . \tag{18b}$$

Table 2. Hipparchus' equinox observations (Alexandria, latitude +31°10'47")

Date		N_δ	JD_{obs}	JD_{mean} [a]	vernal -142 [b]
27 Sep	-161	-3".8	1662522.17	1662520.298	6.848
27 Sep	-158	-6.9	63617.67	63615.795	6.688
27 Sep	-157	-6.4	63982.92	63981.045	6.720
26 Sep	-146	+4.2	68000.42	67998.550	6.816
24 Mar	-145	+3.3	68178.77	68180.678	6.411
27 Sep	-145	+2.2	68365.67	68363.799	6.846
26 Sep	-142	-4.4	69461.17	69459.293	6.684
23 Mar	-134	+0.5	72196.42	72198.328	6.374
23 Mar	-127	+4.5	74753.17	74755.082	6.417

[a] Corrected for refraction, parallax, and nutation.
[b] Mean: 1669276.645 ± 0.065.

3.4 Bernhard Walther's Equinox Observations

Bernhard Walther's observations were made at his private observatory at Nuremberg using parallactic rulers from which the chord of the Sun's zenith distance was read off at noon. Walther's observations certainly range among the most accurate and most reliable positional observations ever made with the naked eye. This is, for instance, proved by a reduction of his solstice observations which has recently been made by Wittmann (1982) for a study of the obliquity of the ecliptic. The equinox data were mainly taken from Riccioli (1665) and Lacaille (1753), they are shown in Table 4. Reduced to the vernal equinox of the year 1490 one has from Walther's observations:

$$\text{vernal eqx. 1490: } JD_{mean} = 2265352.166 \quad . \tag{18c}$$

3.5 Greenwich Equinox Observations

These were made with the Cooke Transit Circle at the Royal Greenwich Observatory at Herstmonceux. The data, already corrected for refraction and parallax, were derived from the tabulations of Buontempo et al. (1973) and Blackwell et al. (1975), they are given in Table 5. Reduced to the vernal equinox of the year 1970 one finally has from the Greenwich observations:

$$\text{vernal eqx. 1970: } JD_{mean} = 2440668.4299 \quad . \tag{18d}$$

A quadratic least-squares fit of the four vernal equinoxes given above yields:

$$JD_{mean}(n) = 1721141.068 + 365.242416\, n - 5.0107 \times 10^{-8}\, n^2 \quad . \tag{19a}$$

Table 3. Arab equinox observations

Date	Lat.	Observer	JD_{obs}	N_δ	JD_{mean} [a]	Vernal 870 [b]
19 Sep 829	33°35	Abi Mansur	2024111.660	-6".8	2024109.654	1.967
19 Sep 830	33.35	-	24476.835	-5.9	24474.829	1.901
16 Mar 831	33.35	-	24655.460	-5.0	24657.469	1.919
19 Sep 831	33.35	-	24842.168	-4.4	24840.163	1.992
16 Mar 832	33.35	-	25020.710	-3.4	25022.720	1.927
18 Sep 832	33.52	Al-Marwarudi	25207.366	-2.6	25205.363	1.949
18 Sep 844	33.35	-	29590.267	-3.7	29588.262	1.941
19 Sep 851	36.22	Al-Makki	32146.837	-1.7	32144.831	1.813
19 Sep 859	34.22	Benu Musa Brothers	35068.878	+4.3	35066.878	1.922
18 Sep 882	36.03	Al-Battani	43469.444	-5.1	43467.435	1.906
18 Sep 888	36.69	Al-Samarqandi	45660.881	-2.1	45658.873	1.890
18 Sep 970	29.61	Al-Sufi	75610.812	+5.4	75608.814	1.962
18 Sep 971	29.61	Al-Sufi	75976.104	+3.8	75974.105	2.010
18 Sep 974	33.35	Al-Buziani	77071.752	-3.0	77069.744	1.923
18 Sep 1019	33.58	Al-Biruṇi	93507.601	-0.6	93505.595	1.869

[a] Corrected for refraction, parallax, and nutation.
[b] Mean: 2038901.926 ± 0.013.

Table 4. Bernhard Walther's equinox observations (Nürnberg, latitude +49°26'25")

Date	UT	Chord	z.d.$_{obs}$	N_δ	δ	JD_{app}[a]	JD_{mean}	Vernal[b] 1490
16 Sep 1475	$11^h07^m07^s$	0.84785	50°09'55".0	+0".6	-2673".9	2260058.074	2260056.109	2.094
16 Mar 1476	11 21 35	0.80394	47 24 10.0	-0.5	+7272.2	60236.851	60238.811	2.206
13 Sep 1476	11 07 50	0.83404	49 17 35.4	-1.7	+ 468.0	60423.295	60421.329	2.075
11 Mar 1477	11 23 15	0.83675	49 27 50.7	-2.7	- 146.4	60602.078	60604.037	2.190
17 Sep 1477	11 05 56	0.85680	50 43 55.7	-3.9	-4710.1	60788.628	60786.663	2.168
11 Mar 1478	11 23 21	0.83820	49 33 20.0	-4.7	- 473.6	60967.308	60969.268	2.179
12 Mar 1478	11 23 01	0.83239	49 11 21.1	-4.7	+ 844.7	60967.379	60969.339	2.250
13 Sep 1478	11 08 00	0.83054	49 04 21.5	-5.7	+1265.9	61153.860	61151.895	2.159
14 Sep 1479	11 07 45	0.83490	49 20 50.6	-6.6	+ 277.7	61519.160	61517.195	2.219
14 Sep 1487	11 07 44	0.83600	49 25 00.4	+4.9	+ 16.4	64440.975	64439.011	2.112
18 Sep 1487	11 06 24	0.85444	50 34 57.1	+4.9	-4180.3	64441.004	64439.039	2.140
10 Mar 1488	11 23 28	0.84090	49 43 33.6	+5.7	-1097.6	64619.748	64621.707	2.192
17 Mar 1488	11 21 14	0.79729	46 59 13.2	+5.7	+8762.8	64619.800	64621.759	2.244
14 Sep 1488	11 07 29	0.84067	49 42 41.3	+6.2	-1045.8	64806.223	64804.259	2.119
15 Sep 1488	11 07 09	0.84695	50 06 30.0	+6.2	-2474.5	64806.212	64804.247	2.108
11 Mar 1489	11 23 14	0.83625	49 25 57.1	+6.7	- 42.1	64985.004	64986.963	2.205
13 Mar 1489	11 22 36	0.82365	48 38 20.6	+6.7	+2814.4	64894.992	64986.950	2.193
11 Mar 1491	11 23 23	0.83925	49 37 18.6	+6.3	- 723.2	65715.484	65717.442	2.200
13 Mar 1491	11 22 45	0.82684	48 50 23.0	+6.3	-2091.8	65715.501	65717.459	2.216
14 Sep 1491	11 07 44	0.83680	49 28 02.0	+5.7	- 166.0	65901.846	65899.882	2.021
15 Sep 1491	11 07 24	0.84237	49 49 07.9	+5.7	-1431.9	65901.950	65899.985	2.125
14 Sep 1493	11 07 33	0.83934	49 37 39.0	+1.9	- 739.2	66632.440	66630.476	2.135
14 Sep 1494	11 07 38	0.83785	49 32 00.5	-0.3	- 398.5	66997.682	66995.717	2.136
14 Sep 1496	11 07 28	0.84125	49 44 53.2	-4.7	-1166.8	67728.138	67726.174	2.111
18 Sep 1498	11 06 57	0.86270	51 06 23.5	-6.7	-6055.1	68458.677	68456.713	2.170
14 Mar 1499	11 22 43	0.82005	48 24 46.1	-6.8	+3642.4	68637.408	68639.366	2.183
10 Mar 1501	11 23 30	0.84234	49 49 01.0	-5.3	-1414.0	69367.971	69369.928	2.260
14 Sep 1501	11 07 32	0.83990	49 39 46.3	-4.6	- 860.0	69554.355	69552.391	2.127
15 Sep 1501	11 07 12	0.85131	50 23 03.3	-4.6	-3457.0	69554.516	69552.552	2.288
12 Mar 1504	11 22 47	0.82750	48 52 52.5	+0.8	+1948.4	70463.602	70465.559	2.163

[a]Corrected for refraction, parallax, and nutation.
[b]Mean: 2265352.166 ± 0.011.

Table 5. Greenwich equinox observations (Herstmonceux, latitude +50°52'18")

Date	JD_{app} [a]	N_{δ}	JD_{mean} [b]	Vernal 1970[c]
21 Mar 1958	2436283.6302	+3"9	2436285.5221	8.4279
23 Sep 1958	36470.0481	+2.8	36468.1415	8.4291
21 Mar 1959	36648.8751	+1.8	36650.7680	8.4317
23 Sep 1959	36835.2981	+0.4	36833.3899	8.4351
20 Mar 1960	37014.1122	-1.5	37016.0026	8.4241
23 Sep 1960	37200.5428	-1.8	37198.6332	8.4360
20 Mar 1961	37379.3552	-2.9	37381.2445	8.4239
23 Sep 1962	37931.0240	-5.6	37929.1121	8.4301
20 Mar 1964	38475.0906	-6.7	38476.9767	8.4296
23 Sep 1964	38661.5118	-6.9	38659.5992	8.4324
23 Sep 1965	39026.7562	-6.3	39024.8442	8.4350
23 Sep 1966	39391.9890	-5.0	39390.0781	8.4265
21 Mar 1967	39570.8162	-4.1	39572.7035	8.4300
20 Mar 1968	39936.0567	-2.3	39937.9451	8.4294
21 Mar 1971	41031.7775	+4.3	41033.6700	8.4278
22 Sep 1972	41583.4416	+6.2	41581.5397	8.4337
20 Mar 1973	41762.2588	+6.6	41764.1526	8.4261

[a]Corrected for refraction and parallax.
[b]Corrected for nutation.
[c]Mean: 2440668.4299 ± 0.0009.

According to this fit the secular decrease of the length of the tropical year is given by

$$Y = 365^{d}05^{h}48^{m}48\overset{s}{.}29 - 0\overset{s}{.}87 \quad T \quad , \tag{19b}$$

where now T is reckoned in Julian centuries from 1900.0.

The length of the tropical year at the epochs AD 364, 1180, and 1730 may also be found directly from the time intervals between the equinoxes given by (18). These are given in Table 6, together with the lengths of the year following from the mean motion of the solar tables analyzed in Sect. 2. For convenience Table 6 also list some additional values taken from the literature, in particular from the tabulation of Houzeau (1882).

A plot of the length of the tropical year as a function of epoch is shown in Fig. 1, where full-drawn dots represent the data obtained from our equinox reduction (plus Newcomb's value), triangles represent the data obtained from longitude comparison (Sect. 2), and crosses represent the supplementary data from the literature. The straight line corresponds to a linear fit of the most reliable data, viz. the large dots:

Table 6. Length of the tropical year according to different authors

Author	Length
Hipparchus (-134)	$365^d05^h55^m02\overset{s}{.}07$
Ptolemy (139)	365 05 55 12.00
Liu Hung and Tshai Yung (206)	365 05 54 30.00
This work (364)	365 05 49 00.99
Chiang Chi (384)	365 05 55 26.80
Tsu Chhung-Chih (462)	365 05 49 39.60
Liu Chhuo (608)	365 05 52 14.39
I Hsing (724)	365 05 51 56.76
Al-Mamun (830)	365 05 46 49.00
Al-Battani (880)	365 05 46 30.00
Ibn Junis (1000)	365 05 48 59.09
This work (1180)	365 05 48 56.67
Alfonso X. Rey de Castilia (1250)	365 05 49 16.00
Kuo Shou-Ching (1278)	365 05 49 12.00
E. Reinhold (1550)	365 05 49 16.39
D. Santbech (1560)	365 05 48 41.71
M. Mästlin (1570)	365 05 49 15.77
J. Kepler (1588)	365 05 48 57.69
T. Brahe (1590)	365 05 48 45.00
P. van Lansberg (1600)	365 05 48 57.04
G. Wendelin (1610)	365 05 49 05.45
C. Severin (1620)	365 05 48 55.00
I. Boulliaud (1640)	365 05 49 04.35
G.B. Riccioli and F.M. Grimaldi (1650)	365 05 48 48.00
P.G. de Lahire (1686)	365 05 48 50.39
J. Flamsteed (1700)	365 05 48 57.50
E. Halley (1719)	365 05 48 54.80
This work (1730)	365 05 48 47.52
J. Cassini (1740)	365 05 48 52.44
N.L. de Lacaille (1749)	365 05 48 46.00
N.L. de Lacaille (1750)	365 05 48 40.00
J.T. Mayer (1760)	365 05 48 42.35
J.J.F. de Lalande (1770)	365 05 48 49.02
J.C. Burckhardt (1800)	365 05 48 49.70
P.S. de Laplace (1800)	365 05 48 50.83
F.X. von Zach (1800)	365 05 48 50.88
G. Piazzi (1800)	365 05 48 49.84
J.B.J. Delambre (1800)	365 05 48 51.61
F. Carlini (1810)	365 05 48 51.39
F.W. Bessel (1830.0)	365 05 48 47.63
U.J.J. Leverrier (1850.0)	365 05 48 46.05
P.A. Hansen and C.F.R. Olufsen (1850.0)	365 05 48 45.46
S. Newcomb (1900.0)	365 05 48 45.97
This work (numerical fit, 1900.0)	365 05 48 46.67

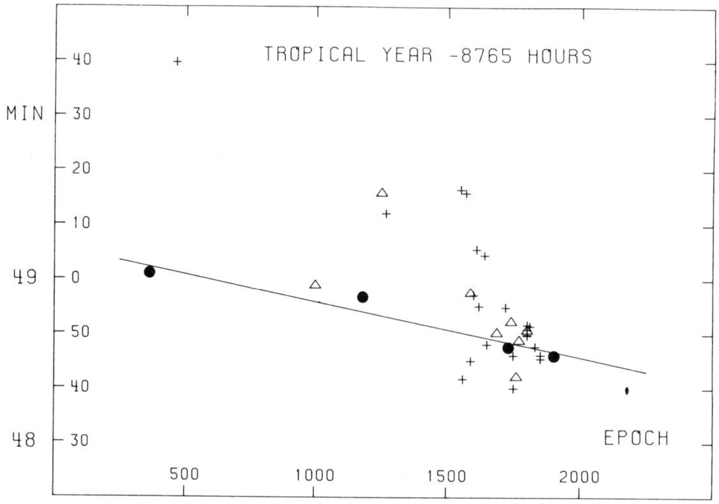

Fig. 1. Length of the tropical year as a function of time. A linear fit to the most reliable data (black dots) is shown by the straight line, the slope of which corresponds to dY/dT = -1.0 s/cy. For details see text

$$Y = (365^{d}05^{h}48^{m}46^{s}.67 \pm 1^{s}.74) - (1^{s}.01 \pm 0^{s}.19) \cdot T \quad . \tag{19c}$$

Note that the latter fit includes Newcomb's value. From (19b) and (19c) we have

$$\dot{Y} = -(0.94 \pm 0.07) \text{ s/cy} \quad . \tag{20}$$

By virtue of (13) we finally get from motion comparison:

$$\dot{n}_{MC} = (1.9 \pm 0.2) \text{ "/cy}^2 \tag{21a}$$

or

$$\dot{n}_{tidal} = 0.8 \text{ "/cy}^2 \quad . \tag{21b}$$

4. Conclusions

Positional observations of the Sun covering a sufficiently large time span may be used to determine the tidal deceleration of the Earth's rotation. Two largely independent methods may be applied, viz. longitude comparison, and mean motion comparison (Goldstein 1981).

The solar tables constructed by Ibn Junis (1008), Alphonsus (1250), Kepler (1627), Lahire (1687), Cassini (1740), Lacaille (1750), Mayer

(1767), Lalande (1771), and von Zach (1804) constitute a numerical fit to a large number of observations, they may thus be compared with a more recent theory of the Sun in order to obtain O-C residuals for the mean longitude and the mean motion. An alternative approach - equivalent to the latter method - is to study the decrease of the length of the tropical year. In the present paper a total of 10 old solar theories (tables) were re-condensed into convenient numerical expressions (Table 1) and compared with Newcomb's (1895) theory. The resulting acceleration coefficient of the mean longitude of the Sun is $\dot{n} = 2.6$ "/cy^2, of which about 1.1 "/cy^2 are due to planetary perturbations of the Earth's orbit, and about 1.5 "/cy^2 are due to spin deceleration. The latter value corresponds to $\dot{\omega}/\omega_0 = -\ 2.31 \times 10^{-8}\ cy^{-1}$.

The length of the tropical year as a function of epoch has been derived not only from the solar tables, but also from an entirely new analysis of equinox observations made by Hipparchus (-161 to -127), Arab astronomers (829 to 1019), Bernhard Walther (1475 to 1504), and by the staff of the Royal Greenwich Observatory (1958 to 1973). The decrease of the length of the tropical year thus found is $dY/dT = -0.94$ s/cy (or $\dot{n} = 1.9$ "/cy^2), of which about -0.54 s/cy are due to planetary perturbations, and about -0.40 s/cy (or 0.8 "/cy^2) are due to spin deceleration. The latter value corresponds to $\dot{\omega}/\omega_0 = -1.23 \times 10^{-8}\ cy^{-1}$.

In conclusion, the net acceleration of the mean Sun, which may be due to tidal deceleration of the Earth's rotation, is found to be

$$\dot{n}_{tidal} = (1.2 \pm 0.3)\ ''/cy^2 \quad ,$$

which is roughly half the value found by Goldstein (1981). Formally, the coefficient of the quadratic term in Newcomb's expression for the mean longitude of the Sun requires a correction of $+\dot{n}_{tidal}$ in order to reconcile his theory with the observations discussed above. With $\dot{\omega}/\omega_0 \simeq -1.8 \times 10^{-8}\ cy^{-1}$ tidal friction alone would lead to an accumulated contribution to ΔT = ET - UT of 28 s in 100 years, 1 h in 1140 years, and half a day (the most inconvenient case) in about 3950 years.

Acknowledgements. It is gratefully acknowledged that this investigation has been stimulated by a preprint version of the paper by Samuel J. Goldstein Jr. (1981), of which a brief account was given at the Bielefeld meeting. Thanks are also due to Drs. Peter Brosche, François Mignard, and Peter Stumpff for very helpful comments. The computations were made with the Univac 1100/82 of the Gesellschaft für wissenschaftliche Datenverarbeitung Göttingen.

References

Buontempo ME, Carey JV, Eldridge P (1973) R Obs Bull 178: R.G.O. Herstmonceux

Blackwell KC, Buontempo ME, Eldridge P, Swifte RHD (1975) R Obs Bull 180: R.G.O. Herstmonceux

Cassini J (1740) Tables Astronomiques. Imprim Royale, Paris

Caussin de Perceval JJA (1804) Le livre de la Grande Table Hakémite, etc. Not Extr Manuscr 7:16-240

Curott DR (1966) Astron J 71:264

Fotheringham JK (1918) Mon Not R Astron Soc 78:406

Fotheringham JK (1920) Mon Not R Astron Soc 81:104

Forbes EG (1971) The Euler-Mayer correspondence (1751-1755). Macmillan, London

Goldstein SJ (1981) Cel Mech (in press)

Hammer F (ed) (1969) J Kepler Gesammelte Werke, Vol X. CH Beck, München

Houzeau JC (1882) Vade-mecum de l'astronome, F Hayez, Bruxelles

Lacaille de NL (1753) His Mem Acad R Sci Paris 1749:40

Lacaille de NL (1754) His Mem Acad R Sci Paris 1750:166 (cf. also p.11)

Lahire de PG (1687) Tabularum Astronomicarum pars prior etc. S Michallet, Paris

Lalande de JJF (1771) Astronomie, Tome 1, Appendix. Desaint, Paris

Manitius K (1963) Des Claudius Ptolemäus Handbuch der Astronomie (Almagest). BG Teubner, Leipzig

Mayer JT (1767) In: Maskelyne N (ed) New and correct tables of the motions of the sun and moon, by Tobias Mayer, to which is added the method of finding the longitude improved by the same author. Nourse, Mount Page, London

Newcomb S (1895) The elements of the four inner planets and the fundamental constants of astronomy. US Gov Print Off, Washington

Newton RR (1970) Ancient astronomical observations and the accelerations of the Earth and Moon. Johns Hopkins, Baltimore

Newton RR (1972) Mem R Astron Soc 76:99

Newton RR (1977) The crime of Claudius Ptolemy. Johns Hopkins, Baltimore

Riccioli GB (1665) Astronomia Reformata tomi duo etc. V Benati, Bologna

Santritter JL (ed) (1492) Tabule Tabularum Celestium Motuum etc. J Hamann, Venezia

Stephenson FR (1978) In: Brosche P, Sündermann J (eds) Tidal friction and the Earth's rotation. Springer Berlin Heidelberg New York, p5

Wittmann A (1982) The obliquity of the ecliptic. II. (in prep); for Part I cf. Astron Astrophys 73:129 (1979)

Zach von FX (1804) Tabulae Novae et Correctae Motuum Solis etc. Suppl Tabulas Motuum Solis anno 1792 editas. Becker, Gotha

Long Time Integration of the Moon's Orbit

F. Mignard[1]

1. Introduction

The problem dealt with in this chapter is not really a new one. Basically the question of the past orbit of the Moon arose from the observation of a secular acceleration in the Moon's mean longitude discovered more than two centuries ago. The presence of a quadratic term in the Moon's longitude implied a secular variation in the Moon's semi-major axis and then a past history of the orbit of the Moon.

Subsequently the observed secular acceleration has been split into a gravitational part (see Appendix I) and a non-gravitational, usually called the tidal acceleration. Today the connection between the secular acceleration and the tidal dissipation is taken for granted for many reasons.

First, the ceaseless motion of the ocean was obviously observed by populations living on the seashore, and the relation between the recurrent motion of the sea and the phases of the Moon was recognized as early as Pytheas of Marseille's travels in the North sea. However, it was not until the mid-eighteenth century that the dissipation of energy in the tides was invoked by Kant. Later Darwin carried out the first modern investigation of the history of the Moon's orbit but his results received little attention, with the notable exception of Poincaré, who gave large excerpts in his Leçons sur les Hypothèses Cosmogoniques (1913).

There is, however, other compelling evidence in the Solar System of the dominant role played by the tidal process in the gradual evolution of the planetary system and its satellites. Tides determine the rotational history of the natural satellites (Peale 1973, 1977), and cause many satellites to be locked in a synchronous rotation, in keeping with the involved timescales in such processes. Tidal process is responsible for the setting up of many of the orbital resonances observed through the solar system (Greenberg 1973, Goldreich 1965, Yoder 1979). The very regular property of the Galilean satellites

---

1 Cerga, Av. Copernic, F-06130 Grasse, France

Tidal Friction and the Earth's Rotation II
ed. by Brosche/Sündermann

and inner Saturnian satellites, which all share a very small eccentricity and lie nearly in the planet's equator, is undoubtedly the result of a tidal evolution which has taken place for several billions of years. In addition the moon of Mars, Phobos, is still evolving and there is observational evidence of a secular acceleration (Sinclair 1978) which amounts to a surprisingly large value of $13^{o}/cy^{2}$, corresponding to a decay of the semi-major axis. To date tidal dissipation seems to be the best candidate to account for this acceleration and the problem has received great attention in the last decade (Burns 1972, 1978, Lambeck 1979, Cazenave et al. 1980, Mignard 1981a,b).

This brief list makes clear that tidal dissipation is far from being an exception within the solar system. Both the observed secular acceleration and the slowing down of the Earth's rotation bear witness to the energy dissipation still being at work .. at home. Consequently it is tempting to take advantage of this fact to seek what the Moon orbit looked like some millions or billions of years ago by reconstructing the tidal dissipation process and thereby the former states of the Earth-Moon system. To be successful in this study we need to figure out how efficient the dissipation is today, in other words a reliable value of the non-gravitational acceleration is required. Various methods have been used to determine $\dot{n}$[1] and recent reviews have appeared by Stephenson (1978), Morrison (1978) and Cazenave (this Volume). Certainly the value of $\dot{n}$ has been erratic for the last two decades, but now seems to be becoming more convergent, and we can conclude with some confidence that the secular acceleration lies in a range from $-20''/cy^2$ to $-30''/cy^2$. The most recent satellite solution by Cazenave and al. (1981) gives $\dot{n} = -21''.3 \pm 1''.25/cy^2$, while the determination based on the record of ancient eclipses leads to $\dot{n} = -30'' \pm 3''/cy^2$ (Stephenson 1978). In addition, if we look at results published around 1970 we find values as large as $-65''/cy^2$ (Van Flandern 1975) or $\dot{n} = -42''/cy^2$ (Newton 1970), but the most recent estimation of Van Flandern (1981), based on extended data, is $\dot{n} = -21''.4 \pm 2''.6/cy^2$. However, it is not possible to be certain that some unmodelled systematic errors do not pervade the present solutions, as all the mean errors quoted come from a standard least-squares estimation and are unfortunately only internal errors reflecting the data dispersion rather than a real accuracy.

---

1 On a quantitative basis this acceleration is well represented as the second derivative with respect to the time of the Moon's mean longitude, namely

$$\dot{n} = \frac{d^2L}{dt^2} .$$

Half of this quantity was formerly used.

2. Physical Processes: A Qualitative Analysis

All these studies of orbital evolution mentioned deal with a number of cumbersome equations that can hide the mechanisms at work. I confine myself to situations that have probably occurred in the history of the Earth-Moon system.

Basically the mechanism which brings about the evolution of the Moon's orbit at present is illustrated in Fig.1. The tidal force distorts the Earth's surface and the equilibrium shape is reached after a certain time Δt. During this time delay the Earth's rotation carries the tidal bulge ahead of the Earth-Moon line while the Moon keeps on moving on its orbit. First assume the Moon's orbit to be circular.

The tangential force F causes the satellite to go farther and farther from the planet and in a circular orbit there is no way to break up the symmetry and the orbit remains circular. However, the tidal force is strongly dependent on the distance between the bodies. The planet's distortion appears as the result of the differential forces between its center and its surface caused by the satellite. Hence the magnitude of this distortion varies as the inverse third power of the distance. On the other hand, a slightly flattened planet gives rise to a field of force decreasing as the fourth power of the distance, superimposed on the usual Newtonian component. The combination of both arguments yields a tidal force that varies as the inverse seventh power of separation of the planet and its satellite. Thus, as soon as the eccentricity becomes appreciable, the major effect is undergone in the vicinity of the periapsis. Let us assume at this point $\omega > df/dt$, where ω is the planet's speed of rotation and f the true anomaly of the satellite. In such a case the satellite receives a kind of impulse and the pericenter distance remains fairly constant while the semi-major axis increases, then we have:

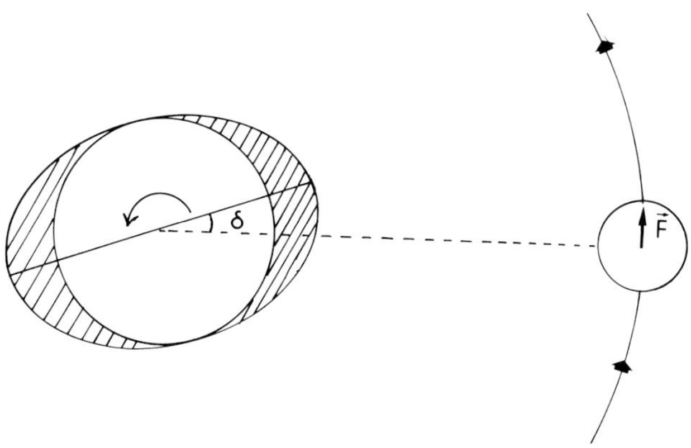

Fig. 1. The tidal bulge due to the time delay. In this situation the planet's speed of rotation is larger than the instantaneous angular velocity of the tide-rising object

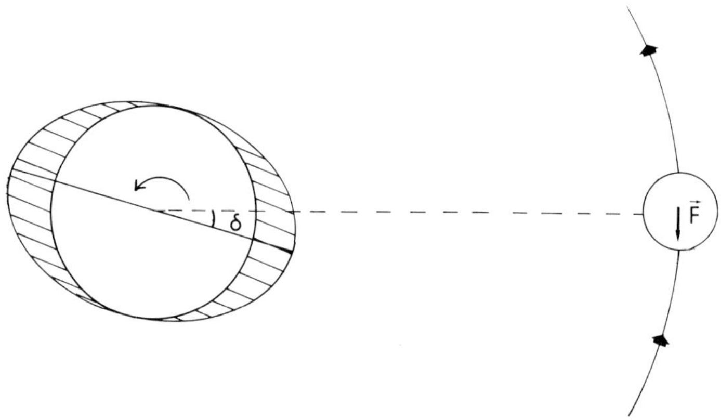

Fig. 2. The tidal bulge due to the time delay. In this situation the planet's speed of rotation is less than the instantaneous angular velocity of the tide-rising object

$$\frac{da}{dt} > 0 \quad \text{and} \quad \frac{d[a(1-e)]}{dt} = 0 \quad ,$$

which implies an increase of the eccentricity. That is the present situation of the Moon's orbit.

Let us consider now an orbit having a very large eccentricity, so that $\omega < df/dt$ at the periapsis. The diagram of this situation is depicted in Fig. 2. Now the tidal bulge is behind the center line and the interacting resulting force looks like a force experienced in a resisting medium. Therefore the orbits shrinks with an invariable pericenter distance and becomes less elliptical. This situation might have occurred in the distant past and this physical explanation will help the reader to accept some not readily interpretable results of the numerical integration that will be presented below.

So far the tide-rising object has been considered as a point mass without any dissipation in it. It is sometimes said that owing to its synchronous rotation, the tidal bulge of the Moon is nothing but a permanent deformation towards the Earth and that no dissipation is taking place. That would be true for a perfectly circular orbit but this result is no longer valid for an elliptical path. The exact synchronism means that the average orbital motion of the Moon is equal to its rotation rate. But the instantaneous orbital angular velocity is more or less than the mean motion according to the longitude. In particular at the pericenter $df/dt > n$ and the Moon's tidal bulge becomes misaligned with the Earth-Moon line. In addition the Earth-Moon distance is continuously changing while the Moon proceeds on its orbit, and so is the magnitude of the Moon's deformation. Accordingly both the radial and eccentricity tides lead to a dissipation of energy within the Moon.

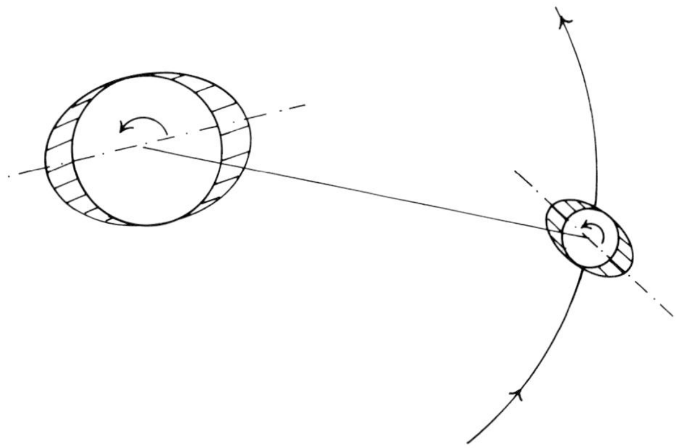

Fig. 3. Location of the Earth's and Moon's tidal bulge in the vicinity of the pericenter. The torque by the Moon tends to slow down the Earth's rotation while the Earth's bulge acts to dilate the Moon's orbit. The torque by the Earth to the Moon tends to shrink the orbit. The maintenance of the synchronism involves an offset of the Moon permanent large axis

However, this description should be modified to allow for the permanent deformation of the Moon. On the average the largest axis of the Moon points towards the Earth but with a slight offset. Inasmuch as the synchronism is maintained the geometrical configuration of the Moon's ellipsoid with respect to the Earth undergoes minor change during the orbital motion. Hence there is a gravitational torque superimposed to the tidal torque. As a rule the result is to lessen the tidal torque and allow the synchronism to be maintained as the semi-major axis of the orbit evolves (Yoder 1982).

What consequences can be drawn from this fact? In Fig. 3 the Moon is pictured close to its pericenter. The Earth's tidal bulge is ahead of the Earth-Moon line ($\omega > df/dt$), while the Moon needs to rotate more about its axis of rotation to reach this line and the Moon's bulge is behind. As in Fig. 1, the Earth's deformation causes the orbit to expend and the eccentricity to increase. The reverse situation arises from the Moon's bulge which tends to shrink the orbit and to make it more circular. Obviously the magnitude of these two effects depends on the relative dissipation in the two bodies and no general conclusion can be drawn on the issue of the competition. All that can be said with confidence is that the angular momentum of the whole system is conserved during the evolution.

At present the Earth's dissipation effect dominates the Moon's and our satellite is receding as its eccentricity increases. But numerous examples in the solar system show that evolved satellites have a tendency to move in a circular orbit. With a simultaneous dissipation within the satellite and the planet it can be shown that the rate of change of the eccentricity of the Galilean satellites is negative and their orbit is observed to tend to a final circular orbit. As soon as the satellite orbit is eccentric it is not legitimate to disregard tidal dissipation within the satellite. For the Moon orbit this effect was predominant in the distant past when the Moon orbited the Earth on a very elongated ellipse. The discovery of volcanoes on Io brought the proof of the efficiency of such a dissipation and what was once an hypothesis is now an observational fact.

Even with the restrictive case $\omega/n > 1$, which excludes the evolution of Phobos or Triton, a great variety of evolutionary tracks have to be considered according to the relative rate of dissipation within the planet and the satellite. Consequently this parameter is going to play a major role in the history of the Moon's orbit.

Comments. The knowledge of the Earth deformation in response to the tidal forces is the keystone of the study of the past orbit of the Moon. The equilibrium distribution of matter within the Earth determines the additional potential acting upon the Moon and which rules its motion. The solution adopted, that we call the standard solution, is the simplest one, and possesses the apppreciable virtue of yielding handlable equations.

The response of a differentiated body like the Earth to the tidal stress is likely more complex than a mere time lag. Kaula (1964) has expanded the Earth's tidal potential in Fourier series which allows him to introduce different phase lags, one for each component. In such a model these unmodelled phases remain unknown and cannot be determined on the basis of a unique observational constant: the secular acceleration. Notwithstanding the fact that this model can closely represent the reality, it is too sophisticated in regard to our current knowledge of the planetary structure and our ability to infer the frequency response to the tidal forces. Kaula was well aware of that, and in his investigation the history of the lunar orbit all the phase lags were kept equal.

The idea of representing the whole dissipation of the Earth by a time lag Δt dates back to Darwin (1880) in his model of a viscous fluid Earth. Given this assumption, Darwin derived a relation tying the relaxation time of the viscous Earth to the angular delay of the bulge of any tidal component.

$$\tan\varepsilon = \nu \cdot \Delta t \quad ,$$

where ν is the frequency of the tidal component and Δt the delay given by $\Delta t = 19\mu/2gR$, where μ denotes the viscosity coefficient g the gravitational acceleration at the planet's surface and R its radius. Only for $\mu < 10^{15}$ Poises is the phase lag small enough to obtain $\tan\varepsilon \simeq \varepsilon$. However, the Earth's mantle viscosity is of the order of 10^{21} Poises (McConnell 1968). In any case, the mantle viscosity cannot be smaller than 10^{17} Poises, a value reached near the ridges and the subduction zones (Vetter and Meissner 1977). The viscosity of the core is everywhere negligible (Gans 1972). The Darwin model does not apply to the Earth but is extensively used to study the tidal evolution of binary stars.

From this short description it turns out that the time-lag model deserves no attention. In fact, however, it is probably better than it seems at first. The Earth is not fluid and if the mantle is plastic it remains a solid with a rigidity $\chi = 1.7 \times 10^{11}$ N m^{-2}. An elastic solid which experiences a periodical driving force vibrates with the same fundamental frequency and shows a phase lag. Outside the resonance zones the involved time lag is independent of the frequency. In short, for the whole Earth the process to reach the equilibrium shape is assimilated to a relaxation with a characteristic damping time Δt.

The tides in the oceans are more complex, but at present the dissipation occurs primarily in the M2 tide, whose equilibrium shape is well represented by an ellipsoid. No longer is this approximation valid in the distant past when the Moon travelled on a fairly elliptical orbit and excited other partial tides much more strongly than today. But the ocean configuration and dissipation was less important at that time. This problem will be discussed later in connection with the time scale problem.

2.1 Forces

Let us proceed now to a more technical matter in this investigation. With the assumptions made in the preceding section concerning the tidal model, the force between a distorted Earth and the Moon takes the following form,

$$\vec{F} = -3Gm^2 k_2 \Delta t R_e^5 \frac{1}{r^{10}} [2(\vec{r}\vec{v})\vec{r} + r^2(\vec{v} + \vec{r}x\vec{\omega})] \quad (1)$$

while the torque comes from $\vec{T} = \vec{r} \times \vec{F}$,

$$\vec{T} = -3Gm^2 k_2 \Delta t R_e^5 \frac{1}{r^8} [(\vec{r}\vec{\omega})\vec{r} - r^2\vec{\omega} + \vec{r}x\vec{v}] \quad . \quad (2)$$

In these two equations the expression of the tidal interaction is restricted to the second harmonic. A more complete expression is given in Mignard (1980). k_2 is the second-order Love number, Δt the time delay, R_e the Earth's radius and m the Moon's mass. The radius vector joining the Earth to the Moon is $\vec{r}$ and $\vec{v}$ is the orbital velocity, whereas $\vec{\omega}$ denotes the rotation vector of the Earth.

Both the force and the torque vanish with ω-n. For example consider a circular and equatorial orbit. In (2) we get the simplifications,

$\vec{r}\vec{\omega} = 0$, $r^2\vec{\omega} - \vec{r}x\vec{v} = a^2(\omega - n)$, and the classical expression of the torque is recovered. This torque disappears at the synchronous state.

The same (1) and (2) can be used in case of a tidal deformation of the Moon, provided we adapt the meaning of the physical parameters. Equation (1) deserves a little more attention when the Moon is considered. When viewed from a Moon point of view the Earth orbits almost in the Moon's equator, the inclination being less than 7°, and the rotational velocity of the Moon is equal to the Earth's mean motion; (1) becomes:

$$\vec{F} \propto 2(\vec{r}\vec{v})\vec{r} + r^2(\vec{v} + \vec{r}x\vec{n}) \quad . \tag{3}$$

Let us consider the two unit vectors $\vec{\alpha}$, $\vec{\beta}$ along $\vec{r}$ and perpendicular to $\vec{r}$ in the orbital plane. After some transformations we obtain,

$$\vec{F} \propto 3\dot{r}\vec{\alpha} - \left(n - \frac{df}{dt}\right)\vec{\beta} \tag{4}$$

where f denotes the true anomaly. Both components in (4) cancel for circular orbits, but both are simultaneously present for an elliptical one and are of the order of the eccentricity. In addition the azimuthal component acts as a retarding force as it could be expected after the qualitative discussion.

2.2 Equations of Evolution

The result of the tidal forces on the orbital elements is computed with the help of the Gauss' equations (Brouwer and Clemence 1961), for example,

$$\frac{da}{dt} = \frac{2}{n(1-e^2)^{\frac{1}{2}}}\left[Q\ e \sin f + (1 - e^2)\ \frac{a}{r}\ S\right] \tag{5}$$

where Q and S are respectively the radial and the azimuthal component of the forces.

Our interest in the history of the Moon orbit only lay in the secular trend of this evolution. Consequently the equations are to be averaged over the mean anomaly. Since the past eccentricity of the Moon may have been fairly large in the past, all the expressions involving the eccentricity must be obtained in a closed form. The detailed derivation has been given in various forms under different assumptions by Gerstenkorn 1969, Kaula 1964, Goldreich 1966, Mignard 1980 and Conway 1982.

Their results are equivalent as long as the dissipation within the Moon is disregarded. From Mignard (1980) the rate of change for the semi-major axis and the eccentricity is written as follows:

$$\frac{1}{a}\cdot\frac{da}{dt} = 2K\left[-(1+A)g_1 + \left(\frac{\omega}{n}\cos I + A\right)g_2\right] \tag{6}$$

$$\frac{de}{dt} = K\left[-(1+A)g_3 + \left(\frac{\omega}{n}\cos I + A\right)g_4\right] \; . \tag{7}$$

In (6) and (7) I stands for the inclination of the Moon's orbit to the Earth's equator.

$K = 12\pi^2 k_2 \Delta t \, \frac{m}{M} \frac{m}{\mu} \frac{1}{P^2} \frac{R_e^8}{a^8}$ where P is the period of an Earth-grazing satellite, and the g's are expressions involving the eccentricity. The parameter A expresses the ratio of the effect of the tides in the satellite and the planet:

$$A = \frac{k_2'}{k_2} \frac{\Delta t'}{\Delta t} \left[\frac{M}{m}\right]^2 \left[\frac{R'}{R_e}\right]^5 \; .$$

The value of A is unfortunately unknown but reasonably ranges between 0 and 20. The g's are as follows:

$$g_1 = \frac{1}{(1-e^2)^{15/2}} \left(1 + \frac{31}{2}e^2 + \frac{255}{8}e^4 + \frac{185}{16}e^6 + \frac{25}{64}e^8\right)$$

$$g_2 = \frac{1}{(1-e^2)^6} \left(1 + \frac{15}{2} + \frac{45}{8}e^2 + \frac{5}{16}e^6\right)$$

$$g_3 = \frac{1}{(1-e^2)^{13/2}} \left(9e + \frac{135}{4}e^3 + \frac{135}{8}e^5 + \frac{45}{64}e^7\right)$$

$$g_4 = \frac{1}{(1-e^2)^5} \left(\frac{11}{2}e + \frac{33}{4}e^3 + \frac{11}{16}e^5\right)$$

2.3 Numerical Values

Equation (6) combined with the observed secular acceleration allows us to obtain $k_2\Delta t$ for the Earth. At present the eccentricity of the Moon's orbit is too small to play a role in the receding of our satellite. Therefore we have,

$$\frac{1}{a}\cdot\frac{da}{dt} = 2K\left[\frac{\omega}{n} - 1\right]$$

and with $\dot{n} = -25''/\mathrm{cy}^2$ we get $k_2\Delta t = 2.8$ min. It must be kept in mind that this figure includes both solid Earth and oceanic dissipation. The value of k_2 for the Earth derived from the model is of the order of 0.3, whereas the observations of Earth's tide lead

$k_2 = 0.27$. Accordingly Δt is about 10 min for the global response of the Earth. Since the M_2 tide dominates the dissipation we can infer a value for the Earth's global ;

$$Q = 1/\omega\Delta t = 20$$

while values around 100 (50 to 400) are reported from the Chandler wobble decay (Smith and Dahlen 1981).

The difference between the two determinations reflects the major contribution of the oceanic dissipation to the receding of the Moon (Lambeck 1975). The present distribution of the dissipation between the hydrosphere and the solid Earth is not known, but given the value of Q about 5% to 15% of the observed secular acceleration can be ascribed to the solid Earth's friction, yielding a time lag of 1 min.

The parameters related to the Moon are still more uncertain in spite of recent determinations derived from lunar laser data (Cappallo et al. 1981). For a constant time lag model we have $k_2' = 0.024 \pm 0.006$. The surprising result, already anticipated by Ferrari et al. (1980) concerns the large dissipation in the Moon since $Q = 22 \pm 6$. However, this figure is model-dependent in that the Moon's rotation is modelled with a solid dissipation factor to reduce the observations. But other phenomena such as a core-mantle interaction, are likely to give in the rotation of the Moon the same signature as the tidal distortion. Such a study is under investigation (Yoder per. comm.) and seems able to account for most of the dissipation. Nonetheless we can compute an upper value of A by taking Q = 22 for the tidal dissipation. At present this hypothesis leads to A = 20 and if only the bodily tides are considered for the Earth we get A = 200! In the numerical investigation of the past history of the Moon we restricted ourselves to $0 < A < 20$. This extreme value is obtained by assuming a value of Q_{tidal} within the Moon comparable to the Earth's Q and by taking into account the tide due to the solid Earth, assumed to have been dominant in the distant past.

2.4 Precessional Equations

In (6) and (7) the orientation of the Moon's orbital plane appears through cos I. We have all learnt that the Moon's node is regressing on the ecliptic plane with a period of 18.6 years and that the Moon maintains a fairly constant inclination to the ecliptic of slightly more than 5°. As a consequence the inclination to the Earth's equator varies with the same period from 18° to 28° and cos I is no longer a constant over this span of time. Therefore the equations of evolution have to be averaged over the precessional period. This work is easy to carry out in the present situation owing to the following relationship,

$$\cos I = \cos\gamma \cos\varepsilon - \sin\gamma \sin\varepsilon \cos\Omega$$

where ε is the obliquity of the ecliptic and γ the inclination of the Moon's orbit to the ecliptic. The node's longitude Ω varies uniformly with the time. Hence the average of cos I is,

$$\cos I = \cos\gamma \cos\varepsilon \quad .$$

Hence γ is a well-suited angle to represent the orientation of the Moon's orbital plane and its tidal evolution can be obtained from the equation

$$\frac{d\gamma}{dt} = \frac{1}{(na(1-e^2)^{\frac{1}{2}}} \frac{r}{a} \cos(\bar{\omega} + f)W \tag{8}$$

which involves W the component of the tidal force normal to the orbital plane and $\bar{\omega}$ the longitude of the pericenter.

But is this solution valid throughout the history of the Moon's orbit? Unfortunately the answer is no.

The Moon is steadily precessing on the ecliptic because the solar distrubances are the major ones. The orbital plane of an Earth's artificial satellite keeps a constant inclination to the equator and not to the ecliptic and a steady precession of its node occurs on the same plane. This precession is well accounted for by introducing the forces due to the Earth's oblateness forces. It appears hence that the precessional plane changes gradually from the Earth's equator to the ecliptic as the semi-major axis increases. The transition zone occurs when both the oblateness and solar disturbances have the same magnitude. The critical semi-major axis a_c at this point is, (Goldreich 1966)

$$\frac{a_c}{R_e} = \left(\frac{2M}{M_\odot} J_2\right)^{1/5} \left(\frac{a_\odot}{R_e}\right)^{3/5} \tag{9}$$

for a circular orbit and becomes larger if eccentric orbits are allowed (Mignard 1981). In this equation R_e is the planet's radius and M and $M_\odot$ respectively denotes the Earth's and the Sun's mass. For the Earth-Moon system (8) yields a value of $a_c/R_e = 10$. An artificial satellite moving in the transition zone would precess on a plane lying between the equator and the ecliptic, called the Laplace plane for the name of the French geometrician who first pointed out that property. An illustration of the three situations is diagrammed in Fig. 4. In (9) $a_\odot$ is the Sun-planet distance.

This representation proves to be good for an artificial satellite, or more precisely each time the orbital angular momentum of the satellite is less than the planet's rotational angular momentum. At present the orbital angular momentum of the Moon is five times that

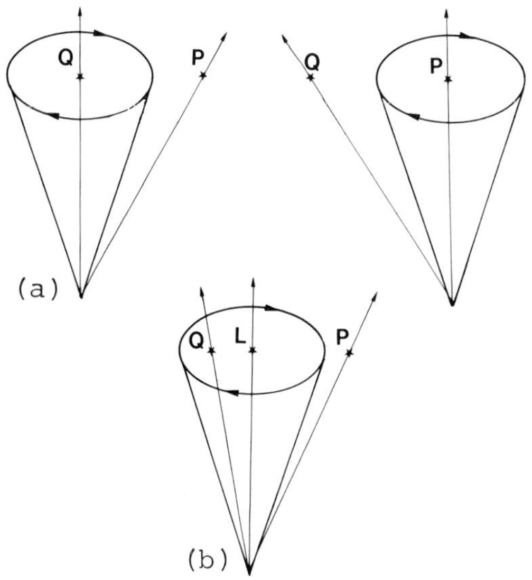

Fig. 4a,b. The precessional motion according to the semi-major axis. (a) A distant satellite precesses about the normal of the planet orbital plane whose pole is Q. (b) The node of a close satellite moves on the planet's equatorial plane whose pole is P. For intermediate semi-major axis the satellite's orbit precesses on the Laplace plane

of the Earth and at 20 R_e both are equal, just at the transition zone. Let us disregard the Sun for a moment and assume the Moon to orbit the Earth at 10 Earth radii. The oblateness of the Earth drives the Moon's node along the equator but as a response, the Moon is massive enough to act upon the equatorial bulge and causes the Earth's spin axis to precess. In this process the total angular momentum remains invariable:

$$\vec{L} = \vec{L}_M + \vec{L}_E$$

where the $\vec{L}$'s are respectively the total, Moon's and Earth's angular momentum. These three vectors are coplanar and the inclination between each pair is constant (Fig. 5). The Laplace plane is no longer the Earth's equator but the plane perpendicular to the total angular momentum. Therefore, as the Moon approaches the Earth, the precessional plane changes gradually from the ecliptic to this maximum area plane. This evolution can be more easily understood by looking at Fig. 4. From a more quantitative point of view, the general equations of precession are derived in Appendix II. As a consequence of this discussion average value of cos I proves to be less simple to calculate than we would like it to be. In fact no general solution of this problem exists so far and in the past different procedures have been used to overcome this difficulty.

Goldreich (1966) made a numerical averaging of the equations of evolution over a precessional period, this number being obtained by a numerical integration of the precessional equations. In principle the method is well suited, but his investigation was limited to circular orbit and is not likely to be extanded to elliptical paths. Consequently the history of the Moon orbit is restricted in the range

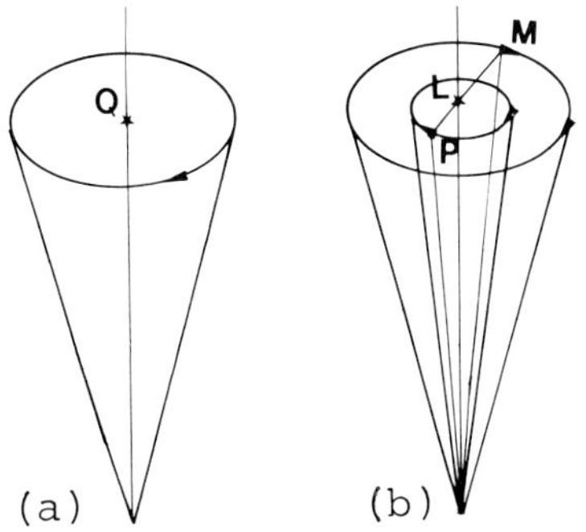

Fig. 5a,b. The current precession of the Moon's orbit occurs on the ecliptic in 18.6 years (a). When the Moon was at 20 Earth radii both the Earth's equator and the orbit of the Moon precessed on a plane normal to the total angular momentum of the Earth-Moon system (b)

from 10 R to 60 R. Rubincam (1975) is limited to a more qualitative discussion about the possibility of originating the Moon in the Earth's equatorial plane and he restricted himself to the geometrical interpretation shown in Figs. 4 and 5. The starting point of his reasoning enlightens the problem and it is worth paying some attention to it. Let us assume the Moon to be in the equatorial plane at 10 R. The three angular momenta, $\vec{L}$, $\vec{L}_M$, $\vec{L}_E$, are aligned and the Laplace plane is the Earth's equator. As the Moon recedes, the Laplace plane leaves its initial position to tilt towards the ecliptic. But the inclination of a precessing satellite to the Laplace plane is an adiabatic invariant and keeps a constant value, here 0°. The orbital plane of the Moon then, will move towards the ecliptic even if the tidal forces lie in the orbital plane. This point makes clear that a good solution of the precessional equations is required not only to obtain a good average value of cos I but to take account properly of the gravitational exchange of angular momentum between the Earth, the Moon and the Sun. In this solution, investigated by Rubincam, there is no way to explain the present inclination to the ecliptic of 5°. On the other hand, the tilting of the orbital plane of a satellite caused by a moving Laplace plane proved to be efficient in the history of Phobos (Cazenave et al. 1980, Mignard 1981b).

For the Moon Mignard (1981a) considered the two degenerate cases, one when the Laplace plane does not differ from the ecliptic between 15 R to 60 R, and the other at the close approach while the two-body approximation can be used. The two different solutions are joined at 10 R. More recently, Conway (1982) has come up with an analytical solution of the linearized precessional equations, an acceptable approximation in the transition zone between 30 R and 10 R. The close approach, when the inclinations may be large, is conducted with the two-body approximation.

2.5 History of the Orbital Elements

From (6), (7), and (8), combined with a solution of the precessional equations and with the loss of the angular momentum of the Earth-Moon system due to the Sun, we can obtain a complete set of five differential equations which governs the evolution of the Moon's orbit and likewise that of the Earth's rotation. The five parameters which describe the state of the system are respectively the semi-major axis, the eccentricity, an angle that gives the orientation of the Moon's orbital plane, the obliquity of the ecliptic and the Earth's rate of rotation.

The numerical integration does not give rise to major difficulty provided that an independent variable is chosen to eliminate the effect of $1/a^6$ in the rate of change of the elements.

All the results are plotted in terms of the semi major axis to separate the question of the timescale from the evolutionary pattern. The most striking feature of the result is seen in Fig. 6. In the recent past the eccentricity was less than today while it was very large at the time of the closest approach. This startling result was first pointed out by Gerstenkorn (1955) and afterwards confirmed by MacDonald (1964), Mignard (1980). Leaving out the time scale, such a result favors an origin of the Moon by capture of a body approaching the Earth with a hyperbolic orbit. In the same figure the comparison of the curves corresponding to different values of the dissipation within the Moon shows that the closest approach is affected by this parameter. The minimal distance remains of the order of 3 R in case of weak and intermediate dissipation when $A < 3$ but increases rapidly with A. If we take $A = 20$ at the time of the closest approach a minimal distance of 10 R is obtained well outside the Roche limit. The rate of variation of the eccentricity with $A = 0$ would be today twice as much as with $A = 20$ but is still too small to be detected in the observations. Unfortunately, the dissipation within the Moon is of little importance in the current receding and cannot be inferred from the observed secular acceleration.

The past history of the Moon's inclination to the ecliptic is given in Fig. 8 and does not show strong variation for $A = 0$. For $A = 20$ the large value of this inclination might be caused by the assumption that the Moon's equator laid in the ecliptic. More interesting is the history of the inclination of the orbit of the Moon to the Earth's equator. Obviously in Fig. 7 there are three stages in this history. For a/R between 10 and 60 the inclination is fairly stable with a minimum of about 16° whatever the dissipation within the Moon. Since during this stage, the Moon precesses mainly on the ecliptic, the inclination to the equator is a mean value over the precessional period. In the vicinity of the closest approach the Moon orbital plane exhibits large changes in its orientation. (The Earth's equator keeps a rather constant orientation with respect to the ecliptic; this fact

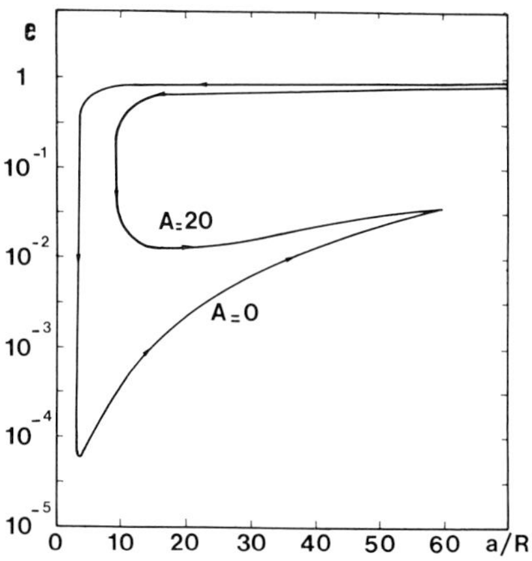

Fig. 6. History of the eccentricity of the orbit of the Moon in term of the semi-major axis. The parameter A is the ratio of the tidal effects of the Moon to that of the Earth

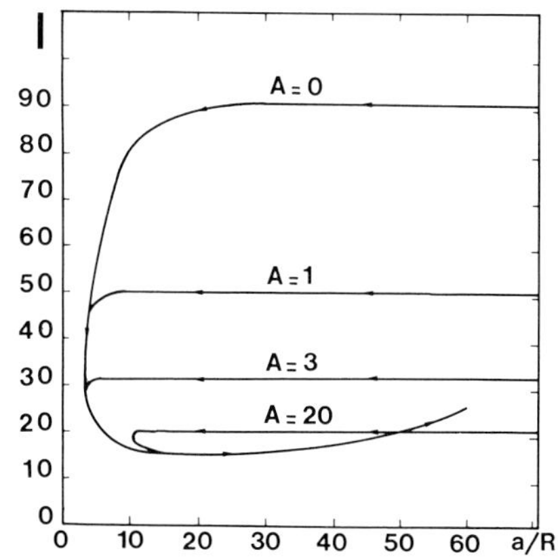

Fig. 7. History of the inclination of the Moon's orbit to the Earth's equator in term of the semi-major axis. The parameter A is the ratio of the tidal effects of the Moon to that of the Earth

is related to the fact that at 10 Earth's radii the Moon's angular momentum is half the rotational angular momentum of the Earth). The amplitude of variation is strongly dependent on the value of A. If the dissipation within the Moon is disregarded, the Moon is found on a polar orbit just before the closest approach, which seems difficult to reconcile with any of the possible origins. The value of the maximum inclination lessens sharply with A and becomes 50° for A = 1, 30° for A = 3 and 20° for A = 20. Once again the relative dissipation in the Moon proves to be a critical parameter in the building up of the past history of the Moon. The obliquity of the ecliptic and the length of the day are depicted in Figs.8 and 9. An initial period of rotation for the Earth of about 4 to 6 h is in keeping with the results obtained from the planetary formation by accretion. Some of the paleodata concern the day-duration via the number of days in a tropical year. To estimate what kind of constraint can be really brought by these investigation we assume that the Moon approached the Earth to within 10 Earth radii 4 billions years ago. Therefore the distance was 55 R 2 billion years ago, 50 R gives 3 billion years and 40 R occurred 3.85 billion years ago. In any case the paleostudies could produce valuable information only on the most quiescent part of Moon history. Even though these figures have been obtained with the assumption of a constant dissipation through the Moon's history, it becomes clear that we cannot rely on the bivalves or on the stromatolites to disclose the origin of the Moon. On the other hand, the strong dependence of the rate of change of the orbital elements with the semi-major axis makes the timescale completely controlled by the evolution which has occurred from 50 to 60 Earth radii. From the point of view of dynamics the past history of the Moon is divided

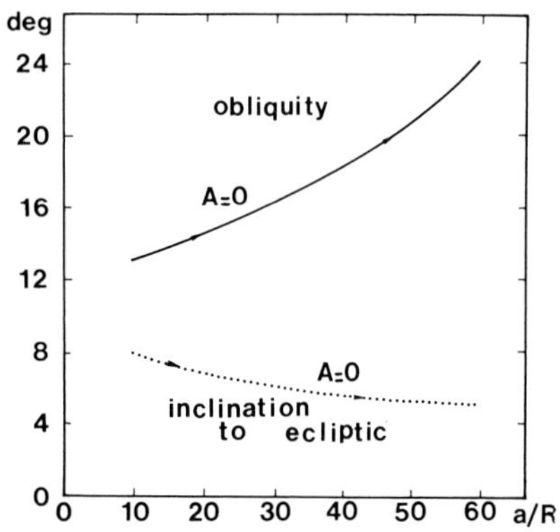

Fig. 8. Variation in obliquity of the Earth's equator (full line) and of the inclination of the Moon's orbit to the ecliptic (dotted line)

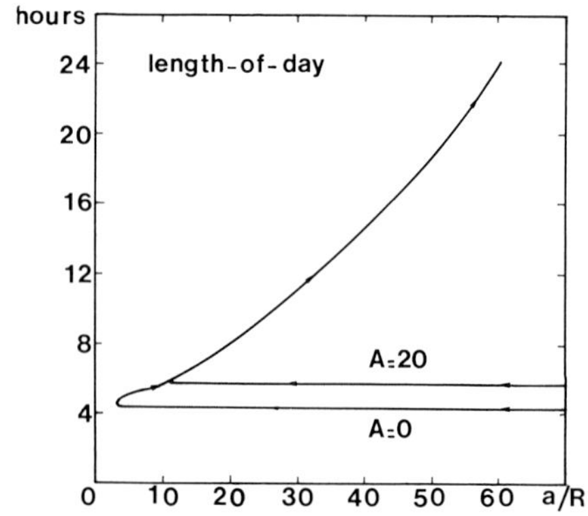

Fig. 9. History of the length of the day in term of the semi-major axis

into two sections: First, the recent past in term of the semi-major axis, during which the tidal model is of little importance for the history of the inclinations or the eccentricity but in return we have a poor idea of the time scale; the second period concerns the closest approach where the evolution is rapid and is affected by the details of the model of interaction that must be cautiously studied.

The Time Scale Problem. So far I have avoided talking about the time scale, even though several opportunities presented themselves in the preceding sections, but now I cannot escape! First let us give the crude result: If the observed secular acceleration, is assumed not to have changed in the past, the Moon was at only 5 R 1.3 billion years ago. Such a time scale has now been proved to be unrealistic. No scars of the important heating which should have occurred are visible either on the Earth or on the Moon. There is strong evidence that tidal bands in stromatolites exist in a 2 billion-year-old specimen (Panella 1975). Likewise the same material indicates, more qualitatively than quantitatively, that the slowing down of the Earth's rotation was an active phenomenon then, and was not as great as it is at present. Accordingly, the Moon has been orbiting the Earth for more than 1.3 billion years. In addition, the direct exploration of the Moon during the Apollo missions has firmly established that nothing but few recent craters has occurred on the Moon's surface for 3.2 billion years, the epoch of the last lava flow. It is likely that our satellite was already steadily receding at that time and had orbited the Earth for more than 1 billion years.

An obvious question arises: what is wrong in the computation of the time scale and how can it be corrected?

The solution to this problem is thought to be a reduced rate of dissipation of energy in the past, which is quite a reasonable assumption. The continental drift has been operating for more than 200 million years, yielding a modification in the repartition of the continents and in the ocean tides. Brosche and Sündermann (1978) modelled a Permian ocean and found a dissipation smaller at this epoch than the observed value. It must be borne in mind that the Moon was at about 58 R 200 million years ago and at the present rate of dissipation 50% of the time of evolution has elapsed while the Moon has moved from 55 R to 60 R, in a region where the model of dissipation (frequency-dependent, constant phase lag...) does not affect the evolution. So the time scale problem concerns mainly the last ten Earth radii in the Moon's journey.

Closer to the Earth we have no problem with the time scale but the model of interaction becomes less reliable. Then the present tide provides us with good indication of what could have happened in the last billion years; farther back the stromatolite data given above, though not conclusive, give a similar trend towards a smaller dissipation in the past than that of the current one. As we have already mentioned the major past of the dissipation takes place in the ocean and roughly 85% of the dissipation occurs in the sea. The Earth hydrosphere is a secondary process and, like the continental crust, has formed from the progressive differentiation of the mantle over geologic time. Hence there is a body of facts in favor of a variable dissipation with time and, even though no undisputable evidence has been put forward so far, the dissipation was probably smaller in the distant past. The coral studies have shown that the rate of dissipation has been fairly constant for the past 600 million years. If we accept this figure, a schematic story of the Moon orbit can be inferred. During the last 600 millions years the Moon has moved away from 55 R to 60 R. If the same dissipation had been kept before, another 700 million years would have been taken to evolve from 5 R to 55 R. But if at that time the dissipation was mainly due to the bodily tides, the time really required was 6 times larger, namely 4 billion years. Given the uncertainty of this estimation, we can locate the time of the closest approach between 3 and 4 billion years ago. This epoch coincides with the extrusion of the maria basalt. The heating source for the melting might have been not only the long-lived radioactive elements, but the tidal dissipation as well. The lack of evidence of young volcanic episodes proved that the heat engine stopped about 3 billion years ago. However, this scenario is not free of difficulties. The flooding of the basins requires only a partial melting during the close approach, since some rocks of the highlands are older than 4 billion years and the date of cristallization of the highland crust is now believed to be 4.45 aeons (S.R. Taylor 1979). As pointed out by Lambeck (1980), if the tidal dissipation is the source of the partial melting, the Moon must have been reasonably close to the Earth, but not too close to cause extended melting. In addition the lava flow lasted 0.7 b.y. (S.R. Taylor 1975) and the orbital evolution model does not allow the Moon to stay in a close approach for such an extended time. Alfven and Arrhenius (1972) called upon the passage of the Moon through resonant states with the Earth's rotation to maintain the Moon in the vicinity of the Earth. This possibility has been recently ruled out by Conway, who proved that there is no significant modification in the time scale in the crossing of resonances. In conclusion, even if we have sound reasons to accept a substantial reduction of the dissipation in the past, we are still lacking evidence of what the Moon's orbit looked like 3 or 4 billion years ago.

4. Conclusion

To conclude this paper I would like to stress desirable improvements that can be made. First the dissipation in the oceans is now being extensively studied with the help of artificial satellites (see for example Cazenave, this Vol.). Nevertheless, the ocean tide perturbations have not yet been applied to the history of the Lunar orbit. Lambeck (1975, 1980) outlined the procedure to be followed and showed the possibility of defining tidal parameters that reflect the total response of both the solid Earth and its hydrosphere. But since the main part of the dissipation takes place in the ocean, it would be valuable to confront the results of the integration presented here with a similar study based on the ocean tide potential alone.

The second point of interest deals with (4). The whole model is not really self-consistent, since the synchronous rotation of the Moon is imposed a priori while the azimuthal component of (4) causes the Moon to spin up. As pointed out by Yoder (1979), the permanent bulge of the Moon is displaced until the resulting torque allows the synchronism to be maintained. This feature is not included in any of the models presented here, and is likely to bring some modifications in case of large dissipations in the Moon and at the epoch when the eccentricity was significant.

Appendix I. The Present Status of the Gravitational Secular Acceleration

The first genuire demonstration of the existence of a secular acceleration in the Moon longitude was brought forward by R. Bunthorne in 1749, in a letter read before the Royal Society. Having compared the observed time for the occurrence of ancient, medieval and modern eclipses to the computed time, he was able to detect a systematic quadratic trend in the discrepancy of the lunar longitude. The secular acceleration found amounted to $\dot{n} = +20''/(cy)^2$. (Throughout this paper the secular acceleration is defined as physicists do, namely the acceleration is twice the coefficient of t^2 in the mean longitude.)

Soon after this breakthrough the result was confirmed by independent studies. It was not until the 19th December 1787 that the origin of such an acceleration was disclosed by the French geometer P.S. Laplace in a paper presented to the French Academy of Sciences. From an investigation in the motion of the Galilean satellites, Laplace noted that the secular variation of the eccentricity of Jupiter's orbit, should produce a secular variation in the mean motion of its satellites and as Laplace said "I immediately transposed this result to the Moon and found out that the secular variation of the Earth's orbital eccentricity yielded the Moon's secular acceleration determined by astronomers....".

Analytically the problem solved by Laplace is in its principle easy to understand. The expression of the disturbing function due to the solar gravitational forces combined with the Lagrange planetary equations enabled him to estimate the variation of the Moon's mean motion:

$$\frac{dn}{dt} = -3m^2 \alpha e_0 n \quad , \qquad (1)$$

the Earth's eccentricity being represented by

$$\begin{aligned} e &= e_0 + \alpha T \\ &= 0.01675 - 4.18 \times 10^{-5} T \quad , \end{aligned}$$

and m is the ratio of the Sun's mean motion to the Moon's. After he had inserted the numerical values in (1), Laplace must have been delighted to get exactly the observed value for $\dot{n}$!

It took nearly one century to improve definitely the computation made by Laplace. Both J.C. Adams and C.E. Delaunay must be credited for having come up with the complete solution of what actually was a difficult question of mathematical analysis which gave rise to decades of controversy and wrangle in the celestial mechanicians' community. Their result is still in use today in the numerical expression of the Moon's mean longitude.

By leaving out the other 37 terms involving the eccentricity and the inclination, the main part of the gravitational secular acceleration takes the following form,

$$\dot{n} = -3m^2 \alpha e_0 n \left(1 - \frac{1257}{32} m^2 - \frac{11349}{32} m^3 - \frac{306865}{144} m^4 - \frac{5701247}{576} m^5 \ldots \right) . \qquad (2)$$

In this expression the first term is the one found by Laplace and all the others are negative and destroyed the perfect agreement between the theory and the observations.

As far as gravitation is concerned, the remaining problem rests upon the accuracy of the knowledge of the Earth's variable eccentricity. To date, for the computation of the Earth's ephemeris, astronomers still trust the theory perfected either by U.J. Le Verrier (1858) or by S. Newcomb (1898), both of them having published a monumental and comprehensive theory of the planetary motion in the solar system. Due to the planetary perturbations the Earth's eccentricity exhibits a periodic variation, whose period is of the order of 100,000 years. Consequently for some ten centuries a Taylor expansion can be used,

$$e = 0.0167498 - 4.258 \times 10^{-5} T \qquad (3)$$

in LeVerrier theory, while Newcomb gives

$$e = 0.0167510 - 4.180 \times 10^{-5} T \quad , \qquad (4)$$

where T is expressed in Julian century from 1900.0. Obviously the secular acceleration involved by this variable eccentricity turns into a periodic change in the Moon's mean motion over duration comparable to 100,000 years.

Accordingly the numerical value of α in (2) is well enough known to warrant a reliable value of the gravitational secular acceleration can be computed. The difference between the two values of e in (3) and (4) stems from the set of planetary mass used by the two astronomers and only a further improvement in the mass of Jupiter and Venus would change the adopted value. In any case the gravitational secular acceleration is far more accurately known than the observed acceleration, which allows us to include it in the theory of the Moon's motion and to talk about the observed secular acceleration as a non-gravitational secular acceleration.

To conclude, it might be interesting to give some explanations about the figure which appears in the mean longitude of the Moon as factor of t^2. From the American Ephemeris I extract the following expression giving the mean longitude of our satellite,

$$L = 270^{\circ}\!.434358 + 13^{\circ}\!.1763965\ d - 4''\!.08 T^2 \quad , \qquad (5)$$

where d designs a day and T a century. Following the convention defined above the secular acceleration in (5) is $-8.16''/(cy)^2$ and is neither purely gravitational nor purely tidal. This number is the result of three effects and was introduced by Clemence (1948) in relation with the E.T.

First we have the gravitational part, $\dot{n}_g = +12.05$

then the acceleration in the motion of the equinox since (5) deals with the tropical longitude $\dot{n}_p = +2.23$

and finally the conventional tidal secular acceleration found by S. Jones (1938) $\dot{n}_t = -22.44$

The whole problem remains the difference between the conventional $\dot{n}_t$ and the true value!

Appendix II. Precessional Equations

The system under consideration to ascertain the precessional motion of the Moon's orbit comprises an oblate planet, the satellite and the Sun. The planet angular momentum is $\vec{L}_1$ and lies along the spin vector. The orbital angular momentum of the satellite is normal to the orbital plane and is denoted $\vec{L}_2$.

Then the total angular momentum can be written as follows,

$$\vec{L} = \vec{L}_1 + \vec{L}_2 \quad .$$

The torque experienced by the planet or the satellite causes their angular momentum to vary; these torques originates from

1. interaction planet's oblateness-satellite mass
2. interaction planet's oblateness-Sun's mass
3. interaction Sun-satellite.

Each of these torques drives the angular momentum in a steady precession about a fixed axis.

For example, an artificial satellite precesses along the Earth's equator so that the orbital angular momentum describes a cone of constant angle, whose axis is the Earth's spin axis. With e = 0 the rate of this motion is,

$$\frac{d\Omega}{dt} = -\frac{3}{2}\, n\left(\frac{R}{a}\right)^2 J_2 \cos I \quad .$$

In term of evolution of the angular momentum the involved equation is,

$$\frac{d\vec{L}_2}{dt} = -\frac{3}{2}\, n\left(\frac{R}{a}\right)^2 J_2(\vec{z}_2\cdot\vec{z}_1)(\vec{z}_1\times\vec{L}_2)$$

where $\vec{L}_2 = L_2\vec{z}_2$ and $\vec{z}_1$ is a unit vector along the Earth's spin axis. When 1, 2, 3 are simultaneously taken into account we can obtain the two following equations:

$$\frac{d\vec{z}_2}{dt} = -\frac{3}{2}\, J_2\left(\frac{R}{a}\right)^2 n(\vec{z}_2\cdot\vec{z}_1)(\vec{z}_1\times\vec{z}_2) - \frac{3}{2}\left(\frac{n_\odot}{n}\right)^2(\vec{z}_2\cdot\vec{z}_3)(\vec{z}_3\times\vec{z}_2) \qquad \text{(II-1)}$$

$$\frac{d\vec{z}_1}{dt} = -\frac{3}{2} J_2 \frac{mn^2}{M\omega} \frac{MR^2}{C} (\vec{z}_1 \cdot \vec{z}_2)(\vec{z}_2 \times \vec{z}_1) - \frac{3}{2} J_2 \frac{n_\odot^2}{\omega} \frac{MR^2}{C} (\vec{z}_1 \cdot \vec{z}_3)(\vec{z}_3 \times \vec{z}_1) \tag{II-2}$$

where $\vec{z}_3$ is the unit vector normal to the ecliptic.

The complete solution of this set of two vectorial equations is required to average the tidal equations over the precessional period. Though there is no simple general solution, the study of some particular cases gives a clue on the general behaviour.

Let us consider the equation,

$$\frac{d\vec{z}}{dt} = A(\vec{z}\cdot\vec{\Omega})(\vec{\Omega} \times \vec{z}) \tag{II-3}$$

where $\vec{\Omega}$ is a given fixed vector.

Clearly we have

$$\frac{d(z^2)}{dt} = 0 \quad \text{and} \quad \frac{d(\vec{z}\cdot\vec{\Omega})}{dt} = 0 \quad .$$

These two relationships express the constancy of the modulus and of the component of $\vec{z}$ along $\vec{\Omega}$. The solution of (II-3) is a uniform precession of $\vec{z}$ around $\vec{\Omega}$ whose rate is $A\Omega$.

This result can be used in (II-1) and (II-2). First we consider a system with only a planet and a satellite, namely we put $n_\odot = 0$. Inserting the expression for the total angular momentum leads to:

$$\frac{d\vec{z}_1}{dt} = -\frac{3n}{2\omega}\left(\frac{R}{a}\right)^2 \frac{J_2}{C} \cos I\cdot L \cdot \frac{\vec{L}}{|\vec{L}|} \times \vec{z}_1$$

$$\frac{d\vec{z}_2}{dt} = -\frac{3n}{2\omega}\left(\frac{R}{a}\right)^2 \frac{J_2}{C} \cos I\cdot L \cdot \frac{\vec{L}}{|\vec{L}|} \times \vec{z}_2 \quad .$$

Clearly both the normal to the orbital plane and the Earth's spin axis precess around the direction of the total angular momentum as represented in Fig. 5b.

If $C\omega > mna^2$, which means the angular momentum embodied in the planet's rotation is much larger than the orbital angular momentum, we can write,

$$\vec{L} = C\vec{\omega}$$

and the rate of precession becomes

$$\Omega = -\frac{3}{2} n \left(\frac{R}{a}\right)^2 J_2 \cos I \quad ,$$

and is similar to the rate of the motion of the node of an artificial satellite. Conversely, if $C\omega < mna^2$ the following expression is obtained,

$$\Omega = -\frac{3}{2} \frac{n^2}{\omega} \frac{C - A}{C} \frac{m}{M} \cos I \quad ,$$

which is not different from the rate of the precession of the Earth's spin axis due to the gravitational torque of the Moon. Present conditions in the Earth-Moon system give:

$$C\omega = 0.17L \quad \text{and} \quad mna^2 = 0.83L \quad .$$

The equality of the two moments took place at about $a/R = 20$, when $C\omega = mna^2$.

Another situation of interest is related to (II-1) when the oblateness perturbation is of the same magnitude as that of the third body. In case of small inclinations we can write

$$\frac{d\vec{z}_2}{dt} = \left[-\frac{3}{2} J_2 \left(\frac{R}{a}\right)^2 n \cdot \vec{z}_1 - \frac{3}{2} \frac{n_\odot^2}{n} \vec{z}_3\right] \times \vec{z}_2 \quad . \tag{II-4}$$

The precession of the node occurs neither on the equator nor on the ecliptic but on an intermediate plane located between them, the direction of which is determined by (II-4). The equality of the two coefficients of $\vec{z}_1$ and $\vec{z}_2$ in (II-4) gives (9) in the main text. In the solar system Iapetus is currently precessing on a Laplace plane inclined by 15° to the equator of Saturn. Equation (II-4) has been extensively used in the study of the past orbit of Phobos by Mignard (1981 b) and Cazenave et al. (1980).

References

Alfven H, Arrhenius G (1972) Origin and evolution of the Earth-Moon Moon 5:210

Burns JA (1972) Dynamical characteristic of Phobos and Deimos. Rev Geophys Space Phys 10:463-482

Burns JA (1978) The dynamical evolution and origin of the Martian moons. Vistas Astron 193-210

Cappallo RJ, Counselman III CC, King RW, Shapiro II (1981) Tidal dissipation in the Moon. J Geophys Res 86:7180-7184

Cazenave A, Daillet S (1981) Lunar tidal acceleration from Earth satellite orbit analyses. J Geophys Res 86:1659-1663

Cazenave A, Dobrovolsky A, Lago B (1980) Orbital history of the Martian satellites with inferences on their origin. Icarus 44:730-744
Clemence GM (1948) On the system of astronomical constants. Astron J 53:169
Conway (1982) The history of the lunar orbit. Icarus (in press)
Darwin GH (1880) On the secular change in the elements of a satellite revolving about a tidally distorted planet. Philos Trans R Soc London 171:713-891
Delaunay CE (1859) C R Acad Sci 48:817
Dunthorne R (1749) A letter concerning the acceleration of the Moon. Philos Trans R Soc London 46:162-172
Ferrari AJ, Sinclair WS, Sjogren WL, Williams JG, Yoder CF (1980) Geophysical parameters of the Earth-Moon system. J Geophys Res 85:3939-3951
Gans RF (1972) Viscosity of the Earth's core. J Geophys Res 77:360
Gerstenkorn H (1955) Über Gezeitenreibung beim Zweikörperproblem. Z Astrophys 36:245-274
Gerstenkorn H (1969) The earliest past of the Earth-Moon system. Icarus 11:189-207
Goldreich P (1965) An explanation of the frequent occurrences of commensurabilities in the Solar System. Mon Not R Astron Soc 130:159-181
Goldreich P (1966) History of the Lunar orbit. Rev Geophys 4:411-439
Greenberg R (1973) Evolution of satellite resonances by tidal dissipation. Astron J 78:338-346
Kaula WH (1964) Tidal dissipation by tidal friction and the resulting orbital evolution. Rev Geophys Space Phys 2:661-685
Lambeck K (1975) Effect of tidal dissipation in the ocean on the Moon's orbit and on the Earth's rotation. J Geophys Res 80:2917-2925
Lambeck K (1979) On the orbital evolution of the martian satellites. J Geophys Res 84:5651-5657
Lambeck K (1980) The Earth's variable rotation. Cambridge Univ Press, Cambridge
LeVerrier UJJ (1858) Annales de l'observatoire de Paris. Memoires vol IV
McConnell (1968) Viscosity of the mantle from relaxation time. J Geophys Res 73:7089
MacDonald GJF (1964) Tidal friction. Rev Geophys Space Phys 2:467-541
Morrison LV (1978) Deceleration of the Earth's rotation deduced from astronomical observations in the period A.D.1600 to the present. In: Brosche P, Sündermann J (eds) Tidal friction and the Earth's rotation. Springer, Berlin Heidelberg New York
Mignard F (1980) The evolution of the Lunar orbit revisited II. Moon Planets 23:185-201
Mignard F (1981a) The evolution of the lunar orbit revisited III. Moon Planets 24:189-207
Mignard F (1981b) Evolution of the Martian satellites. Mon Not R Astron Soc 194:365-379
Newcomb S (1898) Table of the four inner planets. Astron Pap Vol VI, Washington

Newton R (1970) Ancient astronomical observations. John Hopkins Univ Press
Panella G (1975) Paleontological clocks and the history of the Earth's rotation. In: Rosenberg GD, Runcorn SK (eds) Growths rythms and the history of the Earth's rotation. John Wiley, New York, pp 253-284
Peale SJ (1973) Rotation of solid bodies in the solar system. Rev Geophys Space Phys 11:767-793
Peale SJ (1977) Rotation histories of the natural satellites. In: Burns JA (ed) Planetary satellites, chap 6
Poincare H (1913) Leçons sur les hypotheses cosmogonique. Librairie Hermann et fils, Paris
Rubincam DP (1975) Tidal friction and the Early history of the Moon's orbit. J Geophys Res 80:1537-1548
Sinclair AT (1978) Vistas Astron 22:133
Smith ML, Dahlen FA (1981) The period and Q in the Chandler Wobble. Geophys J R Astron Soc 64:223-281
Spencer Jones H (1938) The rotation of the Earth and the secular acceleration of the Sun, Moon and planets. Mon Not R Astron Soc 99:541
Stephenson FR (1978) Pre-telescopic astronomical observations. In: Brosche P, Sündermann J (eds) Tidal friction and the Earth's rotation. Springer, Berlin Heidelberg New York
Sündermann J, Brosche P (1978) The numerical computation of tidal friction for present and ancient oceans. In: Brosche P, Sündermann J (eds) Tidal friction and the Earth's rotation. Springer, Berlin Heidelberg New York
Taylor SR (1975) Lunar science: a post APOLLO view. Pergamon Press, New York
Taylor SR (1979) Structure and evolution of the Moon. Nature (London) 281:105-110
Van Flandern TC (1975) Determination of the rate of change of G. Mon Not R Astron Soc 170:333-342
Van Flandern TC (1981) Is the gravitational constant changing? Astrophys J 248
Vetter UR, Meissner RO (1977) Creep in geodynamic processes. Tecnophysics 42:37
Yoder CF (1979) How tidal heating in Io drives the galilean orbital resonance locks. Nature (London) 279:767-771
Yoder CF (1982) Tidal rigidity of Phobos. Icarus 49:327-346
Yoder CF, Sinclair WJ, Williams JC (1978) The effects of dissipation in the Moon and the lunar physical libration. Lunar Sci 9:1291

The Earth's Non-Uniform Rotation

R. Hide[1]

1. Introduction

In the absence of internal energy sources and mechanical, gravitational or radiative interactions with other bodies, the solid parts of the Earth (i.e., crust, mantle and inner core) and the fluid parts (atmosphere, hydrosphere and outer core) would rotate together as a rigid body at a constant rate about its fixed axis of maximum moment of inertia through its centre of mass. Positional astronomers equipped with perfect telescopes and clocks (and the necessary life-support systems all, together with the astronomers, of zero mass!) would find no variation of the latitude of any station and the longitude of all fixed stars would change at a constant rate.

The successful use of the Earth's rotation as the basis of the earliest attempts to provide a practical time scale attests to the validity of the above picture as a good "zeroth approximation" to the truth (see, e.g., Gaposchkin and Kolaczek 1981). But over the years, as clocks based on other physical phenomena were invented and methods of positional astronomy improved, tiny fluctuations in the "length of the day" (up to a few milliseconds) and very slight movements of the Earth's pole of rotation (up to a few metres) came to light. The interpretation of these variations in the magnitude and direction of the Earth's rotation vector in terms of energetic processes and angular momentum transfer within the Earth-Moon system is one of the most fascinating of scientific problems, for it brings together several otherwise diverse areas of study, including solid Earth geophysics, meteorology, oceanography, hydrology, glaciology, geomagnetism, palaeontology and positional astronomy, as well as certain areas of historical scholarship.

Such studies of the Earth's rotation go back to the last century. The first thorough review of the subject was presented by Munk and MacDonald (1960) and Lambeck (1980) has given an up-to-date discussion taking into account recent improvements in instrumentation, international co-operation and basic ideas in geodynamics.

---

1 Geophysical Fluid Dynamics Laboratory, Meteorological Office, Bracknell, Berkshire RG12 2SZ, United Kingdom

Tidal Friction and the Earth's Rotation II
ed. by Brosche/Sündermann

2. Angular Momentum of the Atmosphere

The magnitude of the principal moment of inertia of the atmosphere I_a is about 1.42×10^{32} kg m^2, which should be compared with that of the whole Earth $I = 8.04 \times 10^{37}$ kg m^2. During the development of the three-dimensional and non-axisymmetric global atmospheric circulation following the imaginary "switching on" of the main source of energy for atmospheric motions, solar heating, the angular momentum of the atmosphere M_a (say) should increase by an amount δM_a (magnitude $\doteqdot 1.5 \times 10^{26}$ kg m^{2}s^{-1}). This increase is a measure of the "super-rotation" of the densest parts of the atmosphere, the troposphere and lower stratosphere, at an average angular speed

$$|\delta M_a|/I_a \doteqdot 10^{-6} \text{ rad s}^{-1} \doteqdot 1.5 \times 10^{-2}|\Omega_s| \tag{1}$$

relative to the underlying planet, the corresponding linear speed being about 10 ms^{-1}. (Here Ω_s is the angular velocity of rotation of the solid Earth, which is very close to the total angular momentum M (magnitude 5.85×10^{33} kg m^{2}s^{-1}) divided by I; thus, $|\Omega_s| \doteqdot 7.28 \times 10^{-5}$ rad s^{-1}.)

Solar radiation provides the kinetic energy of atmospheric motions but it cannot provide the angular momentum. This must come from the underlying planet which, to conserve angular momentum of the whole Earth, would undergo a change $-\delta M_a$, largely manifested as a change

$$|\delta\Omega_s| \doteqdot - |\delta M_a|/I_s \doteqdot - 3 \times 10^{-8}$$

in the magnitude of Ω_s; the corresponding increase in the length of the day (l.o.d.) would be about 3×10^{-3} s. This simple calculation shows, as W.H. Munk and R. Miller were evidently the first to realise (for references see e.g., Munk and MacDonald 1960, Lambeck 1980), that fluctuations in the general atmospheric circulation could produce changes in the length of the day of up to about 10^{-3}s. It also implies that since the biggest of the observed irregular l.o.d. changes — the so-called "decade variations" — are about 5×10^{-3}s, they cannot originate in the atmosphere. The decade variations are now generally supposed to originate largely in the Earth's fluid metallic core, where the main geomagnetic field is produced by the magnetohydrodynamic dynamo process (see Munk and MacDonald 1960, Lambeck 1980).

Length of day fluctuations of meteorological origin are largely associated with displacements in the mean latitude of the major jet-streams in the troposphere. The concomitant vertical transfer of angular momentum at the Earth's surface is due to a slight imbalance between the positive torque on the underlying planet exerted by the average surface westerlies in mid-latitudes and the negative torque exerted by the surface easterlies at low and high latitudes, the net couple being K. Monitoring the angular momentum "budget" of the atmosphere with the aid of improved data on l.o.d. changes is now beginning to interest meteorologists concerned with fluctuations of the atmospheric circulation on time scales upwards of a few days.

Direct determinations of K cannot yet be made with acceptable accuracy. Even with a complete set of surface wind and pressure observations, an understanding of boundary layer processes and airflow over and around mountain ranges that goes well beyond present knowledge would be required (Hide and White 1980, Swinbank 1981). The value of K is best obtained indirectly from the rate of change of the total angular momentum of the atmosphere, which depends only on the distribution of velocity and pressure throughout the atmosphere. But even this quantity is subject to large uncertainties owing to the paucity of regular meteorological data from the southern hemisphere and tropical regions.

During 1979, meteorologists made a special effort to monitor the whole atmosphere, providing for the first time data sets that could be used with confidence in the assessment of fluctuations in M_a. It has been shown from these data sets that observed irregular changes in the l.o.d. during 1979 could be accounted for on the basis of angular momentum exchange with the atmosphere (Hide et al. 1980). This demonstration improves the meteorological usefulness of the data set on rapid changes in the l.o.d. since 1955, when "atomic" clocks were introduced. This data set reveals a pronounced fluctuation on a time-scale of about 7 weeks with an amplitude comparable with the better-known seasonal variations (see, e.g., Feissel and Gambis 1980, Langley et al. 1981). The elucidation of this fluctuation should greatly advance our understanding of the general circulation of the atmosphere and might lead to improvements in long-range weather forecasting by a combination of statistical and dynamical methods.

All components of M_a exhibit fluctuations. The non-axial components give rise to polar motions of a few metres, including a contribution to the so-called Chandler wobble with a period of 14 months (see Lambeck 1980). The excitation and damping of the Chandler wobble are still controversial issues amongst geophysicists but any attempts to infer mantle motions preceding earthquakes from Chandler wobble observations will require in the first instance an assessment of atmospheric effects (Lambeck 1980, Barnes et al. 1982, Wahr 1981). This should be possible soon when a new collaborative programme involving a number of institutions is fully operational.

3. The Decade Variations in the Length of the Day and Core-Mantle Coupling

We have seen that, at several milliseconds in amplitude, the so-called "decade variations" are too great to be accounted for in terms of atmospheric processes and that geophysicists generally agree that, in the absence of a reasonable alternative, the variations must be largely due to angular momentum transfer between the Earth's "solid" mantle of thickness $\doteqdot$ 2900 km and underlying liquid outer core of thickness $\doteqdot$ 2200 km. The distortion and displacement of the geomagnetic field pattern at the Earth's surface, including the well-known westward drift of the non-dipole field at some $3 \times 10^{-4} ms^{-1}$, can be taken as manifestation of core motions, but the accurate determination of all but the broadest features of these motions from geomagnetic observations is an impossible task (for references see Hide 1977). Nevertheless, the quantitative requirement that the time-scale and r.m.s. value of fluctuations in zonal speed of core motions be generally compatible with the amplitude of the decade variations in the length of the day is not particularly restrictive (Munk and MacDonald 1960).

The principal quantitative difficulties emerge when the nature of the horizontal stresses that couple the core to the mantle, across the core-mantle interface, are considered. These stresses must suffice both quantitatively and qualitatively to account for the fluctuating couple at the core-mantle interface implied by foregoing interpretation of the decade variations in the length of the day. We can write (following Hide 1969)

$$F = F_V + F_E + F_T \doteqdot 0.04 \text{ N m}^{-2} \tag{2}$$

where F is the average magnitude of these horizontal stresses, the corresponding magnitude of the unbalanced couple between core and mantle being $10^{19} - 10^{20}$ kg $m^2 s^{-2}$. F_V, F_E and F_T are the contributions to F associated with viscous coupling, electromagnetic coupling and topographic coupling respectively.

If δ is the thickness of the viscous boundary layer at the surface of the core then

$$F_V = C_V \nu \rho \Delta U / \delta \tag{3}$$

where ρ is the density of the core ($\doteqdot 10^4$ kg m^{-3}), ν is the coefficient of molecular or eddy kinematic viscosity, depending on whether or not the boundary layer is turbulent, ΔU is a typical magnitude of fluctuations in the azimuthal flow speed of the core material relative to the mantle and C_V is a 'viscous drag coefficient' which on general grounds should be around unity.

The electric currents responsible for the main geomagnetic field leak out of the metallic core into the weakly conducting lower mantle and these, together with currents induced in the lower mantle by fluctuations in the main geomagnetic field, give rise to electromagnetic coupling between the core and mantle. If B_1 is the fluctuating horizontal component of the magnetic field at the core-mantle interface and B_2 the vertical component, and μ denotes the magnetic permeability (not significantly different from $4\pi \times 10^{-7}$ Hm^{-1}, that of free space) then the term F_E in (2) is given by

$$F_E = C_E B_1 B_2 / \mu \tag{4}$$

where C_E is the "electromagnetic drag coefficient" which, like C_V should be around unity.

The third term F_T in (2) arises if irregular topographic features are present on the core-mantle interface and h, a typical value of the vertical dimensions of these features, exceeds the thickness of the viscous boundary layer δ. F_T is related to h by the expression

$$F_T = C_T \Omega_s \rho \Delta U (h - \delta) \tag{5}$$

where Ω_s is the angular speed of rotation of the mantle ($\doteqdot 10^{-4}$ rad s^{-1}) and C_T is the "topographic drag coefficient". C_T is unlikely to exceed unity but (unlike C_E and C_V) it may in some circumstances take on very small values.

Viscous coupling is almost certainly inadequate but electromagnetic coupling (proposed by Bullard) could be important and so could topographic coupling (proposed by the writer) with the bumps no more than about a kilometre in height (see Jacobs 1975 Hide 1977). But high topography does not necessarily imply strong coupling, owing to the complicated dependence of C_T on h, with C_T decreasing with increasing h when h exceeds a certain critical value h_*, dropping to values at $h \gg h_*$ that are typically very much smaller than those attained when $h < h_*$. The determination of h_* and the investigation of the detailed dependence of C_T on h and other parameters are of considerable importance in the study of the dynamics of atmospheres, oceans and planetary cores, where the influence of topography on fluid flow can in some circumstances be considerable. Preliminary studies of magnetohydrodynamic flow over and around topography in a rapidly rotating fluid have now been made. They indicate that, owing to the presence of magnetohydrodynamic inertial waves produced by interactions involving the magnetic field at the core-mantle interface, the appropriate value of C_T might be comparatively high, around unity (see Hassan and Eltayeb 1982). The corresponding topographic coupling would be adequate to account for the observed "decade" variations in the length of the day.

References

Barnes RTH, Hide R, White AA, Wilson CA (1982) Atmospheric angular momentum fluctuations correlated with length-of-day changes and polar motion (submitted)

Feissel M, Gambis D (1980) La mise en évidence de variations rapide de la durée du jour. C R Acad Sci 291 Ser B:271-273

Gaposchkin EM, Kolaczek B (eds) (1981) Reference coordinate system for Earth dynamics. D Reidel, Dordrecht, Holland

Hassan MHA, Eltayeb IA (1982) On the topographic coupling at the core-mantle interface. Phys Earth Planet Interviews 28 (in press)

Hide R (1969) Interaction between the Earth's liquid core and solid mantle. Nature (London) 222:1055-1056

Hide R (1977) Towards a theory of irregular variations in the length of the day and core-mantle coupling. Philos Trans R Soc London Ser A 284:547-554

Hide R, Birch NT, Morrison LV, Shea DJ, White AA (1980) Atmospheric angular momentum fluctuations and changes in the length of the day. Nature (London) 286:114-117

Hide R, White PW (eds) (1980) Orographic effects in planetary flows, World meteorological Organization (Geneva), Global Atmos Res Programme. Joint Organizing Committee Publ Ser No 23, 449 pages

Jacobs JA (1975) The Earth's core. Academic Press, London New York, pp viii + 253

Lambeck K (1980) The Earth's variable rotation. Cambridge Univ Press, Cambridge, pp xi + 449

Langley RB, King RW, Shapiro II, Rosen RD, Salstein DA (1981) Atmospheric angular momentum and the length of the day: a common fluctuation with a period near 50 days. Nature (London) 294: 730-732

Munk WH, MacDonald GJF (1960) The rotation of the Earth . Cambridge Univ Press, Cambridge, pp xii + 323

Swinbank R (1981) The global atmospheric angular momentum balance inferred from analyses made during the F.G.G.E. and general circulation model experiments (submitted)

Wahr JM (1981) The effects of the atmosphere and oceans on the Earth's wobble: 1 Theory. Geophys J Roy Astron Soc 70:349-372

The Rotation and the Magnetic Field of the Earth*

M. Stix[1]

1. Introduction

The sources of the main geomagnetic field must lie in the Earth's interior. This has been known since the appearance of the Allgemeine Theorie des Erdmagnetismus of Gauß (1839). As a possible mechanism to generate the field Larmor (1919) first had the idea of a self-exciting dynamo. He found this type of field excitation particularly attractive in view of the variability of the Earth's magnetic field, but "... it would require fluidity and residual circulation in deep-seated regions". Later Elsasser (1946), being more confident about the existence of the fluid, electrically conducting core (of radius $a = 3485$ km, see, e.g., Jacobs 1975, Table 1.3), advocated the dynamo hypothesis, which can be described in the following way: An initial magnetic field, $\underset{\sim}{B}$, and a velocity field, $\underset{\sim}{u}$, induce an electric field, $\underset{\sim}{u} \times \underset{\sim}{B}$. If the medium is conducting, a volume current, $\underset{\sim}{j}$, will be driven by this electric field, with an additional magnetic field associated to it. If the latter reinforces the original field, we have a selfexited dynamo. The field then no longer needs to be maintained by external sources. The energy lost through Ohmic dissipation is compensated for by conversion of mechanical into magnetic energy. If the velocity field is given, the dynamo problem is called *kinematic*; it is formulated by means of Maxwell's equations (in the magnetohydrodynamic approximation)

$$\underset{\sim}{j} = \frac{1}{\mu} \operatorname{curl} \underset{\sim}{B} \quad , \quad \frac{\partial \underset{\sim}{B}}{\partial t} = - \operatorname{curl} \underset{\sim}{E} \quad , \tag{1}$$

and Ohm's law

$$\underset{\sim}{j} = \sigma(\underset{\sim}{E} + \underset{\sim}{u} \times \underset{\sim}{B}) \quad . \tag{2}$$

Here $\mu (= 4\pi \times 10^{-7}$ Vs/Am) is the permeability and σ is the conductivity of the medium. These equations are complemented by the boundary requirements that $\underset{\sim}{B}$ has no sources external to the sphere $r = a$ and that $\underset{\sim}{B}$ must not decay to zero for $t \to \infty$. Elimination of $\underset{\sim}{j}$ and $\underset{\sim}{E}$ from (1) and (2) gives the well-known induction equation

$$\frac{\partial B}{\partial t} = \operatorname{curl}(\underset{\sim}{u} \times \underset{\sim}{B}) - \operatorname{curl} \lambda \operatorname{curl} \underset{\sim}{B} \quad , \tag{3}$$

1 Kiepenheuer-Institut für Sonnenphysik, Schöneckstr. 6, D-7800 Freiburg, Fed. Rep. of Germany

* Mitteilungen aus dem Kiepenheuer-Institut Nr. 214

Tidal Friction and the Earth's Rotation II
ed. by Brosche/Sündermann

where $\lambda = (\mu\sigma)^{-1}$ is the electromagnetic diffusivity. For the solution of (3) a number of necessary conditions must be satisfied, the most important of which are: the magnetic Reynolds number $R_m = ua/\lambda$, where u is a typical relative velocity within the fluid, must exceed a critical value of order 1 to 10^3, depending on the particular model (Backus 1958, Childress 1969); the field $\underset{\sim}{B}$ must not be axisymmetric (Cowling 1933; for a recent more general proof see James et al. 1980); and the inducing velocity $\underset{\sim}{u}$ must not be purely toroidal (Bullard and Gellman 1954). Here "toroidal" means tangential to surfaces r = const; in particular, any $\underset{\sim}{u} = (0, 0, u_\varphi)$ is toroidal. None of these conditions poses a serious restriction to the application of dynamo theory to the geomagnetic field, and a large number of solutions of (3) have been derived in recent years and discussed in the geomagnetic context. I shall not review these here since a number of reviews have appeared during the last few years, e.g., those of Gubbins (1974), Braginsky (1976), Stix (1977), Busse (1978), or Soward (1979). Various aspects of the subject are also covered in the books of Moffatt (1978), Parker (1979), and Krause and Rädler (1980).

The role of rotation is a complicated one. First of all, solid body, rotation is irrelevant for (3) since this equation is invariant under a transformation into a rotating frame of reference. On the other hand, *differential* rotation is an important ingredient of many dynamo models, in particular the "strong field" models to be discussed below, as it draws poloidal lines of force (e.g., lines of force lying in meridional planes) around the axis of rotation and thus very effectively generates toroidal field components. Second, it is not presently possible to predict a relationship between the magnitude of the dynamo-generated field and the rate of rotation. Such a relationship could only be obtained from a consideration of the *dynamic* interaction of $\underset{\sim}{u}$ and $\underset{\sim}{B}$ which, in addition to (3), requires a solution of the equation of motion

$$\frac{\partial \underset{\sim}{u}}{\partial t} + (\underset{\sim}{u} \cdot \underset{\sim}{\nabla})\underset{\sim}{u} = -\frac{1}{\rho}\underset{\sim}{\nabla} p + 2\underset{\sim}{u} \times \underset{\sim}{\Omega} + \frac{1}{\mu\rho}\,\mathrm{curl}\,\underset{\sim}{B} \times \underset{\sim}{B} + \nu\Delta\underset{\sim}{u} + \underset{\sim}{f} \quad . \qquad (4)$$

Here ρ is the density, ν the kinematic viscosity (assumed uniform), p is the pressure (including the centrifugal potential), $\underset{\sim}{u}$ is measured in a frame of reference rotating with angular velocity $\underset{\sim}{\Omega}$, and $\underset{\sim}{f}$ represents any volume force, in particular gravity. The interplay between the driving force $\underset{\sim}{f}$, rotation $\underset{\sim}{\Omega}$, and field $\underset{\sim}{B}$ can be demonstrated by means of the homopolar, or disk dynamo (Bullard 1955) depicted in Fig. 1. A conducting disk rotates about its axis, and a circular coil is in contact (by brushes) with the axis and the periphery of the disk. If a field parallel to the axis is applied, a current, I, is induced in the disk and the ciruclar conductor according to

$$L\dot{I} + RI = M\frac{\Omega}{2\pi}I \quad , \qquad (5)$$

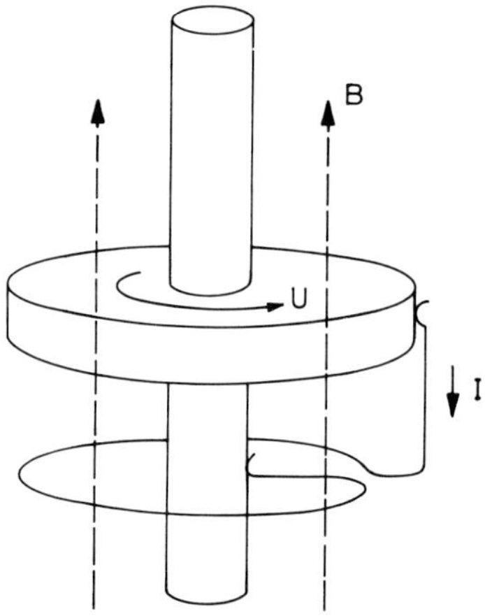

Fig. 1. The homopolar dynamo of Bullard (1955)

where L and R are the self-inductance and the resistance of the circuit, and M is the mutual inductance between the coil and the periphery of the disk. If the direction of the field associated with I is the same as that of the external field, the latter can be turned off; a steady solution of (5) then requires

$$\Omega = 2\pi R/M \quad . \tag{6}$$

The angular velocity, Ω, does however not directly affect the magnitude of I since (5) is linear and homogeneous. If a torque, G, is applied to the disk, then the equation of motion is

$$C\dot{\Omega} = G - MI^2/2\pi \tag{7}$$

where C is the moment of inertia of the disk. The steady solution

$$I = \pm(2\pi G/M)^{\frac{1}{2}} \tag{8}$$

depends on G, and thus also B = B(G); the field depends on Ω only if G itself is a function of Ω.

In the more general dynamo described by (3) and (4) above there must of course exist a relationship between the angular velocity and field strength, since $\underset{\sim}{u}$, and therefore $\underset{\sim}{B}$, explicitly depend on $\underset{\sim}{\Omega}$ in the coupled, non-linear system. In Sect. 2 I shall describe two recent attempts to solve this coupled system, one belonging to the "weak field" class and one to the "strong field" class, depending on the magnitude of the toroidal field in the core.

Is there observational evidence for a relationship $\underset{\sim}{B} = \underset{\sim}{B}(\Omega)$? Tidal friction of course retards the Earth's rotation, and perhaps the field was in fact stronger in the Precambrian: For example, McElhinny and Evans (1968) measured a value of 0.48 ± 0.02 G for the equatorial field strength in 2700 m.y. old samples from South Africa, which is

1.6 times the present value. I am not sure that this is sufficient evidence, and Dr. Piper tells me that values from 2/3 to 3/2 the present field strength have been estimated for the Precambrian field. In any case a relationship between rotation and the terrestrial and planetary fields is suggested by the approximate alignment of their dipole and rotation axes. Moreover, the rapid rotations of Earth, Jupiter, and Saturn seem to favour the presence of fields, although the size and internal planetary constition also control the generation. The observational evidence for *stellar* magnetism and rotation is twofold: Magnetic, or Ap, stars on the average rotate significantly slower than normal A stars (Slettebak 1970, Landstreet et al. 1975); on the other hand, the magnetism of lower main-sequence stars, measured through their chromospheric activity, increases with increasing rate of rotation (Skumanich 1972, Vaughan et al. 1981). As late main sequence stars have outer convection zones their fields are very plausibly explained as dynamo-generated, while the fields of Ap stars could as well be fossil (see, e.g., Mestel 1976). If that is true, then there appears to exist a positive correlation between the rate of rotation and the strength of dynamo-generated fields, planetary and stellar.

There is another context in which the Earth's magnetic field and its rotation must be discussed: The "decade fluctuations" in the length of the day are thought to originate from core motions (Munk and Revelle 1952) and require a mechanical coupling of the core to the mantle. In particular, for the increase of the length of the day by ≈ 3 ms between 1890 and 1900 (see, e.g., Lambeck 1980, p. 73) a couple Γ with fluctuations of order $I_m\dot{\omega} \approx 10^{18}$ Nm is necessary ($I_m = 7.2 \times 10^{37}$ kg m^2 is the moment of inertia of the mantle). Rochester (1960), Roberts (1972) and Hide (1977) have treated and reviewed the subject. It seems that the electromagnetic force, $\underset{\sim}{j} \times \underset{\sim}{B}$, falls short by a factor of 10 or so to provide the coupling. Therefore "topographic" core-mantle coupling has been proposed by Hide (1969, see also Hide 1977, Hassan and Eltayeb 1982). I shall, however, not pursue this alternative in the present paper, but instead — in Sect. 3 — discuss once more the problem of electromagnetic coupling. It will be made clear that the electrical conductivity, σ_m, of the lowest layers of the mantle, which allows the electric current, $\underset{\sim}{j}$, to enter from the core or to be induced by core motions, plays the key role. With a value of σ_m close to the core conductivity at $r = a$, and a very steep drop-off within the mantle, it seems indeed possible to obtain a torque of the order 10^{18} Nm.

2. Weak or Strong Toroidal Fields?

The toroidal magnetic field in the Earth's core corresponds to a poloidal electric current and therefore does not penetrate through the surrounding insulating mantle. For this reason the toroidal field in the core is not accessible to direct measurements. In the "weak-field" model it has the same order of magnitude as the poloidal field, extrapolated down to the core-mantle interface, i.e. a few Gauss, while in the "strong-field" model it may attain values of 100 G or more ($1G = 10^{-4}$ T). It shall first briefly describe the weak-field model of Busse (1975, 1976).

Busse considers thermal convection in a rapidly rotating system, where both the viscous and the advective forces are small compared to the Coriolis force. That is, the Ekman and Rossby numbers,

$$E = \frac{\nu}{\Omega \ell^2} \quad \text{and} \quad Ro = \frac{u}{2\Omega \ell} , \tag{9}$$

both are small (ν is the kinematic viscosity and ℓ is a characteristic length in the core). By assumption, the Lorentz force is small, so that, to leading order, we have the geostrophic equilibrium

$$2\underset{\sim}{\Omega} \times \underset{\sim}{u} = -\underset{\sim}{\nabla} p/\rho \quad . \tag{10}$$

In this case convection occurs in the form of rolls parallel to the axis of rotation, see, e.g., Fig. 1 of Busse (1970). However, due to the inclination, $\tan\eta$, of the core-mantle interface at the northern and southern ends of the rolls, the motion is not strictly steady, but time-dependent. Its frequency, ω, is given by a dispersion relationship, viz.

$$\omega = - \frac{2\eta a_y}{a^2(1 + P)} , \tag{11}$$

where P is the Prandtl number, $a^2 = a_x^2 + a_y^2$, and a_x and a_y are wave numbers in the radial and azimuthal directions. Time is measured in units of Ω^{-1} and length in units of ℓ, which Busse (1976) takes as half the total length of his rolls.

In the presence of a mean magnetic field, $\underset{\sim}{B}$, additional modes of oscillation become possible. The frequencies then are the roots of a third order equation,

$$\omega(1 + P) + 2\eta a_y a^{-2} + a_y^2 C \frac{\omega(\lambda/\varkappa - 1)}{\omega^2 + \tau^2 a^4} = 0 \quad ; \tag{12}$$

here $\varkappa$ is the thermal diffusivity, $\tau = \lambda/\Omega\ell^2$, and

$$C = \frac{B^2}{\Omega^2 \ell^2 \rho\mu} \quad . \tag{13}$$

The dispersion relation (12) is qualitatively shown in Fig. 2. In the case $\omega^2 \gg \tau^2 a^4$, real frequencies become impossible if C exceeds the critical value

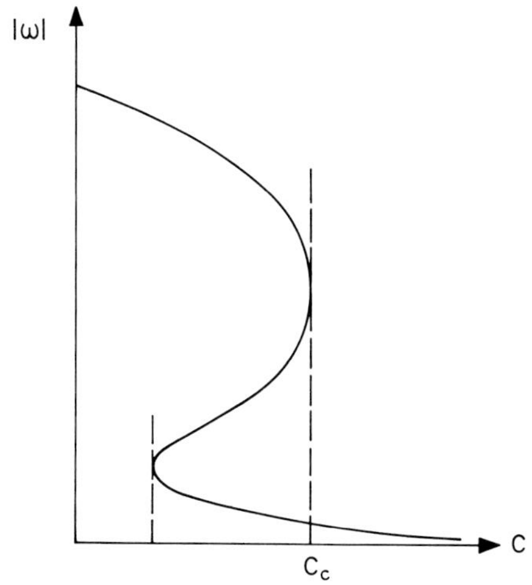

Fig. 2. Frequencies, ω, of thermal convection modes, according to (12), as a function of dimensionless field energy, C. The high frequency modes which support the field become impossible for $C > C_c$. (After Busse 1979)

$$C_c = \frac{\eta^2}{a^4(1 + P)(\lambda/\varkappa - 1)} \quad . \tag{14}$$

Busse now identifies the field $\underset{\sim}{B}$ with the field generated by the dynamo process to be discussed presently, and proposes that this field grows until C reaches the critical value. At this amplitude the field destroys the flow pattern which originally gave rise to its amplification.

The amplification process itself depends on velocity components parallel to the axis of rotation. These are caused by the inclination of the end-surfaces of the convection rolls, and by Ekman pumping (the former effect being the stronger of the two). Their direction is such that the flow becomes helical, that is all convection rolls in the nothern hemisphere are left-handed, and all rolls in the southern hemisphere are right-handed. As a consequence, toroidal lines of force of the mean field $\underset{\sim}{B}$ are deformed in such a way that a net contribution to the poloidal field is generated, and poloidal lines of force regenerate the toroidal field by the same mechanism. The equations

$$D \cdot B_{pol} = \alpha \cdot B_{tor} \tag{15a}$$

$$D \cdot B_{tor} = \alpha \cdot B_{pol} \tag{15b}$$

symbolically describe this mechanism, which has been termed the "α^2-dynamo". The operator D describes transport and diffusion of the mean field, while α denotes the regeneration mechanism just described. From the symmetry of (15) it is clear that B_{tor} and B_{pol} will have the same order of magnitude, in consistance with the original assumption of a "weak" toroidal field. A quantitative evaluation of the field amplitude given by (13) and (14) is, however,

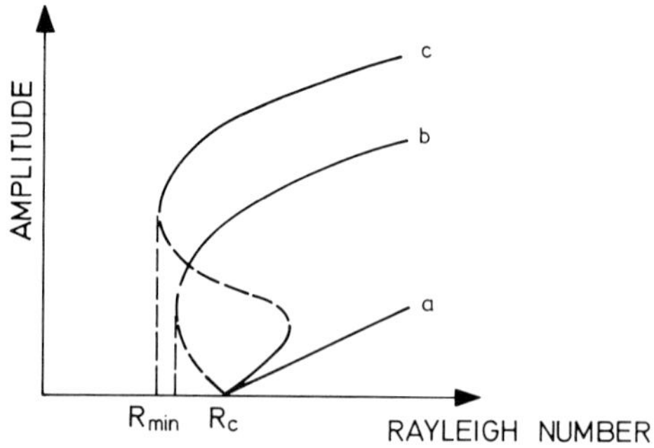

Fig. 3. Amplitudes of thermal convection, as functions of the Rayleigh number (non-dimensional temperature gradient). Case (a) stable small amplitude, at or above the critical Rayleigh number, R_c. Cases (b) and (c) subcritical finite amplitude, at $R_{min} < R_c$. (After Roberts 1978)

difficult. Busse (1976, 1979) takes a ~10 in order to adapt the field to values observed in the Earth and other planets. On the other hand, his approximation of *small* amplitude convection and the condition that the mean field described by (15) does not decay to zero are in accord only if $(\varkappa/\lambda)^2 a_x a_y E^{\frac{1}{2}} \gg 1$ (Busse 1975, p. 450; according to our present notation, I have replaced Busse's γ by a_x and α by a_y, and added the missing exponent of $\varkappa/\lambda$). Such a condition would require values of a very much in excess of 10! Numerical modelling might eventually render the approximation of small amplitude convection as well as a number of geometrical approximations unnecessary and thus resolve the difficulty.

Roberts (1978) has criticized the weak-field assumption on different grounds. He argues that convection in the Earth's core occurs as a subcritical finite amplitude instability. Here, "subcritical" means that the temperature gradient required to drive the flow can be *smaller* than the critical temperature gradient which is needed in order that convection of infinitesimally small amplitude first occurs. Figure 3 illustrates the situation. The point then is of course that, plausibly, the finite amplitude motion generates a stronger field than the motion of small amplitude.

The most prominent example of a "strong-field" geodynamo has been developed by Braginsky since 1964, and a review of his "model Z" is available (Braginsky 1976). The idea of a strong toroidal field rests on non-uniform rotation in the core, with differences in angular velocity, ω, of order 10^{-10} s^{-1}, corresponding to the observed westward drift of the non-dipole geomagnetic field of 0.2° longitude per year. The magnetic Reynolds number, $R_m = \mu\sigma\omega a^2$, based on such a non-uniform rotation, the radius, a, and conductivity, σ, of the core, is of order 10^2. The azimuthal component of (3) above says that a toroidal field, of order $R_m B_p$, will be generated from a given poloidal field; and the latter is 4 G, obtained by downward extrapolation from the surface. A field of several hundred Gauss is even more suggestive as it yields a Lorentz force per unit mass, $B^2/\mu\rho a$, of order 10^{-8} ms^{-2}, comparable to the Coriolis acceleration (Roberts and Soward 1972). The balance between these two forces (plus the pressure gradient) is called the magneto-geostrophic equilibrium).

In order to obtain approximate solutions of (3) and (4) above, Braginsky divides the motion $\mathbf{u}$, and the field, $\mathbf{B}$, into axisymmetric means, $\bar{\mathbf{u}}$ and $\bar{\mathbf{B}}$, and asymmetric parts (assumed small), $\mathbf{u}'$ and $\mathbf{B}'$. The former are further divided into their poloidal and toroidal parts, $\bar{\mathbf{u}}_p$, $\bar{u}_\varphi$ and $\bar{\mathbf{B}}_p$, $\bar{B}_\varphi$ respectively, where $\mathbf{B}_p = \mathrm{curl}\ (\mathbf{e}_\varphi A)$ can be derived from a toroidal vector potential. All quantities are now expanded in powers of $R_m^{-\frac{1}{2}}$, and the leading terms are related by the assumptions (consistent with the large toroidal field)

$$\bar{u}_p/\bar{u}_\varphi \sim \bar{B}_p/\bar{B}_\varphi \sim (u'/\bar{u}_\varphi)^2 \sim (B'/\bar{B}_\varphi)^2 \sim R_m^{-1} \quad . \tag{16}$$

Averaging (3), and separating its poloidal and toroidal components then yields, to leading order

$$\frac{\partial A}{\partial t} + s^{-1}\mathbf{u}_p \cdot \nabla(sA) - \lambda\Delta_1 A = \lambda\alpha B_\varphi \quad , \tag{17a}$$

$$\frac{\partial B_\varphi}{\partial t} + s\mathbf{u}_p \cdot \nabla(B_\varphi/s) - \lambda\Delta_1 B_\varphi = s\mathbf{B}_p \cdot \nabla\omega \quad , \tag{17b}$$

where $s = r\sin\theta$, and $\omega = \bar{u}_\varphi/s$ is the angular velocity, relative to the rotation, Ω, of the mantle. The operator Δ_1 is $\Delta - s^{-2}$. The overbars are omitted for brevity; moreover, the symbols A, $\mathbf{u}_p$, and $\mathbf{B}_p$ now mean "effective" fields, including contributions from $\mathbf{u}'$. Equations (17) completely correspond to (15) above, except for two differences. First, the generation coefficient, α (Braginsky's Γ) on the right hand side of (17a) is now based on small-scale motions, $\mathbf{u}'$, which develope around the basic magnetogeostrophic balance. In particular, Braginsky has in mind wave-like motions for which magnetic (M), buoyancy (A, from Archimedean) and Coriolis (C) forces play the dominant parts, and which, therefore, have been termed MAC waves. Secondly, the generation of the toroidal field from the poloidal now does *not* depend on small-scale motions. Instead, the right hand side of (17b) describes the generation of B_φ from $\mathbf{B}_p$ by nonuniform rotation. In contrast to the α^2-dynamos of (15), dynamos described by equations of the type (17) often are called "$\alpha\omega$-dynamos".

In Braginsky's model, the coefficient α is considered as a given quantity although it is recognized that, of course, the MAC waves depend on the field generated, and therefore, in principle $\alpha = \alpha(B)$. On the other hand, the effect of the magnetic force upon the angular velocity, ω, is taken into account explicitly. To this end, the total magnetic torque acting on a cylindrical surface, of radius s, parallel to the axis of rotation,

$$\frac{2\pi s}{\mu}\int (\mathrm{curl}\ \mathbf{B}\times\mathbf{B})_\varphi\ dz = \frac{2\pi}{\mu s}\frac{\partial}{\partial s}\left(s^2\int B_s B_\varphi\ dz\right) \quad , \tag{18}$$

where the integral is over the entire length of the cylinder, is considered. As observed by Taylor (1963), this expression must be zero

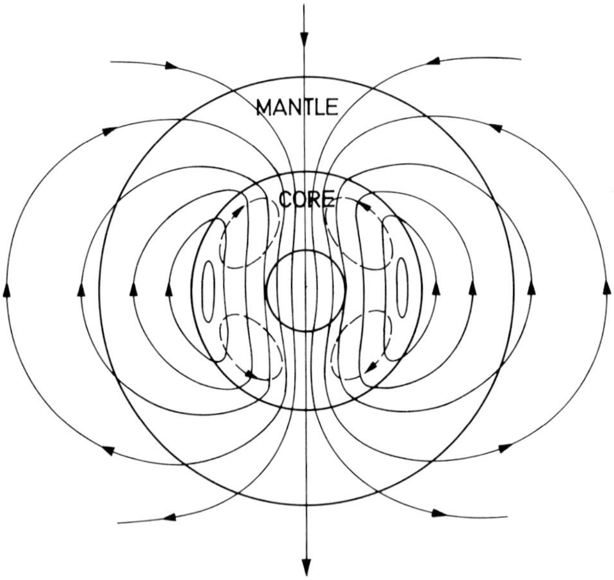

Fig. 4. Poloidal lines of force of "model Z": the field in the core is nearly parallel to the axis of rotation, with a rather abrupt transition into the potential field in and above the mantle. Dashed curves indicate meridional circulation in the core. (After Braginsky 1976)

in the absence of viscosity, since the buoyancy and Coriolis forces do not contribute to the total torque. Braginsky allows for a non-zero, although small, torque, which he balances by viscous torques in an Ekman boundary layer, of thickness $\delta = (\nu/\Omega)^{\frac{1}{2}}$, and by electromagnetic core-mantle coupling. The magnetic torque is kept small through a small B_s, with the consequence that the poloidal lines of force in the core are largely parallel to the axis of rotation, cf. Fig. 4, which explains the name "model Z". Since the field in the mantle is a potential field, the lines of force are sharply bent near the core-mantle boundary, thus forming a current layer.

Braginsky (1975, 1978) has elaborated many more details of his model. Here I mention only the perhaps most conspicuous property, which it has in common with all "strong-field" models, and which differentiates it from the "weak-field" models: Its energy expenditure is considerable, exceeding 10^{12}W in two examples calculated numerically. Much of it is dissipated in the current layer and in the Ekman layer. The difference to the energy needs of "weak-field" dynamos is several orders of magnitude, and is plausible since the dissipated energy

$$Q = \int j^2/\sigma \, dV \tag{19}$$

is proportional to the square of the magnetic field strength.

I have discussed the energy budget of the Earth's magnetic field in my review on planetary dynamos (Stix 1977), and Gubbins (1977) and Gubbins et al. (1979) comprehensively treat the subject. Of course, the 4×10^{12}W tidal dissipation do not help since the source is needed in the core. Two other sources, decay of radioactive nuclei and cooling (plus release of latent heat) have been suggested in the past. A power of order 10^{12}W could be obtained, but, as Bragisnky (1964b) points out, a dynamo operating on such a basis would be a thermodynamic engine of very low efficiency, so that only a few percent would be dissipated electrically — too little for a "strong-field" model. A more capable energy source is chemical differentiation, as already described by Urey (1952), in connection with the formation of the solid inner core. If the liquid outer core consists of a heavy (iron and nickel) and a lighter constituent (e.g., FeS or FeSi), and if these constituents separate continuously during the freezing at the surface of the inner core, a density discontinuity $\Delta\rho$ will result. The light residue in the liquid is buoyant, and "compositional convection" sets in. This process is not, in contrast to thermal convection, subject to a small efficiency because the energy release is gravitational. The available energy is $\Delta\rho$ Vga, where V is the Volume of the inner core, a the core radius, and g an average value of the gravitational acceleration. With $\Delta\rho = 0.75 \times 10^3$ kg/m^3 (Gubbins et al. 1979) we obtain 10^{29} Joule. Assuming that the inner core has grown during the Earth's history (10^{17}s) we can secure the required 10^{12}W.

The details of compositional convection, in particular the thermodynamic aspects, have been studied recently by Loper (1978a,b) and Loper and Roberts (1978, 1980, 1981, 1982). Their results, which support the gravitationally powered dynamo of Braginsky, are summarized in the last of these papers: The convective instability is indeed dominated by compositional rather than by thermal buoyancy. The available power is

$$P = \beta\Delta\rho - 4.5 \times 10^{10} \text{W} \qquad (20)$$

where $\beta = 8.4 \times 10^8$Wm^3/kg, and the second term is the part of the gravitational energy release which is useless due to diffusion of the light constituent in the core. It is also found that the fluid near the freezing interface is supercooled (below the freezing temperature); Loper and Roberts suggest that this results in the growth of dendrites extending from the interface into the adjacent liquid.

As an alternative, chemical differentiation and concomitant gravitational energy release could also occur at the core-mantle boundary. For example, Urey (1952) considered the reaction Si + 2 FeO = = SiO_2 + 2 Fe, and its possible effect of upward transport of Si in "pockets" of iron in the lower mantle. Professor Wänke suggests that such a process could continuously replace Si from the core by Fe from the mantle, and quotes in support the low iron content of the Earth's mantle, of approx. 6%, in comparison to the moon's 10%. The difference leads to an estimated release of $\sim 10^{29}$ to 10^{30} J of gravitational potential energy, quite sufficient to drive a strong-field dynamo over 10^{17}s. The early formation of the core is not affected by this speculation; we must only require that the mantle's original chemical constitution was that of the present moon.

Reversals of the geomagnetic polarity have occurred frequently in the Earth's history, and Brosche (1981) has speculated that such abrupt changes might be caused by the gradual decrease of the Earth's rotation (due to the tidal torque), with a sudden switch from one magnetic mode to another when a critical value of the angular velocity is passed. It is true, rotation influences the generation coefficients of (15) and (17) above, α and $\underset{\sim}{\nabla}\omega$. Changes in these coefficients may indeed cause mode switching, as, e.g., demonstrated by Deinzer et al. (1974) who find transitions from steady to oscillatory modes. It is also true that magnetically active late stars can be divided into two groups, of rapid rotators with intense activity and slow rotators with less intense activity (Vaughan and Preston 1980); magnetic mode switching has been suggested as an explanation (Durney at al. 1981). On the other hand, the terrestrial rotation has changed only by a few percent in the last 10^8 years, while many reversals did occur during this period. It thus appears more natural to assume that the system fluctuates (as evidenced in the secular variation) around a steady magnetic state. A particularly large excursion from this state could then settle down to the state with the opposite sign as well as to the original state, which both equally solve the basic equations (3) and (4) above. A model system, consisting of two coupled disk dynamos of the type shown in Fig. 1, does exhibit such reversals; it is described in the book of Rikitake (1966), and discussed further by Cook and Roberts (1970).

3. Electromagnetic Core-Mantle Coupling

In order to study the hydromagnetic terrestrial dynamo most authors have considered the idealized model of a core with electrical conductivity σ_c, surrounded by an insulating mantle. Even Braginsky, who in principle includes the possibility of electric currents in the lower mantle into his dynamo, later neglects this effect in order to simplify his numerical solution (Braginsky 1978).

Mantle conductivity, σ_m, is however essential to the problem of angular momentum transfer from the core to the mantle. A poloidal current, corresponding to a toroidal field, B_φ, in the core will leak into the mantle, causing there a field $B_\varphi \sigma_m/\sigma_c$. The associated Maxwell stress is $B_r B_\varphi \sigma_m/\sigma_c \mu$, and the "leakage" couple (Roberts 1972),

$$\Gamma_L = 4\pi a^3 B_r B_\varphi \sigma_m / \sigma_c \mu \quad ,$$

is of order 10^{18}kg m^2/s^2 if $B_r \sim 4$G, $B_\varphi \sim 100$G, and $\sigma_m/\sigma_c = 10^{-3}$. If there are core motions, $\underset{\sim}{u}$, relative to the mantle, there will, in addition, be an induced mantle current of order $\sigma_m u B_r$, resulting in a volume force $\sigma_m u B_r^2$ and a couple

$$\Gamma_F = 4\pi a^3 B_r^2 u \sigma_m L \quad ,$$

where L is the thickness of the conducting layer in the lower mantle. In the notation of Roberts (1972), the subscript "F" stands for frozen field: the induction is caused by the convection of magnetic flux in the (almost perfectly) conducting core. For $L \approx 10^6$m Γ_F is also of order 10^{18}Nm, and the two couples must balance each other in the mean if core and mantle rotate together (of course other couples, such as that caused by undulations of the core-mantle interface (Hide 1969) must be included into this balance). Now we need instantaneous violations of this balance of 10^{18}Nm, as I have mentioned in the introduction, i.e., of the same order as Γ_L and Γ_F themselves. As Roberts has observed, the situation is even worse since a fluctuation of Γ_F is more plausibly explained by a *local* change in $\underset{\sim}{u}$ rather than by a uniform acceleration of the core relative to the mantle. For a source of size $10^{12}m^2$ the couple would be reduced by a factor ~ 50, and should, therefore, exceed 10^{18} Nm by this factor if computed on the assumption of uniform core rotation.

The calculation of Γ_F below intends to demonstrate that such large couples are indeed possible, if only a large conductivity is allowed at the mantle's *base*. The model is essential that of Bullard et al. (1950), except that their constant mantle conductivity is replaced by

$$\sigma_m(r) = \sigma_a (a/r)^\alpha \quad . \tag{21}$$

The poloidal field is assumed to be a pure dipole, with a radial component

$$B_r = 2g_1^o (R/r)^3 \cos\theta \quad , \tag{22}$$

where R is the Earth's radius and $g_1^o = -0.301$ G is the strength of the dipole field at the equator. A conductivity $\sigma_m(r)$ of the type (21) has been considered by Braginsky and Fishman (1976) in a more general case, including non-dipole fields. It has the convenient property that the induction equation (3) for the field in the mantle (where $\underset{\sim}{u} = 0$),

$$\frac{\partial \underset{\sim}{B}}{\partial t} = - \operatorname{curl}\lambda \operatorname{curl} \underset{\sim}{B} \tag{23}$$

allows a simple analytical solution, at least in the approximation of slowly variable $\underset{\sim}{B}$, where the left hand side of (23) is neglected. When the expression for the unknown toroidal field,

$$\underset{\sim}{B} = \text{curl } (T_2(r)P_2(\theta),\ 0,\ 0) \tag{24}$$

(in spherical polar co-ordinates, r, θ, φ), is substituted into (23), we obtain

$$\frac{d^2T_2}{dr^2} + \frac{\alpha}{r}\frac{dT_2}{dr} - \frac{6}{r^2}T_2 = 0 \quad , \tag{25}$$

where $\alpha = r\lambda^{-1}d\lambda/dr$ is the exponent in (21). The boundary conditions for $T_2(r)$ are that the toroidal field must be regular at r = 0 and vanish at the surface (r = b) of the conducting part of the mantle, $a \le r \le b$, and that T_2 and the tangential (θ-) component of the electric field must be continuous at the core-mantle boundary. Restricting the attention to the case of solid body rotation, $\underset{\sim}{u} = (0, 0\ \omega r \sin\theta)$, of the core relative to the mantle, the latter condition is

$$\frac{1}{\sigma_a}\frac{dT_2}{dr}\bigg|_{a^+} - \frac{1}{\sigma_c}\frac{dT_2}{dr}\bigg|_{a^-} = \frac{2g_1^0\omega R^3}{3a} \quad . \tag{26}$$

In the core (α = 0) (25) is solved by $T_2 \sim r^3$, while in the mantle we have

$$T_2 = C_1r^{\beta_1} + C_2r^{\beta_2} \tag{27}$$

and

$$\beta_{1,2} = \frac{1-\alpha}{2} \pm \left(\left(\frac{1-\alpha}{2}\right)^2 + 6\right)^{\frac{1}{2}} \quad . \tag{28}$$

The boundary conditions determine C_1 and C_2. Once the field is known, we can calculate the current density, $\underset{\sim}{j} = \mu^{-1}$ curl $\underset{\sim}{B}$, the volume force, $\underset{\sim}{j} \times \underset{\sim}{B}$, and the couple, acting from the core upon the mantle

$$\Gamma_F = \int (\underset{\sim}{j} \times \underset{\sim}{B})_\varphi r \sin\theta \, dV \quad , \tag{29}$$

where the integral is over the conducting part of the mantle. Using the solution (27), we can readily evaluate (29) and obtain

$$\Gamma_F = \frac{32\pi\sigma_a R^6 (g_1^0)^2\left(1 - (a/b)^{\beta_1-\beta_2}\right)\omega}{45a\left[\frac{\sigma_a}{\sigma_c} - \frac{\beta_2}{3} + \left(\frac{a}{b}\right)^{\beta_1-\beta_2}\left(\frac{\beta_1}{3} - \frac{\sigma_a}{\sigma_c}\right)\right]} \quad . \tag{30}$$

A second term, corresponding to Γ_L above, is obtained in a similar way if a toroidal field (e.g. the dynamo-generated field) is specified in the core, or if, as in the model of Bullard et al. (1950), an additional discontinuity in angular velocity is postulated at some depth within the core. The present discussion will be restricted to Γ_F. As Bullard et al., Rochester (1960), and others, I use $\omega = 10^{-10}\ s^{-1}$, a value suggested by the westward drift of the geomagnetic field. The thickness, b-a, of the conducting lower mantle will be taken as 2000 km; this is not of much consequence since in all interesting cases we will have $(a/b)^{\beta_1-\beta_2} \ll 1$. For constant mantle conductivity, $\bar{\sigma}_m = \sigma_a$, we have $\alpha = 0$, $\beta_1 = 3$, $\beta_2 = -2$, and recover the result of Bullard et al.; with the conductivity used by Rochester, $\bar{\sigma}_m = 10^2$ A/Vm, this is

$$\Gamma_{FB} = \frac{16\pi\bar{\sigma}_m R^6 (g_1^o)^2 \omega}{15a} = 6\times 10^{17}\ \text{Nm} \quad , \tag{31}$$

where $\bar{\sigma}_m \ll \sigma_c$ has been used.

A number of attempts have been made to obtain larger torques than the Γ_{FB} of (31). Rochester considered the contributions from the non-dipole harmonics up to degree n = 6 and found an increase by a factor of 1.6. Using the harmonic coefficients of Barraclough et al. (1975), I have extended this to n = 12; but the gain is a factor 2 at most. Rochester also included the time dependence of $\underset{\sim}{B}$ in (23), but a comparison of his results with my own computation (with $\partial\underset{\sim}{B}/\partial t = 0$) shows that the difference is small. Roden (1963) first considered depth-dependent mantle conductivity. In two-layer models, such as his model 9 indicated in Fig. 5, he approximated the conductivity found by McDonald (1957) and obtained torques up to 2.5×10^{18}Nm. A general recipe to obtain Γ_F, from arbitrary core motions $\underset{\sim}{u}$, has been given by Roberts (1972), although he confined his consideration to the case of constant σ_m and slowly variable $\underset{\sim}{B}$. A variable $\sigma_m(r)$, of type (21), was taken into account by Braginsky and Fishman (1976). They observed that the relative contribution of non-dipole field harmonics becomes more significant when $\sigma_m(r)$ declines rapidly outwards from the core-mantle boundary; this is plausible since the effect of higher harmonics in any case is more concentrated towards that boundary.

In order to illustrate that even the dipole alone can produce a rather large torque when $\sigma_m(r)$ is large near r = a, I shall now take the extreme, advocated by Kolomiytseva (1972). She found that the morphology of the 60-year geomagnetic field variations would not allow a large discontinuity of σ at r = a, and suggested a distribution (21), with $\sigma_a = \sigma_c$ and $\alpha = 30$. In this case $\beta_1 - \beta_2 = 29.4$, and $(a/b)^{\beta_1-\beta_2} \ll 1$ even when b is not much larger than a. Thus, (30) reduces to

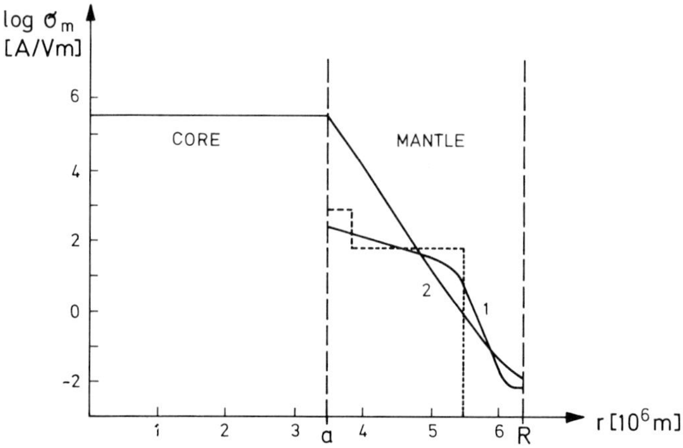

Fig. 5. Electrical conductivity in the Earth's interior. The two solid curves (1) and (2) in the mantle represent results of McDonald (1957) and Kolomiytseva (1972), respectively. The *dotted step* function is the conductivity adopted by Roden (1963) in his model 9

$$\Gamma_F \approx \frac{32\pi\sigma_c R^6 (g_1^o)^2 \omega}{45a(1-\beta_2/3)} = \frac{2\sigma_c \Gamma_{FB}}{3\bar{\sigma}_m(1-\beta_2/3)} \quad . \tag{32}$$

Using $\beta_2 = -29.2$, $\bar{\sigma}_m = 10^2$ A/Vm, and $\sigma_c = 5\times10^5$ A/Vm, we have $\Gamma_F/\Gamma_{FB} = 3.1\times10^2$. Even a reduction of Γ_F by a factor 50 (due to the local rather than uniform character of $\underset{\sim}{u}$) would leave us with a couple of order 10^{19} Nm. Moreover, some of the less extreme estimates of lower mantle conductivity (e.g., $\sigma = 10^4$ A/Vm, and a correspondingly smaller α) still appear sufficient to produce torques capable of generating the observed fluctuations in the length of the day. For a recent review of mantle conductivity studies I refer to the book of Lambeck (1980, p. 254).

I have not discussed in this contribution the time scales involved in electromagnetic core-mantle coupling, and refer to the treatment of Roberts (1972). He pointed out that the response of the mantle to a fluctuating torque does not only depend on the electromechanical time scale, $\tau_I = I/K$, where $K = \Gamma_F/\omega$ and I ($= 7.7\times10^{36}$ kg m^2) is the reduced moment of inertia of the core-mantle system. If the diffusion time

$$\tau_\eta = \mu L \int_a^b \sigma_m(r)dr \tag{33}$$

exceeds τ_I, the response is slower, its time-scale being $\tau_c = \tau_I^{2/3}\tau_\eta^{1/3}$. For the thickness, L, of the conducting layer at the base of the mantle I take here 10^5 m since at this distance the $\sigma_m(r)$ of (21), with

α = 30, has dropped to less than $\sigma_a/2$. Thus, Kolomiytseva's $\sigma_m(r)$ yields $\tau_\eta \sim 250$ years, while τ_I, using $\Gamma_F = 10^{19}$ Nm, obtains 2.5 years. Hence $\tau_c \sim 10$ years, probably short enough to explain the decade fluctuations. Other distributions of $\sigma_m(r)$ probably give similar results, since a reduction of τ_a leads to an increase of τ_I, and a decrease of τ_η. Thus, it appears that electromagnetic core-mantle coupling can be saved as a cause of the decade fluctuations of the length of the day. The assumptions concerning mantle conductivity must — as Professor Hide puts it — be pushed to their limits, but not beyond.

Acknowledgements. I am grateful to Professors F.H. Busse, J.D.A. Piper, M.G. Rochester, and P.H. Roberts for very helpful and encouraging correspondence.

References

(Russian journals are quoted by their English translations)

Backus GE (1958) A class of self-sustaining dissipative spherical dynamos. Ann Phys 4:372

Barraclough DR, Harwood JM, Leaton BR, Malin SRC (1975) A model of the geomagnetic field at epoch 1975. Geophys J R Astron Soc 43:645

Braginsky SI (1964a) Self-excitation of a magnetic field during the motion of a highly conducting fluid. Sov Phys JETP 20:726

Braginsky SI (1964b) Magnetohydrodynamics of the Earth's core. Geomagn Aeron 4:698

Braginsky SI (1975) Nearly axially symmetric model of the hydromagnetic dynamo of the Earth. Geomagn Aeron 15:122

Braginsky SI (1976) On the nearly axially-symmetrical model of the hydromagnetic dynamo of the Earth. Phys Earth Planet Inter 11:191

Braginsky SI (1978) Nearly axially symmetric model of the hydromagnetic dynamo of the Earth. Geomagn Aeron 18:225

Braginsky SI, Fishman VM (1976) Electromagnetic interaction of the core and mantle when electrical conductivity is concentrated near the core boundary. Geomagn Aeron 16:443

Brosche P (1981) Geomagnetic reversals and tidal friction. Naturwissenschaften 68:139

Bullard EC (1955) The stability of a homopolar dynamo. Proc Cambridge Philos Soc 51:744

Bullard EC, Gellman H (1954) Homogeneous dynamos and terrestrial magnetism. Philos Trans R Soc London Ser A 247:213

Bullard EC, Freedman C, Gellman H, Nixon J (1950) The westward drift of the Earth's magnetic field. Phil Trans R Soc London Ser A 243:67

Busse FH (1970) Thermal instabilities in rapidly rotating systems. J Fluid Mech 44:441

Busse FH (1975) A model of the geodynamo. Geophys J R Astron Soc 42:437
Busse FH (1976) Generation of planetary magnetism by convection. Phys Earth Planet Inter 12:350
Busse FH (1978) Magnetohydrodynamics of the Earth's dynamo. Annu Rev Fluid Mech 10:435
Busse FH (1979) Some new results on spherical dynamos. Phys Earth Planet Inter 20:152
Childress S (1969) Théorie magnétohydrodynamique de l'effet dynamo. Rep Dep Mech Fac Sci, Paris
Cook AE, Roberts PH (1970) The Rikitake two-disc dynamo system. Proc Cambridge Philos Soc 68:547
Cowling TG (1933) The magnetic field of sunspots. Mon Not R Astron Soc 94:39
Deinzer W, v Kusserow H-U, Stix M (1974) Steady and oscillatory $\alpha\omega$-dynamos. Astron Astrophys 36:69
Durney BR, Mihalas D, Robinson RD (1981) A preliminary interpretation of stellar chromospheric CaII emission within the framework of stellar dynamo theory. Publ Astron Soc Pac 93:537
Elsasser WM (1946) Induction effects in terrestrial magnetism, 1. Theory. Phys Rev 69:106
Gauß CF (1839) Allgemeine Theorie des Erdmagnetismus. Resultate aus den Beobachtungen des magnetischen Vereins im Jahre 1838, Leipzig
Gubbins D (1974) Theories of the geomagnetic and solar dynamos. Rev Geophys Space Phys 12:137
Gubbins D (1977) Energetics of the Earth's core. J Geophys 43:453
Gubbins D, Masters TG, Jacobs JA (1979) Thermal evolution of the Earth's core. Geophys J R Astron Soc 59:57
Hassan MHA, Eltayeb IA (1982) On the topographic coupling at the core-mantle interface (Preprint)
Hide R (1969) Interaction between the Earth's liquid core and solid mantle. Nature (London) 222:1055
Hide R (1977) Towards a theory of irregular variations in the length of the day and core-mantle coupling. Philos Trans R Soc London Ser A 284:547
Jacobs JA (1975) The Earth's core. Academic Press, London New York
James RW, Roberts PH, Winch DE (1980) The Cowling anti-dynamo theorem. Geophys Astrophys Fluid Dyn 15:149
Kolomiytseva GI (1972) Distribution of electric conductivity in the mantle of the Earth, according to data on secular geomagnetic field variations. Geomagn Aeron 12:938
Krause F, Rädler K-H (1980) Mean-field magnetohydrodynamics and dynamo theory. Akademie, Berlin, and Pergamon Press, Oxford
Lambeck K (1980) The Earth's variable rotation: Geophysical causes and consequences. Cambridge Univ Press, Cambridge
Landstreet JD, Borra EF, Angel JRP, Illing RME (1975) A search for strong magnetic fields in rapidly rotating Ap. stars. Astrophys J 201:624
Larmor J (1919) The magnetisation of a rotating body. Rep Br Assoc Adv Sci, p 159

Loper DE (1978a) The gravitationally powered dynamo. Geophys J R Astron Soc 54:389
Loper DE (1978b) Some thermal consequences of a gravitationally powered dynamo. J Geophys Res 83:5961
Loper DE, Roberts PH (1978) On the motion of an iron-alloy core containing a slurry. I. General theory. Geophys Astrophys Fluid Dyn 9:289
Loper DE, Roberts PH (1980) On the motion of an iron-alloy core containing a slurry. II. A simple model. Geophys Astrophys Fluid Dyn 16:83
Loper DE, Roberts PH (1981) A study of conditions at the inner core boundary of the Earth. Phys Earth Planet Inter 24:302
Loper DE, Roberts PH (1982) Compositional convection and the gravitationally powered dynamo. Geophys Astrophys Fluid Dyn (in press)
McDonald KL (1957) Penetration of the geomagnetic secular field through a mantle with variable conductivity. J Geophys Res 62:117
McElhinny MW, Evans ME (1968) An investigation of the strength of the geomagnetic field in the early Precambrian. Phys Earth Planet Inter 1:485
Mestel L (1976) Ap-star models. In Weiss WW, Jenkner H, Wood J (eds) Physics of Ap. stars. IAU-Coll. No 32, Wien, p 1
Moffatt HK (1978) Magnetic field generation in electrically conducting fluids. Cambridge Univ Press, Cambridge
Munk W, Revelle R (1952) On the geophysical interpretation of irregularities in the rotation of the Earth. Mon Not R Astron Soc Geophys Suppl 6:331
Parker EN (1979) Cosmical magnetic fields. Clarendon Press, Oxford
Rikitake T (1966) Electromagnetism and the Earth's interior. Elsevier, Amsterdam
Roberts PH (1972) Electromagnetic core-mantle coupling. J Geomagn Geoelectr 24:231
Roberts PH (1978) Magneto-convection in a rapidly rotating fluid. In Roberts PH, Soward AM (eds) Rotating fluids in geophysics. Academic Press, London New York, p 421
Roberts PH, Soward AM (1972) Magnetohydrodynamics of the Earth's core. Annu Rev Fluid Mech 4:117
Rochester MG (1960) Geomagnetic westward drift and irregularities in the Earth's rotation. Philos Trans R Soc London Ser A 252:531
Roden RB (1963) Electromagnetic core-mantle coupling. Geophys J 7:361
Skumanich A (1972) Time scales for CaII emission decay, rotational braking, and lithium depletion. Astrophys J 171:565
Sletteback A (1970) Stellar rotation. Reidel, Dordrecht-Holland
Soward AM (1979) Convection driven dynamos. Phys Earth Planet Inter 20:134
Stix M (1977) Planetarische dynamos. J Geophys 43:695
Taylor JB (1963) The magnetohydrodynamics of a rotating fluid and the Earth's dynamo problem. Proc R Soc London Ser A 274:274
Urey HC (1952) The planets: Their origin and development. Yale Univ Press, New Haven

Vaughan AH, Preston GW (1980) A survey of chromospheric CaII H and K emission in field stars of the solar neighborhood. Publ Astron Soc Pac 92:385

Vaughan AH, Baliunas SL, Middelkoop F, Hartmann LW, Mihalas D, Noyes RW, Preston GW (1981) Astrophys J 250:276

Gravitational Heating of Jovian Satellites by Tidal Friction

Z. Kopal[1]

Until quite recently, astronomical problems associated with tidal friction have arisen mainly in connection with the dynamics of the Earth-Moon system. However, since 1979, when the existence of phenomena thought to be of volcanic origin were detected by the Voyager spacecraft on the Jovian satellite Io, tidal friction was regarded also (among other possibilities) as providing, the requisite source of heat. Peale et al. (1979) or Yoder (1979) invoked to this end the dissipation of dynamical tides raised by Jupiter on Io considered as a solid globe of imperfect elasticity – an interesting idea even if it was not quite new; for the present writer proposed the same idea many years before (cf. Kopal 1963) in connection with our Moon.

The approach followed by Kopal and Peale et al. differed, to be sure, in several respects. In 1963, Kopal regarded, for this purpose, our Moon as behaving as a fluid globe of high viscosity; whereas Peale et al. considered Io as a solid globe of imperfect elasticity. But inasmuch as in both approaches the respective mechanism is activated by the same type of dynamical tides, their description of dissipative phenomena is essentially the same; and differs mainly in numerical values of the constants adopted.

The aim of the present paper will, however, be to demonstrate that – whatever values we adopt for these constants – the gravitational heating produced by internal dissipation of dynamical tides in the Jovian system – if significant – would not be limited to Io, but should be operative even more effectively in the interiors of the remaining Galilean satellites (Jupiter II - IV) at which no traces of volcanic activity were reported by the Voyagers.

In order to demonstrate this, let us avail ourselves of the viscous-fluid approach, for the details of which the reader can be referred to Kopal's 1963 paper; and only the results need to be invoked here. These showed that the temperature T produced in the interior of a viscous globe by dissipation of dynamical tides of second-harmonic symmetry (the most important partial tide under the circumstances) can be expressed as

1 Department of Astronomy, University of Manchester, Manchester, United Kingdom

Tidal Friction and the Earth's Rotation II
ed. by Brosche/Sündermann

$$T = (\mu/k)C\tau(r,t) \text{ degrees} \quad , \tag{1}$$

where μ stands for the effective viscosity of the respective configuration; k, its coefficient of heat conduction; and

$$C = 3\left\{12\, r_{\mathfrak{C}} \frac{eh_2}{P} \frac{m_J}{m_{\mathfrak{C}}} \frac{R_{\mathfrak{C}}^3}{A_{\mathfrak{C}}^3}\right\}^2 \tag{2}$$

is a constant, in which A stands for the semi-major axis of the orbit of the respective satellite; e, its eccentricity; P, its orbital period; m_J/m_C, the mass-ratio Jupiter: satellite; $R_{\mathfrak{C}}$, the radius of the satellite; and h_2, the radial Love number for which (by analogy with our Moon) a constant value of 0.76 will hereafter be adopted. Lastly, the function $\tau(r,t)$ represents the distribution of heat produced by friction in radial distance r from the satellite's centre, and the time t which elapsed since heat production by tidal friction became operative.

For the explicit form of this function and its tabulations the reader is referred to our previous paper (Kopal 1963). After a time which, in our case, should be close to 4.5 billion years, and for plausible values of thermal conductivity of the constituent material, the numerical values of $\tau(R_{\mathfrak{C}},\ 4.5\times 10^9 \text{ yr})$ turn out to lie between 0.0015 and 0.0025. The corresponding rate of heat flow across the surface $r = R_{\mathfrak{C}}$ is then given by the equation

$$\frac{\partial\tau}{\partial r} = \frac{\mu C}{K}\left(\frac{\partial\tau}{\partial r}\right)_{R_{\mathfrak{C}}} \quad . \tag{3}$$

In our paper already referred to, applications of this theory were carried out for the Earth-Moon system as well as for the effects of the tides raised on Mercury and Venus by the Sun. In the case of our Moon, gravitational heating produced by dissipation of Earth-raised tides turned out, in fact, sufficient for plausible values of the parameters involved to raise its internal temperature by several hundred degrees over and above that produced by radiogenic heat; and subsequent discovery of recurrence of moonquakes in the period of one synodic month underlined the significance of bodily tides raised on the Moon by our Earth. In the Sun-Mercury and Sun-Venus systems these phenomena proved, however, to be insignificant; and conclusion arrived at that..." the Earth-Moon system appears to be the only case in the solar family where gravitational heating due to dissipation of bodily tides could be of cosmogonic significance" (Kopal 1963, p. 420).

The recent discovery of (supposedly) thermal phenomena on the surface of Io in 1979 prompt us to re-open this question de novo, on the basis of our 1963 theory, for all Galilean satellites of the Jovian system; and the basic data are collected in Table 1: Let us consider,

Table 1. Basa data (see text)

Satellite	P (in days)	$R_{\mathfrak{c}}$ (in km)	$R_{\mathfrak{c}}/P$ (in cm s^{-1})	$10^3(R_{\mathfrak{c}}/A)$	e	$10^{-4}(m_J/m_{\mathfrak{c}})$	$10^7 C$ (in cm^2 s^{-2})
Io	1.769	1820	1191	4.315	0.00001	2.149	1.055
Europa	3.551	1540	502.1	2.295	0.00013	3.807	2.250
Ganymedes	7.155	2635	426.2	2.462	0.00139	1.279	31.86
Callisto	16.689	2500	173.4	1.328	0.00736	1.791	7.136

in their light, what the theory leads us to expect. Io — the satellite nearest to Jupiter — possesses a mass equal to 0.83 that of our Moon, and a radius only 1.05 times as large. If, therefore, we adopt for μ a value of the order of 10^8 g cm^{-1} s^{-1} estimated for the Moon, and $k \sim 5 \times 10^5$ erg cm^{-1} s^{-1} deg^{-1} appropriate for silicate rocks (cf. Birch 1952), it would follow from (1) that

$$T \simeq 10^{18}(2 \times 10^{-6}) \times 2.1 \times 10^{-7} \times 0.002 \simeq 840 \text{ K} \quad ;$$

not very far from that established in 1963 for our Moon (for which $C = 1.79 \times 10^{-7}$ cm^2 s^{-2}).

However — and this is evident at a glance at the values of C for the Jovian satellites as listed in the last column of our tabulation — if the internal structures of the remaining Galilean satellites of Jupiter do not differ greatly from each other, the gravitational heat production in each should be proportional to C; and its value for Ganymede turns out to be 30 times as large for Io. For Callisto it is 6 3/4 times as large; and even for Europa it is more than twice as large as for Io, which is characterized by the smallest value of C of the form.

The reason is the fact that the height of the second-harmonic dynamical tides is proportional to the orbital eccentricity e of the respective satellite; and, consequently, the value of C is proportional to e^2 which between Io and Callisto varies by a factor of more than 500,000. This is the principal reason why the C's for Jupiter III and IV are so large, and their disparity can be offset neither by their larger masses, nor by their greater distances from the central planet.

It is, of course, possible that, for these latter satellites (as well as Europa), the ratios μ/k factoring C on the right-hand side of (1) may be much less than for our Moon; and as long as we do not know their values with greater accuracy, it is impossible to predict the absolute values of internal temperatures in these satellites which can be produced by gravitational heating. But it is difficult to escape the conclusion that if such a process is significantly operative in the interiors of the first four Jovian satellites, the warmest of them should be Ganymedes, followed by Callisto and Europa; with Io holding the last place.

Observations made by the Voyagers (or, indeed, by any other method) failed, however, to note any thermal anomaly for Ganymedes or Callisto. The presumption is, therefore, strong that gravitational heating plays no significant role in Jovian satellites; and that internal heat needed to account for the phenomena observed in Io and ascribed to sub-surface volcanism must be produced by some other process (chemical, or electromagnetic) which does not affect other satellites; electromagnetic heating (cf., e.g., Drobyshevski 1979, or Gold 1979) for Io being the likeliest of the three.

In conclusion, one retrospective remark may be mentioned in this place. Much of the belief in the current volcanic activity on Io is based on the observed discharge of gas from its surface recorded by the Voyagers in March and July 1979, and regarded as volcanic plumes dispersing into vacuum of empty space in which that satellite revolves. No measurements of the actual temperatures of these plumes have been made by the Voyagers at the time of their fly-bys; so that we do not know if the escaping material was hot or cold. If, however, the velocity with which these gases escaped from the satellite could be established from evidence as yet unpublished, their absolute temperatures could be inferred indirectly – by virtue of the fact that the velocity of gas discharge into vacuum is known to be one-third of that of sound; and the latter is, in turn, proportional to the square-root of the absolute temperature. But if the requisite data on the velocity of dispersal of the plumes are available, their utilization for estimates of their temperatures remains still a task for the future.

References

Birch F (1952) J Geophys Res 57:227
Drobyshevski EM (1979) Nature (London) 282:811
Gold T (1979) Science 206:1071
Kopal Z (1963) Icarus 1:412
Peale SJ, Cassen P, Reynolds RT (1979) Science 203:892
Yoder CF (1979) Nature (London) 279:767

Balance Problems in Tidal Computations

H.-R. Baader[1]

1. Introduction

1.1 Observational Facts

Empirical evidence suggests the following secular changes in the Earth-Moon system:

- a slowing down of the Earth's rotation with an approximate rate of 5.5×10^{-22} rad s^{-2},
- an increase of the Moon's geocentric distance by nearly 1.4×10^{-9} m s^{-1}, and
- a deceleration of the lunar orbital mean motion by about 1.5×10^{-23} rad s^{-2}.

This knowledge is obtained from paleontological (growth increments of corals, molluscs, stromatolites), astronomical (solar eclipses, Mercury passages), and quite recently "geodetical" (lunar laser ranging, orbit perturbations of artificial satellites) information.

The dominating *lunar tidal interaction with the oceans* mainly via the M_2 partial tide (in comparison with solar, planetary and atmospheric effects) is regarded as the most plausible physical reason (e.g., see Bartels 1957, Munk and MacDonald 1960, Lambeck 1977, 1980, Brosche 1980). An example for another more exotic proposal is the decrease of the gravitational constant (e.g., Wesson 1980).

1.2 Theoretical Approach

The mentioned changing quantities are connected by the explicit conservation equations for mass, momentum, angular momentum and (total) energy in celestinal-mechanics and hydrodynamics (or respectively elasto-mechanics). These laws describe the interactions (i.e., flows of the corresponding quantities) between the individual fluid (or solid) volume units and (in the formulation of spatially integrated

1 Sternwarte der Universität Bonn, Auf dem Hügel 71, D-5300 Bonn, Fed. Rep. of Germany

Tidal Friction and the Earth's Rotation II
ed. by Brosche/Sündermann

balance equations) apply to the transfers and conversions within the Earth-Moon system on a time scale of tidal periods. Their celestial-mechanical versions ultimately determine the system's evolution on a billion year scale, with time (or Earth-Moon distance) functions of the lunar-terrestrial angular momentum transfer or the total dissipation (i.e., conversion to internal energy) being the crucial key (e.g., Lambeck 1980). The (extended Kepler-) problem of finding reasonable solutions for our changing evolutionary quantities and especially the question of a lunar near-by position (Gerstenkorn event) are still open. The reason is the up to now at best fragmentary (or rather too approximate) "construction" of these key functions (e.g., Sündermann and Brosche 1978, Krohn and Sündermann 1982, this Vol., see also the rough estimate of Sect.2.5 based on a drastic assumption).

For that purpose the evaluation of the corresponding period scale balances is necessary. Due to the lack of appropriate and sufficient measurements and owing to the absence of realistic analytical solutions the missing data (for the volume units or "parcels") must be generated numerically: as for the ocean by the normal three-dimensional (i.e., independent of the third, radial space coordinate) models of the (linearized, periodic) Laplace – or the (non-linear) hydrodynamic-numerical (HN) type (e.g., Hendershott 1972, Pekeris and Accad 1969, or Zahel 1970, Sündermann and Brosche 1978). This procedure requires given internal parameters as elasticity or viscosity constants. (On the other hand, conclusions about the internal parameters are possible if measured short-time data of the evolution quantities are at hand. As in a puzzle, the pieces must be moved to and fro, in order to reach a better understanding).

The basic balances also constitute the "threads" of this study. Its subjects are the angular momentum and mechanical (i.e., kinetic plus potential) energy interactions in a simplified Earth-Moon system consisting of a point-mass Moon in an equatorial planar orbit, a rigid solid Earth and a radially homogeneous ocean.

1.3 Motivation and Aims

This study was motivated by the apparently missing complete and critical synopsis of the theoretical and numerical problems associated with the angular momentum and energy balances. As for the *theory*, astronomers are only interested in the evolutionary aspects of the Earth-Moon system and are not familiar with the implications of hydrodynamics in general and of the ocean models in particular. On the other hand, oceanographers determine quantities which do not belong to their discipline and therefore are not fully understood in their interrelations. Their models are constructed only to depict the tidal currents and water elevations, but are possibly not accurate enough for our purpose. Thus a presentation of the theoretical points with all their aspects, especially of the individual transfers, conversions and constraints seems necessary in our interdisciplinary discussion. (See Sect.2 plus the extra, analogically organized appendix containing the mathematical details.)

As for the *numerical* aspect of the budgets, only estimates of individual terms have been published, but not complete evaluations. The latter would allow for conclusions about the accuracy and reliability of the results. This will be dealt with in Sect.3. An (extensive) list of symbols is provided at the end of the paper.

2. Theoretical Considerations

2.1 General Physical and Celestial-Mechanical Remarks

2.1.1 On the Terminology of Balance Equations

At the beginning it seems convenient to premise some remarks about the general form and about the terminology of balance equations because of their important role in the following. Balance equations for continua (e.g., Glansdorff and Prigogine 1971) can be formulated in two ways [see Sect.A.1.1: Eqs. (1,2)]:

- for an *extensive quantity* (depending on the volume of the continuum) with the resulting statement: "the total change in time corresponds to the production per time unit inside the volume plus the flux of that quantity through the surface"
- or for the adequate *intensive quantity* (depending on the density) in local form: "the local time change in this quantity consists of the source per time and volume unit and the divergence of the flux density".

Here the second formulation is preferably used: as much as possible should be subsumed in the flux divergences. The conservation property is expressed in a zero source term.

2.1.2 Special Balance Equations

The balance equations of mass, momentum, angular momentum and of the kinetic, potential and internal energy forms for a continuous medium are useful in this study. They are compiled with their sources and fluxes in Sect.A.1.2. The system of hydrodynamic differential equations from which the formulation of the ocean model starts is consistent with the balances of the first two quantities. Apparent forces — centrifugal-, Coriolis- and effects due to the acceleration of translation and rotation — were taken into account in order to allow for non-Newtonian coordinate frames. In the potential energy equation this quantity is interpreted as a quantity of matter advected or transported away by the motion. Potentials which directly depend on time, and consequently non-conservative forces in the strict sense have also been admitted.

2.1.3 "Absolute" Balances of the Earth-Moon System

Owing to the assumption of a mechanically isolated Earth-Moon system, i.e., of vanishing exterior mass fluxes, forces and torques, its center-of-mass (hereafter: c.o.m.) favourably serves as origin of the *absolute* coordinates. (The problem-dependent specification of the axes is not necessary at the moment.) Their advantages are the exact and rather simple-looking formulation of the whole system's balances and accelerations free of apparent forces. Unfortunately they are unsuitable on evaluation grounds: data, if measured or model-generated at all, will refer to *relative* (or accelerated) coordinates. A "natural" origin (in the case of the Earth) is the planet's c.o.m. or — preferably in the case of a rigid solid Earth (e.g., because of the time-fixed continental edges) — that of the solid body alone. Then, according to the first case, the (absolute) distances, velocities and accelerations are split (for Earth and Moon respectively) into c.o.m.-, rotational and relative parts (see Sect.A.1.3). In the (absolute balance) equations of angular momentum and kinetic energy, the corresponding *relative balance* terms plus contributions by the reduced mass, by the tensor of inertia, by angular velocity and its acceleration appear. The second possibility leads to additional contributions depending on the changes in distance between the oceanic and solid Earth centres-of-mass.

A direct evaluation would be possible if the accelerations were known besides distances and velocities which is not the case. Thus they must be inferred from the forces (see Sect.A.1.2); this involves the calculation of the relative balances with their independence from the collective (i.e., c.o.m.-) dynamics. This again indicates the completely different meaning of the absolute and the relative balance equations (e.g., Brosche and Sündermann 1971).

2.1.4 Interaction of the Subsystems

Forces also imply angular momentum and kinetic energy interactions of the now simplified lunar, solid Earth, and oceanic subsystems. This consideration is based on their individual *relative* balances:

- All lunar terms vanish trivially (due to point-mass assumption).
- As for the solid Earth's angular momentum, the assumed rigidity implies a zero sum of apparent, body and surface forces or the statement of the Euler equation, as it should be. Here, the torque exerted in the spin direction especially by the oceanic surface

stress forces produces an acceleration of the Earth's rotation and, consequently, a changing rotational energy. The (relative) kinetic energy balance has only trivial zero terms.
- The oceanic balances are considered in more detail in Sect.2.2 and show the disappearance of the corresponding relative changes (storages) in average during the period. Especially the surface stress torque (with reversed sign as compared to the solid Earth's budget) is then balanced by the interaction with the Moon via the tidal forces.

With this information and that about the disappearance of the period-averaged additional terms (only partly in the case of the presumably less important ones depending on the ocean's c.o.m., in spite of its elliptic path; see also Brosche and Hövel 1982) these two (formal) equivalences in the absolute balances of the whole system are confirmed:

- *that of the torque transferred from the oceans to the Moon with the changing orbital angular momentum of its reduced mass,*
- *and the one of the (reduced) Moon's mechanical energy gain with the decay of (solid) Earth's rotational energy minus the oceanic conversion to internal energy.*

The "instantaneous" picture is more complicated, but all terms can be evaluated by the results of the ocean model.

2.1.5 The Gerstenkorn Event

The basic (absolute) balances of the conservative quantities are used in a rough attempt to answer the question about the so-called Gerstenkorn event, i.e., about the instant of minimum distance between the Earth and the Moon. Essentially, this is a "time-scale problem". Starting from the actual estimate of the secular angular momentum transfer to the Moon, our trabant would have coincided with the position of the Earth 1.3 billion years ago. Thus this event should have happened within the age of the system. There is, however, no evidence for it. Therefore, something is perhaps wrong with the assumptions of this estimate. In the hypothesis of Sect.A.1.5 the Moon's orbital (radial and longitudinal) motion and the change of its angular momentum are expressed by a power relation with respect to the Earth-Moon distance. This is suggested by the *momentum* equation in cylinder coordinates, respectively by the minus three power dependence of tidal potential and of water elevation amplitudes (which can easily be seen in belt channel considerations). As modifications of powers seem to be of little effect, the proportionality "constants" must be strong time functions to shift this event further into the past. Oceanic processes of the time scale considered seem to work. They must be treated explicitly and may comprise a modified frictional and viscous behaviour as well as transits through resonance situations, induced by changing tidal frequencies and continental drifts (Brosche and Hövel 1982, Krohn and Sündermann 1982, this Vol.).

2.2 Oceanographical Remarks

2.2.1 The Consideration of Barotropic Tides and Its Implications

The rigorous consideration of the oceanic angular momentum- and energy-conservation and the sound interpretation demand a recapitulation of the manipulation-, assumption- and omission-steps prior to the budget evaluation (Sect.3). For the time order of tidal periods the acceleration of the Earth's rotation and of its c.o.m.- translation can be neglected (Sect.A.2.1). The centrifugal force is subsumed in the effective gravity (e.g., Holton 1973). The shape of the Earth and the aim of a global computation compel us to use spherical coordinates. In the Navier-Stokes equations the curvature terms which are manifestations of the direction changes of the local orthonormal unity vectors along the water parcel's trajectory must be retained as well as the full advection terms. Their normal omission (e.g., Sündermann and Brosche 1978) would produce spurious sources in the budgets, especially in that of the angular momentum (as for the magnitude see Sect.3). The radial component equation is substituted by the relation of hydrostatic equilibrium. This leads, by radial integration, to a formula for the pressure and to the introduction of an elevation-dependent pressure gradient force in the equations of the two other components.

This equilibrium assumption is legitimate only for macroscale processes. In this context the termini macroscale and, consequently small-scale, have a disposable, problem-dependent meaning; they are not used in the sense of a conventional classification of the oceanic fluctuations (in which tidal phenomena would belong to the mesoscale; e.g., Monin et al. 1977). Thus a discrimination is conceivable according to the "validity" of the Boussinesq and hydrostatic equilibrium hypotheses, but also according to the chosen resolution (grid constants) of the computation (e.g., for depth-averaged, global tidal models the processes, then defined macroscalar, must be relevant in space-time domains of, say, $>10^5$ m horizontal, $>10^3$ m radial, $>10^3$ s temporal range). In any case, the orginally kinematic viscosity must be reinterpreted as eddy-viscosity: This *Austauschparametrization* (e.g., Holton 1973) tries to account for the momentum interaction with smaller scales by normally two parameters because of the different exchange intensity with respect to radial and horizontal directions. Let us imagine full four-dimensional spectral expansions of the oceanic variables in order to explain the eddy-viscosity (see e.g., Baader 1978). A cut-off at the limits of the particular space-time domain considered leaves the macroscale parts. In the momentum equation, however, the smaller scale "amplitudes" give rise to correlation products which belong to the macroscale and may be separated by some

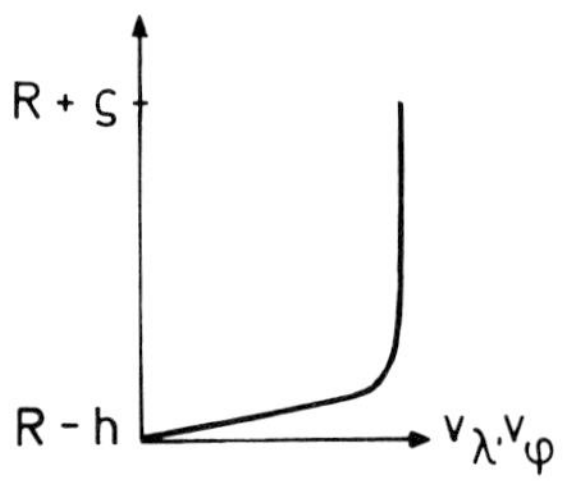

Sketch 1:
Velocity profile of a barotropic tide

Fig. 1. Radial dependence of the tidally induced horizontal velocities: constant generation throughout nearly the entire depth of the ocean. (see e.g., Monin et al. 1977)

filtering process (see also Fortak 1968). These correlations are equivalent to a force – to the eddy-viscosity – in the case of a formulation in macroscale quantities according to the mixing-length-hypothesis. In this hypothesis a fluid "parcel" carrying the average momentum of its environment for some characteristic distance (analogous to the mean free path of the molecular viscosity), produces a turbulent fluctuation at the new position: its effect, presuming isotropy, depends on this distance and on the mean velocity shear. Apart from the frequently questionable legitimation of those assumptions, evidently an eddy-viscosity parameter is not a material "constant", but is a function of the time- and space scalings and, moreover, of the dynamical state of the fluid (see Schwiderski 1980, Krauss 1980). Thus, the eddy-viscosity (its formula as well as the value of the "Austauschkoeffizient") is more or less at the modeler's disposition (see Schwiderski 1980). He will presume that all unresolved effects are lumped together and reasonably treated according to his law. The criteria for its verification, then, are the demonstration of numerical stability and the generation of realistic velocity- and elevation patterns in his model (e.g., Zahel 1970, Schwiderski 1980).

According to personal preference, the horizontal eddy-viscosity of a "barotropic" macroscale could then be applied in the usual formulation [Zahel 1977, Sündermann and Brosche 1978, A.2.1: Eq. (4)] or in one with the additional terms [perhaps reducing the eddy-viscosity effects in the balances, in favour of the bottom-friction; see Eq. (5)] which was found by means of the velocity profile (Fig. 1) as well as of suitable kinematic and dynamic boundary conditions. Because of the radial averaging (see Fortak 1962) such an Austauch-like (internal) viscosity and a further parametrization, the quadratic Newtonian friction at the sea-bottom interface, appear in the hydrodynamic differential equations of a barotropic fluid (Sündermann and Brosche 1978). These equations represent the basis of our thus closed oceanographic model.

Of course the "macroscale" balances which are the subject of the following three sections, are not identical with the complete budgets. For example, in the angular momentum balance the dissipative tensor should normally result in a zero surface integral (Weizel 1962) by means of suitable boundary conditions; the introduction of different radial and horizontal eddy-viscosities leads to a contribution indeed. Consequently, budget parts exist due to the smaller scales (Figs. 2,3; stressed quantities) which cannot be resolved, but, with appropriate assumptions, can be interpreted reasonably.

2.2.2 The Budget of Angular Momentum

The oceanic budget of the (relative macroscalar) angular momentum in the direction of Earth rotation is deduced from the equation of zonal motion. There are five contributions (Fig. 2) to the storage:

- the well-known transfer to and from the (reduced) Moon;
- the normal bottom-friction torque;
- the often disregarded eddy-viscosity contribution describing a transfer to and from the smaller scales (see Sect.2.2.1), which (especially in the period average and neglecting small-scale transfer to the Moon) may be largely attributed to friction- and pressure interactions at the sea-bottom interface (i.e., the bottom-friction does not comprise the total pressure and stress effects there);
- the torque by the pressure gradient force which, like its meteorological analogon, the so-called "mountain-torque" (e.g., Holton 1973; Oort and Bowman 1974) stems from different hydrostatic pressures at both sides of meridional sea bottom variations, e.g., ridges, valleys, coastal slopes etc., thus pushing the solid Earth around;
- and a term originating in the Coriolis force which can be transformed in a time change of the so-called Ω-momentum (i.e., absolute angular momentum of parcels fixed to the Earth's surface, Holton 1973) and which may be subsumed in the storage.

This may be compressed in the statement:

The net result of the solid Earth's feeding torque mediated by bottom friction, lateral eddy-viscosity (via smaller scales) and pressure gradient forces minus the transfer to the Moon causes an identical increase of stored angular momentum.

It implicitly anticipates the flow directions in the time mean, when all storage terms should have zero values (which is exact at least in the case of a single partial tide, because of periodicity). Even then the internal viscosity- and pressure gradient force terms must not simply by ignored a priori (as it was still done by Sündermann and Brosche 1978, and Brosche and Sündermann 1979, and others). In general, the first term will contribute (except for the zero "Austauschkoeffizient") as will the other (except for harmonic and at shore zero-valued water elevations) thus leading to a "three-mode exchange" with the solid Earth (also Baader 1979, pers. Commun. to J. Sündermann).

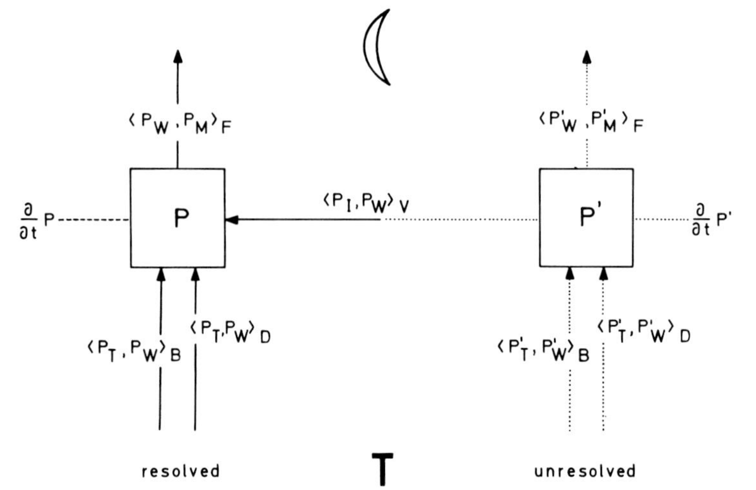

Sketch 2:

Tentative schematic box diagram of relative angular momentum transfer

Fig. 2a

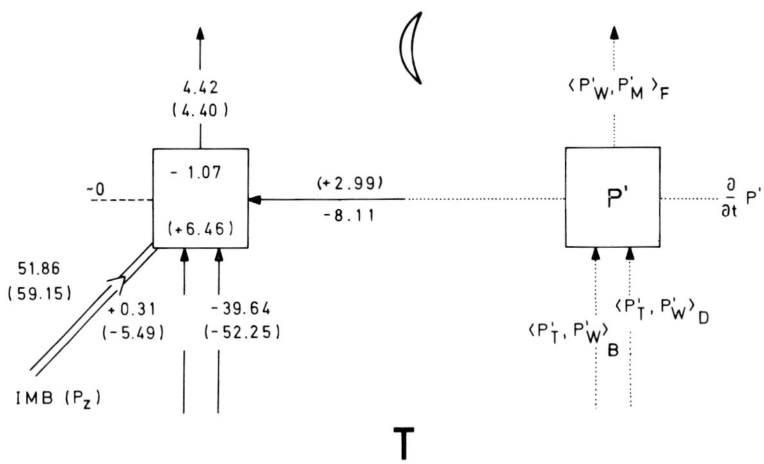

Box diagram of sketch 2:

Transfers in angular momentum: 10^{16} kg m^2 s^{-2}

Contents in angular momentum: 10^{22} kg m^2 s^{-1}

Values after 500 (400) periods

Fig. 2b

Fig. 2a,b. Oceanic budget of (relative macroscalar) angular momentum (P). (a) Theoretical budget: storage ($\partial/\partial t$ P, containing the Coriolis contribution) balanced by the transfer ($<P_W, P_M>_F$) to the Moon (☾), by flows from solid Earth (T) due to bottom friction ($<P_T, P_W>_B$) and pressure gradient force ($<P_T, P_W>_D$) and by internal exchange ($<P_T, P_W>_V$) with the unresolved smaller scale reservoir (stressed quantities, analogous flows); flow direction as expected in time average. (b) Budget

2.2.3 The Budget of Kinetic Energy

The oceanic budget of (relative macroscalar) kinetic energy is deduced in Sect.A.2.3. It means (Fig. 3):

The storage of kinetic energy is the net result of conversion from potential energy, i.e., the work done by the water "parcels" against tidal and pressure gradient forces minus the conversion to what may be called the internal energy of smaller scales via bottom friction and (lateral) eddy-viscosity.

The latter conversion is not completely identical with the *dissipation* which is rather a heat transfer to the internal energy of the statistical, unordered motions connected to temperature. The kinetic energy reservoir of the smaller-scale collective motions, for its part in interaction with a small-scale potential energy, is inserted. Since in time (period) mean a negligible generation of this energy is expected and all energy storages should not contribute (in analogy to the angular momentum budget, Sect.2.2.2), equivalence of the potential-to-kinetic energy conversion, with that to small-scale internal energy, with dissipation and ultimately, with the heat flow out of the ocean then exists.

2.2.4 The Budget of Potential Energy

The oceanic budget of (relative macroscalar) potential energy is also indicated in Fig. 3. This energy form is defined in a somewhat new, but convenient manner (Sect.A.2.4). It consists of two parts: one originating from the Earth's gravity field, the other from the tidal potential. Balance considerations require:

The net result of the input, i.e., the generation, minus the output, i.e., the conversion to kinetic energy via tidal and pressure gradient forces serves to raise the potential energy reservoir contents.

In the time (period) mean the storage and the gravity-potential energy generation as well as, in consequence, the work against the pressure gradient forces should be zero. For the M_2-tide, the generation of the other potential energy form proves to be identical with the losses in the solid Earth's rotational plus the (reduced) Moon's mechanical energies.

Legend to Fig. 2a,b (cont.)

estimates for (relative macroscalar) angular momentum (z-component in direction of the Earth's rotation) in the ocean mediated by the M_2-tide: period averages. Flows as indicated by arrows. Exchange with solid Earth is not reasonable. (Imbalance IMB is too large: see Sect.3.2.1)

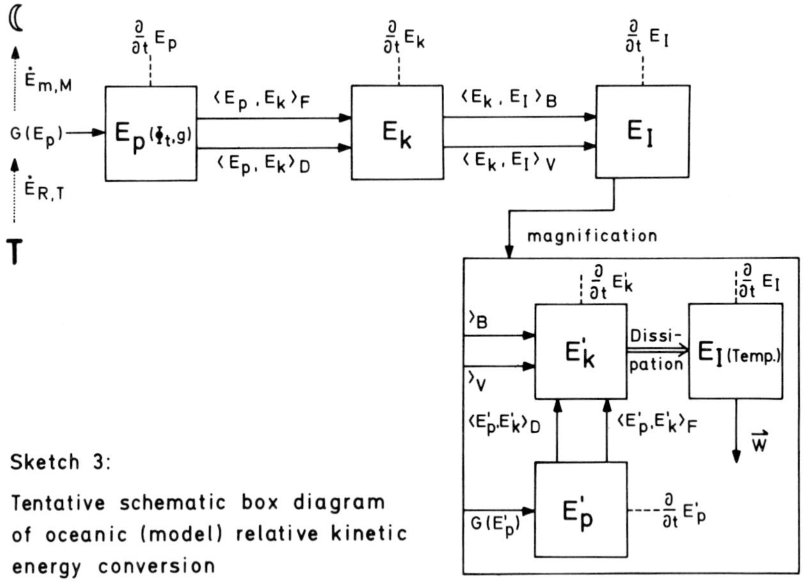

Sketch 3:

Tentative schematic box diagram of oceanic (model) relative kinetic energy conversion

Fig. 3a

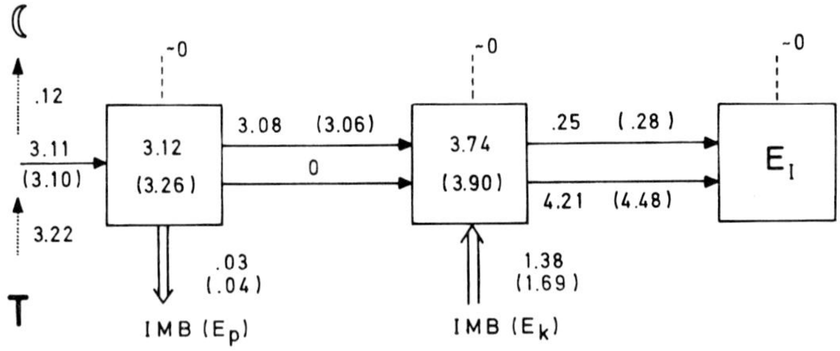

Box diagram of sketch 3:

Conversions: 10^{12} kg m^2s^{-3}

Contents of the reservoirs: 10^{17} kg m^2s^{-2}

Values after 500 (400) periods

Fig. 3b

Fig. 3a,b. Oceanic budget of (macroscalar relative) kinetic energy (E_k) connected with the corresponding budgets of potential (E_p) and internal (E_I) energy. (a) Theoretical budget (E_I (Temp.) that of the unordered fluctuations; stresses refer to small-scale quantities). Storages ($\partial/\partial t$) balanced by conversions (direction as expected in time mean): between potential and kinetic energy mediated by tidal ($\langle E_p, E_k\rangle_F$) and pressure gradient forces ($\langle E_p, E_k\rangle_D$); between kinetic and internal energy by bottom friction ($\langle E_k, E_I\rangle_B$) and lateral eddy-viscosity ($\langle E_k, E_I\rangle_V$). The generation of potential energy is fed by the changes of the Earth's rotational ($\dot{E}_{R,T}$) and the Moon's mechanical energy ($\dot{E}_{m,M}$). The always positive-definite dissipation partly changes

3. Numerical Estimates

3.1 General Remarks

Prior to the evaluation of the individual terms of the oceanic angular momentum-and mechanical energy balances (originating from the relative motion field, see Sects.2.2 and A.2) zonal and meridional velocities as well as tide elevations must be generated numerically, because of the scarcity of directly measured information. This is achieved by a modified version of Zahel's nonlinear $4^o \times 4^o$-HN-model (1970; including his eddy-viscosity and his formal allowance for solid Earth tides by the Love number $\gamma_L = 0.69$; see also Brosche and Hövel 1982, this Vol.). Subject here is the M_2-tide. In this model, velocity component- and elevation data only exist in their own, dislocated sets of grid points. Thus the actual evaluation program starts by a linear interpolation procedure with the result of a uniformly covered $2^o \times 2^o$ grid at each of the 128 moments during one period. For mass conservation the tidal elevation is slightly corrected to a zero oceanic average. Of course, these manipulations do not guarantee the exact validity of the continuity equation. As the following results indicate, this, especially, seems to play a role. As "good-natured" HN-models gradually converge to a stable, in our case periodic, solution, the generation is continued for several hundred simulated periods and after some intervals the balance equation program is repeated in order to gain some insight into the behaviour of the system. The results of 400 and 500 periods are reported. To the author's knowledge, other modelers usually stopped the generation process after some tens of periods, when a sufficient accuracy (2 digits) in the angular momentum transfer to the Moon had been reached.

In the numerical practice balance equations are not fulfilled exactly. The remainder is responsible for the so-called *imbalance* (IMB): a spurious contribution defined to leave no net budget flow. In our case it may stem from:

- not met constraints (e.g., theoretically zero contributions);
- lack of periodicity, presumed in our balance equations;
- discretization effects with respect to time- and space coordinates

---

Legend to Fig. 3a,b (cont.)

into the heat flux out of the seas. (b) Budget estimates for kinetic and potential energies of M_2-tide period averages. The theoretical value (A.2.4; as for the numerical estimate see Table 2.2) is given to the conversion due to the pressure gradient force. Energy flows as indicated by arrows. (Balance remainders, i.e., the imbalances IMB are acceptable.) Potential energy value due only to gravity (during the procedure of generation as well as of evaluation, e.g., during the mentioned manipulations at the beginning of the program.

Its magnitude should illustrate the significance and reliability of each estimated budget contribution. The "time-resolved" budgets given also, for which no other comparable information seems to exist in literature, serve the same purpose.

3.2 Angular Momenta with Respect to the Earth's Rotation and Budget Quantities

3.2.1 Period Mean

Whereas the energy balances (Sect.3.3) are in a quite satisfactory condition, this is unfortunately not the case for the angular momentum with respect to the Earth's rotation axis and its balance, at least in the period average. This is obvious at first glance at Table 1 or Fig. 2b, with their deterrent data composition. The most striking fact is the fluctuation of the majority of terms in magnitude and even sometimes in sign. That starts with the angular momentum itself and ends with a substantial Coriolis contribution, where theory demands zero. Significantly, the imbalance is dominating. Nonperiodicity, i.e., an insufficient damping-out of the oceanic eigenmotions, and discretization effects are chiefly responsible for its huge values. An example for the latter faults is quoted in the last line of Table 1A. The last two lines show that the full metric and convective terms as well as *nonlinearized* mass fluxes have to be considered for a reasonable angular momentum balance. Thus a Laplace model where even these terms are omitted leaves no chance for such a balance because of the inevitable spurious sources. Of course, the genuine transfers do not entail the same amount of errors. In analogy to the potential energy generation (Sect.3.3), the estimate of the transfer to the Moon is quite plausible. The reason for this is that the potential force is an exact quantity per definition, and derivatives do not have to be estimated by difference-quotients. Our model generates the velocity- and elevation data with a relative accuracy of only $O(10^{-2})$ (grid constants relative to total domain: $\Delta\lambda/360 = \Delta\phi/360 = 4/360$; $\sigma\Delta t/2\pi = 1/128$). The oceanic lunar transfer loses 5 significant digits on its way from a local value to an oceanic and period mean, but only 1 or 2 digits in the essential part depending on water elevation. (In the period mean the undisturbed water depth does not contribute because of its simple correlation with a periodic function.)

The Coriolis- and pressure gradient terms which comparatively lose seven significant digits are determined essentially by the time mean, i.e., $O(10^{-3})$ parts of the model data (see also Brosche and Hövel 1980). Thus these contributions cannot reliably be figured out by our model, but the transfer to the Moon can. Satisfying results can be expected

Table 1A,B. Angular momentum balance quantities for the world ocean relative to the Earth's rotation axis: M_2-period averages after 400 and 500 simulated periods. Attention: Domination of imbalance! Definitions: see A.2.2 and list of symbols

A) Reservoir contents in 10^{22} kg m^2 s^{-1}

Symbol	$t_{sim}(\sigma/2\pi) = 400$	$t_{sim}(\sigma/2\pi) = 500$
$P_{z,rel}$	6.46	- 1.07
$P^{**}_{z,rel}$	5.00×10^{-2}	4.37×10^{-2}

B) Transfers in 10^{16} kg m^2 s^{-2}

Symbol	$t_{sim}(\sigma/2\pi) = 400$	$t_{sim}(\sigma/2\pi) = 500$
$-\langle P_W, P_M\rangle_F$	- 4.40	- 4.42[a]
$\langle P_T, P_W\rangle_B$	- 5.49	0.31
$\langle P_T, P_W\rangle_V$	2.99	- 8.11
$\langle P_T, P_W\rangle_D$	- 52.25	- 39.64
$IMB(P_{z,rel})$	59.15	51.86
Coriolis = 0	- 35.76	4.40
metric term	1.23	1.17
$[2u\,(h+\zeta)\,\frac{\Delta}{\Delta\lambda}\,u + u^2\,\frac{\Delta}{\Delta\lambda}\,(h+\zeta)] = 0$	11.71	12.37

[a]This corresponds to a deceleration of Earth's rotation $\dot{\Omega} = -5.44 \times 10^{-22}$ rad s^{-2}, and to a decrease of solid Earth's rotational energy $\dot{E}_{R,T} = -3.22 \times 10^{12}$ kg m^2 s^{-3} (A.1.4). The following changes can be deduced as well (see A.1.5): of the Earth-Moon distance $\dot{a} = 1.52 \times 10^{-9}$ m s^{-1}, of the Moon's radial motion $\ddot{a} = -2.99 \times 10^{-27}$ m s^{-2}, of its longitudinal motion $\dot{n} = -1.58 \times 10^{-23}$ rad s^{-2}

when temporal and spatial "steps" are reduced to one tenth of the present ones. On the other hand, the basic continuity and motion equations are exact to at best 1 (relative part) per 10^{+4} of the estimated omissions (Sect.A.2.1). Moreover, detrimental "eigenfrequencies" enter a HN-model via the required fields of initial values. The choice of zero velocities and elevations, as in Zahel's model does not seem optimal: Analytical considerations in a linear belt channel with typical bottom friction and eddy-viscosity suggest decay times of 10^3 periods for resulting (10^{-3} - 10^{-4})-contributions of the eigensolutions. Therefore, the use of Laplace-model solutions (e.g., Pekeris and Accad 1969) for initial values in a HN model is proposed.

3.2.2 The "Time-Resolved" Angular Momentum Budget

Things look very much better in time resolution, as Fig. 4 reveals. The essential counterparts for the time change of *relative* angular momentum, i.e, storage, are the contributions due to the tidal potential, pressure gradient and Coriolis forces. The Coriolis torque may be interpreted as a time change of Ω- or rotational angular momentum (see Sects.2.2.2 and A.2.2). The terms mediated by eddy-viscosity and bottom-friction are of minor importance (please pay attention to the orders of magnitude). Surprisingly, the pressure gradient force and not the bottom-friction is the most relevant agent for the exchange with the solid Earth. The latter is responsible for the only term with a significant non-harmonic behaviour (i.e., the undisturbed water depth dominates in the other terms). Just as unexpected is the relatively small value of the angular momentum transfer to the Moon with only ~10% of the storage amplitude. The imbalance (not shown here) has the same order of magnitude.

3.3 Mechanical Energies and Budget Quantities

3.3.1 Period Means

Table 2 shows a compilation of some relevant (relative) kinetic energy quantities. For an easier survey, especially as far as the directions of the energy flows are concerned, most of this information is also condensed in Fig. 3b. Kinetic and potential energies, not surprisingly, compare reasonably with the few specifications found in the literature (Zahel 1977, Hendershott 1972). Roughly, they are equipartitioned, as are the kinetic energies of the zonal and meridional motions.

Only contributions of the undisturbed water depth are relevant. The values are quite stable, indicated by small differences after 400 or 500 simulated periods. The last property is also exhibited by the conversion terms. Surprisingly small is the contribution of the bottom

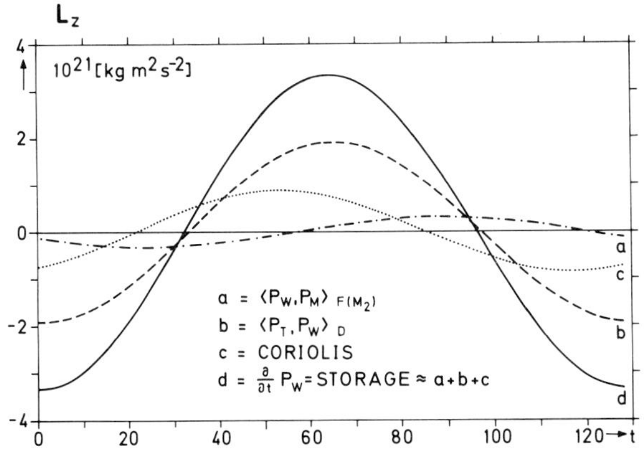

Fig. 4A

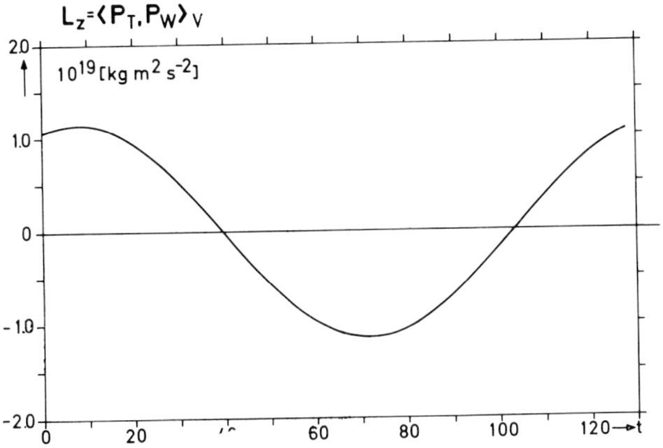

Fig. 4B

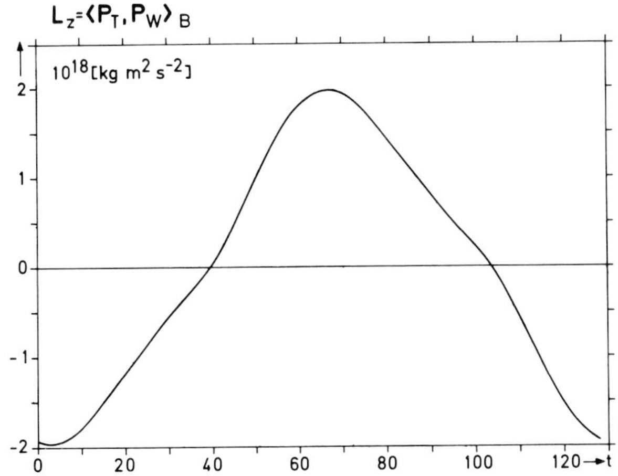

Fig. 4C

Fig. 4A-C. Angular momentum balance for the M_2-tide of the model world ocean: torques as time functions [t in $(2\pi/128 \times \sigma^{-1})$-units]. (A) Essential terms in 10^{21} kg m^2 s^{-2}: the transfers to the Moon (a), the transfer from solid Earth, due to the pressure gradient force (b); the storage of the Ω-momentum via inertia tensor changes (Coriolis torque; c); the storage of the relative angular momentum (d). (B) The angular momentum transfer from smaller scales to the macroscale via lateral viscosity. Units: 10^{19} kg m^2 s^{-2}. (C) The angular momentum transfer from solid Earth to the macroscale ocean due to bottom-friction. Units: 10^{18} kg m^2 s^{-2}

Table 2A,B. Some (relative) kinetic energy balance quantities for the world ocean: M_2-period averages after 400 and 500 simulated periods. Definitions: see A2.3 and list of symbols

A) Reservoir contents in 10^{17} kg m^2 s^{-2}

Symbol	$t_{sim}(\sigma/2\pi)$ = 400	$t_{sim}(\sigma/2\pi)$ = 500
E_{ku}	1.77	1.70
E_{kv}	2.13	2.05
$E_k = E_{ku} + E_{kv}$	3.90	3.74
E_k^{**}	1.85×10^{-5}	2.89×10^{-6}
E_{pg}	3.26	3.12

B) Energy conversions in 10^{12} kg m^2 s^{-3}

Symbol	$t_{sim}(\sigma/2\pi)$ = 400)	$t_{sim}(\sigma/2\pi)$ = 500
$\langle E_p,E_{ku}\rangle_F$	1.97	1.98
$\langle E_p,E_{ku}\rangle_D$	2.46	2.31
$-\langle E_{ku},E_I\rangle_B$	- 0.14	- 0.13
$-\langle E_{ku},E_I\rangle_V$	- 1.85	- 1.70
$-\langle E_{ku},E_{kv}\rangle_C$	- 1.80	- 1.71
$IMB(E_{ku}) = -\Sigma..$	- 0.64	- 0.75
$\langle E_p,E_{kv}\rangle_F$	1.10	1.10
$\langle E_p,E_{kv}\rangle_D$	- 0.54	- 0.47
$\langle E_{kv},E_I\rangle_B$	- 0.13	- 0.12
$-\langle E_{kv},E_I\rangle_V$	- 2.64	- 2.51
$\langle E_{ku},E_{kv}\rangle_C$	1.80	1.71
$IMB(E_{kv}) = -\Sigma..$	0.41	0.29
$\langle E_p,E_k\rangle_F$	3.06	3.08[c]
$\langle E_p,E_k\rangle_D$ (= 0)	1.93	1.84
$-\langle E_k,E_I\rangle_B$	0.28	- 0.25[b]
$-\langle E_k,E_I\rangle_V$	- 4.48	- 4.21[a]
$IMB(E_k) = -\Sigma..$	- 0.23 (+1.70)	- 0.46 (+1.38)

[a,b,c]This is equivalent to Q_V = 23,0, Q_B = 387.5, Q_F = 31.5, respectively (Q = 21.7, for the definition of the Q-factors, see list of symbols).

friction in comparison to that of the lateral eddy-viscosity (see Zahel 1977). In particular, it is the only term which is below the imbalance level. So far, no reliability can be attributed to it. Theoretically, kinetic energy input of work done against the potential forces should equal the corresponding output, i.e., the conversion to internal energy via friction forces.

As far as only information on kinetic energy balance is regarded, a fine agreement can be ascertained. But considering the additional information from the potential energy budget, that the work against the pressure gradient force should have equal posi- and negative zonal, respectively meridional parts and therefore a zero value, the total "dissipation" of 4.5 units is revealed as an overestimate of ~30% (Table 2B; normal values: (2.2 - 6.0) units: see Parke 1978, Pekeris and Accad 1969). This is a consequence of the expected approximate equality of the work against the tidal forces and the potential energy generation of 3.1 units (see Munk and MacDonald 1960, Hendershott 1972). The latter can be figured out from the angular momentum transfer to the Moon (see A.2.4). It represents the net energy transfer from the Earth's rotational and the (reduced) Moon's mechanical energy. This example shows that the imbalance must not be uniformely redistributed to the single budget terms. The possible neglect of any metric, advection, or non-linear mass-flux term would lead to a spurious contribution of $\sim 10^{-5}$ units. Thus the kinetic energy balance seems very insensitive to the linearization of the basic equations (Laplace model). The kinetic and potential energies (in the Earth's gravitiy field) as well as the dissipation rates due to bottom friction and internal friction were used to estimate oceanic Q-factors (definition: see list of symbols or Munk and MacDonald, 1960). The total dissipative effect seems to be reasonably represented (Table 2B), especially by the Q = 31.5 estimated by the generation term (compared to Q = 21.7 for the total dissipation and normal values Q~27-30; the peak mechanical energy was 6.9×10^{17} kg m^2 s^{-2}).

3.3.2 The "Time-Resolved" Kinetic Energy Budget

The relative importance of each contribution and of the errors become more obvious with the aid of Fig. 5, where the evaluated kinetic energy conversions for our M_2-model ocean are depicted as time functions during one period. Keeping in mind that the work against the pressure gradient force is equivalent to the negative storage of the potential energy in the Earth gravity field [see A.2.4: Eqs. (1.2)] two particular quasi-balances are evident:

- a quasi-conservation of that special potential plus kinetic energy (curves d,e: d ~ e);
- a quasi-equivalence of potential force- and lateral viscosity contributions, since the dissipation via bottom friction is negligible in a first approximation [curves a,b,c: a ~ -c; not shown Imb. ~ 0 (b)].

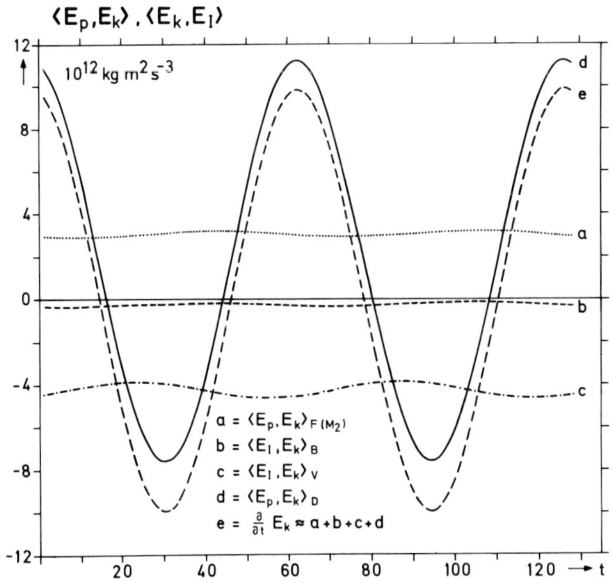

Fig. 5. Kinetic energy conversions for the M_2-tide of the model world ocean in 10^{12} kg m^2 s^{-3} as time function [t in $(2\pi/128 \cdot \sigma^{-1})$ -units]: from potential energy due to tidal-(a) and pressure gradient forces (d); to internal energy due to bottom-friction (b) and eddy-viscosity (c); the storage term (e)

In the period mean these particular balances should be fulfilled, the first with exact zero constituents, the second approximately. Thus we infer a *systematic* overestimation of the (gravitational) potential energy conversion (from: $d > e$; maxima versus minima) and of the internal viscosity effect (from: $a > c$; formulation of a: with the analytical derivative, of c: with the error-loaded double difference quotients). The essential causes are assumed to be mass conservation errors and spurious "eigensolutions" in the first case, in the second case the differing discrete formulation of the Laplace-operator in the generating, respectively evaluating program due to the grid interpolation. All curves, with the exception of the bottom-friction contribution, show a double-frequency behaviour. This is a clear indication that only the undisturbed depth is dominating in the terms [with metric and convective contributions of $O(10^{-2})$ units] thus being reasonably estimated by linear (Laplace) model data.

4. Conclusions

The topics of this study are the tidal interactions of a punctiform M_2-Moon, a (quasi-)rigid solid Earth, and a frictionally (at the bottom and internally) decelerated ocean. *Theory*, condensed of the conservation properties of angular momentum and energy, states for the period average:

- That amount of the angular momentum which is lost to the orbit of the ("reduced") Moon, is gained by the ocean from the decelerating solid Earth's spin via three mechanisms depending on: *bottom friction*, *small-scale exchange*, and *pressure gradient force*.
- Simultaneously the resulting diminution of the Earth's rotational energy serves to increase the Moon's orbital mechanical energy as well as, mainly, the (suitably defined) oceanic tidal potential energy, thus maintaining the kinetic energy of the motion field against the dissipation by bottom-friction and lateral eddy-viscosity.

Numerical computation provided fine unprecedented "time-resolved" results. The estimates of the period averages were also satisfying except for the solid Earth-ocean angular momentum exchange. Its failure, in the individual terms as well as in their sum, was not unexpected in view of the discretization errors. It would be the same in the case of the Laplace-model data. Owing to angular momentum conservation we must simply attribute the value of the oceanic-lunar transfer $\sim 4.4 \times 10^{16}$ kg m^2 s^{-2} to that exchange as well. Here the pressure gradient force mechanism seems to dominate (according to time resolution), leaving the oceanic system quasi-closed to the exterior tidal torque. Bottom-friction plays only a minor role, also in comparison to the internal friction. The same feature is obvious in the kinetic energy balance. There, a plausible secular value of $\sim 3.1 \times 10^{12}$ kg m^2 s^{-3} is estimated for (essentially internal) dissipation and potential energy generation. During the period two approximate subbalances appear:

Gravitational and potential kinetic energy are quasi-conservative, and the work against the tidal forces nearly corresponds to the internal dissipation. The small value of the bottom-friction contributions casts some doubts on foundation and results of the energy-flux method where dissipation is estimated via the energy transports from the deep ocean into shelf seas. As long as such a differing, possibly distorted relative importance of bottom-and internal friction is found, the question about a linear or a quadratic bottom-friction law in a model is secondary. It is more important to ask about the formula and coefficient of eddy-viscosity. A new formula is proposed. As internal (and bottom-) friction influences all torques and energy conversions, directly (e.g., dissipation) or indirectly (e.g., oceanic-lunar angular momentum transfer via the amplitudes and phases of the water elevations, see e.g., Pekeris and Accad 1969, Krauss 1973) it should be more than a disposable, stability producing quasi-numerical effect. Thus the associated problems of parametrization, calibration and "verification" demand further study.

The somewhat unclear treatment of frictional physics has been the weak point of all the models up to now. Until this situation improves, it is recommendable to use:

- the simplest (bottom- and internal) friction laws, i.e., linear ones (perhaps with some dependence on depth; e.g., Schwiderski 1980) and, consequently
- at best the (fully linearized) Laplace-formulation.

Laplace models require less computational effort (computer time and capacity) thus allowing for a relatively greater spatial resolution than the HN-models, and the time behaviour of their solutions (free of the undesired "eigenmotions" per definition) is easier to understand. Although physically reduced, Laplace models generate comparable and sufficiently exact data to evaluate our essential "interaction integrals" of the Earth-Moon system, namely, the secular oceanic-lunar angular momentum transfer, respectively the (new and proportional) potential energy generation in the seas.

Appendix

A.1 General Physical and Celestial-Mechanical Considerations

A.1.1 Balance Equations (General Form)

Two formulations are to be distinguished:

- for an extensive quantity X:

$$\frac{d}{dt} X = \left(\frac{d}{dt} X\right)_i + \left(\frac{d}{dt} X\right)_e$$

$$= Q[X] + \Psi[X]$$

$$= \int \sigma[X]dV + \int j_n[X]dO \qquad (1)$$

- for the intensive quantity $x = \delta X/\delta V$

$$\frac{\partial}{\partial t} x = \sigma[X] + \nabla\cdot\vec{j}[X] \quad . \qquad (2)$$

A.1.2 Special Balance Equations

In the following the general conservation laws of mass, momentum, angular momentum and energy formulated for the intensive quantities (see A.1.1) and further information, interesting in this context and with regard to our tidal problems (in which the absolute, bodily forces in question are only gravitational or tidal ones) are collected (sources: Glansdorff and Prigogine 1971, De Groot 1960, Landau and Lifschitz 1966b).

1. Continuity equation (mass conservation: $\sigma[X] = 0$; $x = \rho$)

$$\frac{\partial}{\partial t}\rho + \nabla \cdot \rho\vec{v} = 0 \tag{1}$$

2. Momentum equation (for an individual component; sum-convention)

$$(\rho v_i)_{,t} + (\rho v_i v_j - \Pi_{ij})_{,j} = \rho(F_i + A_i) = \sigma[M_i] \tag{2}$$

which is a special form of the equation of motion:

$$\frac{d}{dt}\vec{v} = (\vec{F} + \vec{A}) + \nabla \cdot \underline{\Pi} \tag{2'}$$

$\vec{A}$:	
centrifugal force	$-\vec{\Omega} \times (\vec{\Omega} \times \vec{r})$
Coriolis force	$-2\vec{\Omega} \times \vec{v}$
due to accelerated rotation	$-\dot{\vec{\Omega}} \times \vec{r}$
due to accelerated translation	$-\ddot{\vec{a}}$

$\Pi_{ij} = \varepsilon_{ij} + \eta_{ij}$ (i.e., elastic and dissipative parts).

Fluids: $\varepsilon_{ij} = -p\delta_{ij}$ (reduction to scalar hydrostatic pressure)

Hooke's elasticity: $\varepsilon_{ij} = \lambda\vartheta_{nn}\delta_{ij} + 2\mu\vartheta_{ij}$ (stress-strain relation)

(in general:) $\eta_{ij} = \eta_1(v_{i,j} + v_{j,i} - \frac{2}{3}\delta_{ij}v_{n,n}) + \eta_2\delta_{ij}v_{n,n}$

if: η_1, η_2 = const; $\nabla \cdot \vec{v} = 0$ (incompressibility)

then $\eta_{ij,j} = \eta_1 \Delta v_i$

and (2') turns over to the NAVIER-STOKES-eqs.

3. Angular momentum equation (for an individual component)

$$\frac{\partial}{\partial t}(\vec{r}\times\vec{v})_i + \nabla\cdot\vec{v}(\vec{r}\times\vec{v})_i = \rho(\vec{r}\times(\vec{F}+\vec{A}))_i + (\vec{r}\times(\nabla\cdot\underline{\Pi}))_i = \sigma[P_i] \tag{3}$$

4. Energy equations

(4)

a) kinetic: $$(\rho e_k)_{,t}+\left(\rho\frac{v_i^2}{2}v_j-\Pi_{ij}v_i\right)_{,j}=\underset{\text{I}}{\rho(F_i+A_i)v_i}\underset{\text{II}}{-\Pi_{ij}v_{i,j}}=\sigma[E_k]$$

b) potential: $$(\rho e_p)_{,t}+(\rho v_j e_p)_{,j}=\underset{\text{I}}{-\rho F_i v_i}+\rho(e_p)_{,t}=\sigma[E_p]$$

c) internal: $$(\rho e_I)_{,t}+(\rho e_I v_j+W_j)_{,j}=\underset{\text{II}}{\Pi_{ij}v_{i,j}}=\sigma[E_I]$$

d) all forms: $$\left[\rho\left(\frac{v_i^2}{2}+e_p+e_I\right)\right]_{,t}+\left[W_j+\Pi_{ij}v_i+\left(\frac{1}{2}v_i^2+e_p+e_I\right)\rho v_j\right]_{,j}=\sigma[E]$$

$$\sigma[E]=\begin{cases}0 & \text{inertial system + conservative forces}\\ \rho A_i v_i+\rho(e_p)_{,t} & \text{in the other case}\end{cases}$$

Terms having reversed signs in different energy equations (marked I and II) represent *conversions*.
These conservation equations are suitable to "absolute" ($\vec{A}=0$) and "relative" reference systems.

A.1.3 Conservation Relations in the "Absolute" System

Before formulating the respective conservation equations for the Earth-Moon system it is convenient to separate the individual center-of-mass parts:

$$\vec{d}=\vec{a}_*+\vec{r} \qquad \left(\text{and: } \int\rho\vec{r}dV=0\ ;\quad \int\rho\dot{\vec{r}}dV=0\ ;\quad \int\rho\ddot{\vec{r}}dV=0\right)$$

$$\dot{\vec{d}}=\dot{\vec{a}}_*+\dot{\vec{r}}+\vec{\Omega}\times\vec{r} \qquad \text{(rigid rotation)}$$

$$\ddot{\vec{d}}=\ddot{\vec{a}}_*+\ddot{\vec{r}}+\dot{\vec{\Omega}}\times\vec{r}+\vec{\Omega}\times(\vec{\Omega}\times\vec{r})+2\vec{\Omega}\times\dot{\vec{r}}$$

$$=\text{I}+\text{II}+\text{III}+\text{IV}+\text{V}$$

I - V: contributions due to accelerations of translation, of relative motions, of rotation, due to the centrifugal-, to the Coriolis force, respectively.

There is conservation in the whole system (l = E,M) of:

1. *mass* $\left(= \sum_l \int_{V_l} \rho_l \, dV\right)$: $\frac{d}{dt} \sum_l \int_{V_l} \rho_l \, dV = 0$ (1)

2. *momentum* $\left(= \sum_l \int_{V_l} \rho_l \dot{\vec{d}}_l \, dV\right)$: $\frac{d}{dt} \sum_l \int_{V_l} \rho_l \dot{\vec{d}}_l \, dV = \sum_l \int_{V_l} \rho_l \ddot{\vec{d}}_l \, dV = 0$ (2)

3. *angular momentum* $\left(\vec{P} = m\vec{a} \times \dot{\vec{a}} + \sum_l \left(\int_{V_l} \rho_l \, dV \, \vec{r}_l \times \dot{\vec{r}}_l + \underline{\Theta}_l \cdot \vec{\Omega}_l\right)\right)$:

$$\frac{d}{dt} \sum_l \int_{V_l} \rho_l \vec{d}_l \times \dot{\vec{d}}_l \, dV = \sum_l \dot{\vec{P}}_l = \sum_l \vec{L}_l = 0 = m\vec{a} \times \ddot{\vec{a}} + \sum_l (\dot{\underline{\Theta}}_l \cdot \vec{\Omega}_l + \underline{\Theta}_l \cdot \dot{\vec{\Omega}}_l$$

$$+ \vec{\Omega}_l \times (\underline{\Theta}_l \cdot \vec{\Omega}_l) \tag{3}$$

$$+ \underbrace{\int_{V_l} \rho_l \, dV(\vec{r}_l(\dot{\vec{r}}_l \cdot \vec{\Omega}_l) - \dot{\vec{r}}_l(\vec{r}_l \cdot \vec{\Omega}_l) + \vec{r}_l \times \ddot{\vec{r}}_l))}_{I}$$

4. *total energy:*

$$\frac{d}{dt} \sum_l \int_{V_l} \rho_l \frac{\dot{\vec{d}}_l^2}{2} \, dV + \sum_l \left(\int_{V_l} dV(\nabla \cdot \vec{W} - \rho_l \frac{\partial}{\partial t} e_p) + \frac{d}{dt} \int_{V_l} \rho_l \, dV(e_p + e_I)\right)_{abs}$$

$$= 0 \tag{4}$$

[A.1.2: Eq.(4)]

resulting in the "absolute" balance relation for the

4a. *kinetic energy* $\left(E_k = \frac{m}{2} \dot{\vec{a}}^2 + \sum_l \left(\frac{1}{2} \Theta_{ij} \Omega_i \Omega_j + \int_{V_l} \rho_l \, dV(\vec{\Omega}_l(\vec{r}_l \times \dot{\vec{r}}_l) + \frac{\dot{\vec{r}}_l^2}{2})\right)\right)$:

$$\frac{d}{dt} E_k = \sum_l \int_{V_l} (\rho_l \dot{\vec{d}}_l \cdot \vec{F}_l - (\underline{\Pi}_l \cdot \nabla) \cdot \dot{\vec{d}}_l) dV = m\dot{\vec{a}}\ddot{\vec{a}} + \sum_l (\vec{\Omega}_l \cdot (\underline{\Theta}_l \cdot \dot{\vec{\Omega}}_l)$$

$$+ \frac{1}{2} \vec{\Omega}_l \cdot (\dot{\underline{\Theta}}_l \cdot \vec{\Omega}_l) + \underbrace{\int_{V_l} \rho_l \, dV(\dot{\vec{\Omega}}_l \cdot \vec{r}_l \times \dot{\vec{r}}_l + \vec{\Omega}_l \cdot \vec{r}_l \times \ddot{\vec{r}}_l + \dot{\vec{r}}_l \cdot \ddot{\vec{r}}_l))}_{II} \tag{4a}$$

(for the definition of the tensor of inertia: e.g., Landau and Lifschitz 1966a, for the vector products: e.g., Bronstein and Semendjajew 1966)

I,II-marked terms: time changes of the "relative" quantities, $\dot{\vec{P}}_{1,rel}$ and $\dot{E}_{k,1,rel}$ (involving the relative balances).

If the terrestrial *relative* quantities ($\vec{r}_E$, $\dot{\vec{r}}_E$, $\ddot{\vec{r}}_E$) refer to the solid Earth's barycenter, additional terms appear in Eqs. (3) and (4a):

$$
{}_*\vec{L} = -m_W\left(\frac{m_W}{m_E}\right)\Big(({}_*\vec{a}\times\vec{\Omega}_E)(\vec{\Omega}_E\cdot{}_*\vec{a})+\dot{\vec{\Omega}}_E\,{}_*\vec{a}^{\,2} - \vec{a}(\dot{\vec{\Omega}}_E\cdot{}_*\vec{a})
+ 2\vec{\Omega}_E({}_*\vec{a}\cdot{}_*\dot{\vec{a}})-2\,{}_*\dot{\vec{a}}({}_*\vec{a}\cdot\vec{\Omega}_E)+{}_*\vec{a}\times{}_*\ddot{\vec{a}}\Big)
$$

respectively

$$
{}_*\dot{E}_k = -m_W\left(\frac{m_W}{m_E}\right)\Big(((\vec{\Omega}_E\times{}_*\vec{a})\cdot(\dot{\vec{\Omega}}_E\times{}_*\vec{a})+(\vec{\Omega}_E\times{}_*\dot{\vec{a}})\cdot(\vec{\Omega}_E\times{}_*\vec{a})+\vec{\Omega}_E\cdot({}_*\vec{a}\times{}_*\ddot{\vec{a}})
+ \dot{\vec{\Omega}}_E({}_*\vec{a}\times{}_*\dot{\vec{a}})+{}_*\dot{\vec{a}}\cdot{}_*\ddot{\vec{a}}\Big)
$$

depending on the dynamics of the distance between the centers-of-mass of the ocean and of the solid Earth.

A.1.4 Interactions of the Subsystems and Simplifications

The evaluation of the individual M,T,W-contributions in the angular momentum and kinetic energy relations of Sect.A.1.3 requires the consideration of the interactions between the subsystems, i.e., the application of the special balance equations [A.1.2: Eqs. (3) and (4a)].

Moon: It is assumed punctiform, therefore: $\vec{r} = 0$; $\dot{\vec{r}} = \vec{v} = 0$; $\ddot{\vec{r}} = \dot{\vec{v}} = 0$; $\underline{\Theta} = 0$; $\dot{\underline{\Theta}} = 0$; $\vec{\Omega}_M = 0$; $\dot{\vec{\Omega}}_M = 0$ (also zero heat flux and internal energy quantities) and thus the "M"-sums on the right sides of Eqs. (3), (4a) of A.1.3 do not contribute.

Solid Earth: Because of its assumed rigidity: $\vec{\Omega}_T = \vec{\Omega}_E$; $\dot{\vec{r}} = \vec{v} = 0$; $\ddot{\vec{r}} = \dot{\vec{v}} = 0$; $\dot{\underline{\Theta}} = 0$; $\dot{\vec{P}}_{T,rel} = 0$ leading [via A.1.3: Eq.(3)] to the EULER relation, a relation for $\dot{\vec{\Omega}}_E$:

$$\frac{d}{dt}(\underline{\Theta}_T\cdot\vec{\Omega}_E)+\vec{\Omega}_E\times(\underline{\Theta}_T\cdot\vec{\Omega}_E) = \vec{L}_e$$

$$= \int_{V_T} \rho_T\vec{r}\times\vec{F}dV + \int_{O_T} \vec{r}\times(\underline{\Pi}\cdot d\vec{O}) \qquad (1)$$

$$\text{(formally)} = (\langle P_M,P_T\rangle_t+\langle P_W,P_T\rangle_g)+\langle P_W,P_T\rangle_c$$

since: $\vec{F} = -\nabla e_p$; $e_p = \frac{m_M}{a} + \Phi_t$ + terms $({}_*a)$ + gravitational contributions of the oceans

(in this context, see also Bursa 1982, this Vol.).
In the period mean:

$$\overline{L_{T,e,z}} = \overline{\langle P_W,P_T\rangle_c} = \Theta_{T,zz}\,\overline{\dot{\Omega}_E} = \overline{\dot{E}_{R,T}\Omega_E^{-1}} \simeq \overline{\dot{E}_{R,T}}\,\Omega_E^{-1} \qquad (2)$$

(because of periodicity and the normal neglect of any non-radial Earth gravity), and only $\langle P_W,P_T\rangle_c$ respectively

$$\overline{\dot{E}_{R,T}} = \vec{\Omega}_E\cdot(\underline{\Theta}_T\cdot\vec{\Omega}_E) \simeq \Omega_E\,\overline{\langle P_W,P_T\rangle_c} \qquad (3)$$

contribute in the "T"-sums of the absolute relations.

Ocean: For times of order of one period, the apparent forces originating from the acceleration of rotation respectively translation can be neglected (and *must* be neglected in order to start the generation of u, v, ζ and the subsequent calculation of even $\ddot{\vec{\Omega}}$, $\vec{a}$ according to our strategy; rel. error $\lesssim O(10^{-3} - 10^{-4})$):

$$\dot{\vec{\Omega}} \simeq 0;\; m_M(m_E + m_M)^{-1}\,\ddot{\vec{a}} \simeq 0;\; m_W m_E^{-1}\,{}_*\ddot{\vec{a}} \simeq 0 \qquad (4)$$

Because of periodicity, the period averaged terms:

$$\overline{L_{W,e,z}} = \overline{\langle P_T,P_W\rangle_c} - \overline{\langle P_W,P_M\rangle_F} = 0 \qquad (5)$$

(as well as $\overline{\dot{P}_{W,C,z}} = 0$; $\dot{\vec{P}}_{W,C}=\dot{\underline{\Theta}}_W\cdot\vec{\Omega}_E+\int_{V_W}\rho_W dV(\vec{r}(\dot{\vec{r}}\cdot\vec{\Omega}_E-\dot{\vec{r}}(\vec{r}\cdot\vec{\Omega}_E)))$

respectively: $\overline{\dot{E}_{k,W,rel}} = 0 = \overline{\langle E_p,E_k\rangle} - \overline{\langle E_k,E_I\rangle}$ (6)

(and also the other quantities transformable into total time derivatives according to the $\vec{\Omega}$ = const-assumption) are zero.

Summing up the individual contributions of the subsystems to the absolute balances [A.1.3: Eqs. (3), (4a)] these (secular) equivalences in the (whole model) Earth-Moon system result [most easily via Eqs. (1), (2) of A.1.5 as for the kinetic energy changes]:

$$m(\vec{a} \times \ddot{\vec{a}})_z = \overline{\langle P_W, P_M \rangle_F} \tag{7}$$

and

$$\frac{d}{dt}(m\dot{\vec{a}}^2 - \gamma m_E m_M a^{-1} + E_{R,T}) = -\overline{\langle E_k, E_I \rangle} = (n - \Omega_E)\overline{\langle P_W, P_M \rangle_F} \quad . \tag{8}$$

A.1.5 Estimate for the Gerstenkorn Event

In cylinder coordinates the conservation of the radial *momentum* (i.e., in direction of $\vec{a}$) for the "reduced" Moon leads to

$$\ddot{a}_r - \frac{\dot{a}_\lambda}{a} = -\gamma(m_E + m_M)a^{-2} \quad . \tag{1}$$

The orbital angular momentum is:

$$P_{M,z} = ma\dot{a}_\lambda = ma^2 n \quad ; \quad n = \frac{\dot{a}_\lambda}{a} \quad . \tag{2}$$

The assumption that the reduced mass is always forced into a quasi-circular orbit according to

$$n := n_o\left(\frac{a}{a_o}\right)^{-3/2} \quad ; \quad \dot{a} := \dot{a}_o\left(\frac{a}{a_o}\right)^{-1/2}$$

fulfilling (1), produces:

$$\left(\frac{\dot{n}}{n}\right) = -\frac{3}{2}\left(\frac{\dot{a}}{a}\right) \tag{3}$$

$$\left(\frac{\ddot{a}_r}{\dot{a}}\right) = -\frac{1}{2}\left(\frac{\dot{a}}{a}\right) \tag{4}$$

$$\left(\frac{\dot{P}_{M,z}}{P_{M,z}}\right) = \frac{1}{2}\left(\frac{\dot{a}}{a}\right) \quad . \tag{5}$$

On the other hand:

$$\dot{P}_{M,z} = \overline{\langle P_W, P_M \rangle_F} := \chi_p \, a^{-6} \tag{6}$$

(see A.1.4 and $\Phi \sim a^{-3}$; $\zeta \sim \Phi$ e.g., for a belt channel). With the drastic assumption, that $\chi_p = \text{const}$:

$$a^{13/2} = a_o^{13/2} + \frac{13}{2}\dot{a}_o^{11/2} a_o t \tag{7}$$

results and $t(a \sim 0) = -1.34 \times 10^9$ y.

The implications of this assumption cannot be estimated; in general, however, $\chi_p = \chi_p(h, A_H, F_B, \Omega, n)$, i.e., on this time scale χ_p will depend on the changing ocean depth topography, friction parameters, on varying Earth's rotation and Moon's mean motion. Therefore, $\langle P_W, P_M \rangle_F$ has to be computed explicitly for different times in the past with regard to [A.1.4: Eq. (2)]:

$$\dot{\Omega} = -\Theta_{T,zz}^{-1} \ \overline{\langle P_{\vec{W}}, P_M \rangle_F} \ ; \ \left(\Omega = \Omega_o - \Theta_{T,zz}^{-1} \ m a_o^2 n_o \left(\frac{a^2 n}{a_o^2 n_o} - 1\right)\right)$$

and to:

$$\dot{a} = 2(man)^{-1} \ \overline{\langle P_W, P_M \rangle_F} \ . \tag{8}$$

A.2 Oceanic Considerations

A.2.1 Derivation of the Model Equations

The hydrodynamic equations in spherical coordinates are (e.g., Landau and Lifschitz 1966b):

1. *Equation of continuity*:

$$\rho^{-1} \frac{d}{dt} \rho = \nabla \cdot \vec{v} = 0 = r^{-2}\left(\frac{\partial}{\partial r} r^2 v_r\right) + (r\mu)^{-1}\left(\frac{\partial}{\partial \lambda} v_\lambda + \frac{\partial}{\partial \phi} \mu v_\phi\right) \tag{1}$$

(incompressible continuum)

2. *Equation of motion*:

$$\frac{\partial}{\partial t} \vec{v} + (\vec{v} \cdot \nabla)\vec{v} = -\rho^{-1} \nabla p + \nu \Delta \vec{v} + \vec{g} + \vec{F}_t - 2\vec{\Omega} \times \vec{v} \tag{2}$$

(NAVIER-STOKES: centrifugal force is subsumed in $\vec{g}$)

longitudinal component:

$$\frac{d}{dt} v_\lambda = \frac{\partial}{\partial t} v_\lambda + \underbrace{\left[v_r \frac{\partial}{\partial r} + (r\mu)^{-1} v_\lambda \frac{\partial}{\partial \lambda} + r^{-1} v_\phi \frac{\partial}{\partial \phi}\right] v_\lambda}_{\mathrm{I}}$$

$$= \underbrace{-r^{-1} v_\lambda (v_r - v_\phi \mathrm{tg}\phi)}_{\mathrm{II}} \underbrace{-(\rho r \mu)^{-1} \frac{\partial p}{\partial \lambda}}_{\mathrm{III}} \underbrace{-(r\mu)^{-1} \frac{\partial}{\partial \lambda} \Phi_t}_{\mathrm{IV}} + 2\Omega(v_\phi \sin\phi - v_r \cos\phi)$$

$$\underbrace{+\nu r^{-2} \frac{\partial}{\partial r}\left(r^2 \frac{\partial}{\partial r} v_\lambda\right)}_{\mathrm{V}} \underbrace{+\nu r^{-2}\left[\left(\frac{\partial^2}{\partial \phi^2} + \mu^{-2} \frac{\partial^2}{\partial \lambda^2} - \mathrm{tg}\phi \frac{\partial}{\partial \phi} - \mu^{-2}\right) v_\lambda\right.}_{\mathrm{VI}}$$

$$+2\mu^{-1}\left(\frac{\partial v_r}{\partial\lambda} - tg\phi\,\frac{\partial v_\phi}{\partial\lambda}\right)\Big]$$

meridional component:

$$\frac{d}{dt}v_\phi = \frac{\partial}{\partial t}v_\phi + \underbrace{\left[v_r\frac{\partial}{\partial r} + (r\mu)^{-1}v_\lambda\frac{\partial}{\partial\lambda} + r^{-1}v_\phi\frac{\partial}{\partial\phi}\right]v_\phi}_{I}$$

$$= \underbrace{-r^{-1}(v_r v_\phi - v_\lambda^2\, tg\phi)}_{II} - \underbrace{(\rho r)^{-1}\frac{\partial p}{\partial\phi}}_{III} - \underbrace{r^{-1}\frac{\partial}{\partial\phi}\Phi_t}_{IV} - 2\Omega v_\lambda \sin\phi$$

$$+\underbrace{\nu r^{-2}\frac{\partial}{\partial r}\left(r^2\frac{\partial}{\partial r}v_\phi\right)}_{V} + \underbrace{\nu r^{-2}\Big[\left(\frac{\partial^2}{\partial\phi^2} + \mu^{-2}\frac{\partial^2}{\partial\lambda^2} - tg\phi\frac{\partial}{\partial\phi} - \mu^{-2}\right)v_\phi}_{VI}$$

$$+2\mu^{-1}\left(\mu\frac{\partial}{\partial\phi}v_r + tg\phi\frac{\partial}{\partial\lambda}v_\lambda\right)\Big]$$

radial component:

$$\frac{d}{dt}v_r = \frac{\partial}{\partial t}v_r + \underbrace{\left[v_r\frac{\partial}{\partial r} + (r\mu)^{-1}v_\lambda\frac{\partial}{\partial\lambda} + r^{-1}v_\phi\frac{\partial}{\partial\phi}\right]v_r}_{I}$$

$$= \underbrace{r^{-1}(v_\lambda^2 + v_\phi^2)}_{II} - \underbrace{\left(\rho^{-1}\frac{\partial p}{\partial r} - g\right)}_{VII} - \underbrace{\frac{\partial}{\partial r}\Phi_t}_{IV} + 2\Omega v_\lambda\cos\phi$$

$$+\underbrace{\nu r^{-2}\frac{\partial}{\partial r}\left(r^2\frac{\partial}{\partial r}v_r\right)}_{V} + \underbrace{\nu r^{-2}\Big[\left(\frac{\partial^2}{\partial\phi^2} + \mu^{-2}\frac{\partial^2}{\partial\lambda^2} - tg\phi\frac{\partial}{\partial\phi} - 2\right)v_r}_{VI}$$

$$- 2\left(\left(\frac{\partial}{\partial\phi} - tg\phi\right)v_\phi + \mu^{-1}\frac{\partial}{\partial\lambda}v_\lambda\right)\Big]$$

I, II : advection, curvature terms

III, IV : pressure gradient, tidal forces

V, VI : radial, horizontal contributions to viscosity

VII = 0 : relation of hydrostatic equilibrium

VII leads to: $$p(r) = p|_{R+\zeta} + \rho g(r - h + \zeta) \qquad (3)$$

and normally $p|_{R+\zeta} = 0$ (i.e., the atmospheric pressure is not regarded). As the author found no reference about the order of magnitude of the error imported by the oceanic version of the hydrostatic assumption, let us take the atmospheric analogon: $0(10^{-3} - 10^{-4})$ (see Holton 1973).

Before averaging radially, Eqs. (1) and (2) are multiplied by some positive power of r in order to remove the negative ones and to reduce the influence of v_r. After the averaging procedure the hydrodynamic equations for a "barotropic" tide follow (the meaning of I,.., IV is as before, V leads to the bottom friction):

$$\frac{\partial}{\partial t}(h+\zeta) + \tilde{\nabla}\cdot\tilde{v}(h+\zeta) := 0 \qquad (= O(10^{-3} - 10^{-4})) \tag{1'}$$

$$\underset{\text{I}}{\left(\frac{\partial}{\partial t} + \tilde{\nabla}\cdot\tilde{v}\right)u} - \underset{\text{II}}{fv - uvR^{-1}tg\phi} + \underset{\text{III IV}}{(R\mu)^{-1}\frac{\partial}{\partial\lambda}(g\zeta+\Phi_t)} = \underset{\text{V}}{-r_B(h+\zeta)^{-1}(u^2+v^2)^{1/2}u} + \underset{\text{VI}}{\eta_\lambda}$$

$$\left(\frac{\partial}{\partial t} + \tilde{\nabla}\cdot\tilde{v}\right)v + fu + u^2R^{-1}tg\phi + R^{-1}\frac{\partial}{\partial\phi}(g\zeta+\Phi_t) = -r_B(h+\zeta)^{-1}(u^2+v^2)^{1/2}v + \eta_\phi \tag{2'}$$

with the (lateral) eddy-viscosity forces $\eta_1 = R_1 + T_1$, $1=\lambda,\ \phi$.

Zahel (e.g., 1977) used:

$$R_\lambda = A_H(R\mu)^{-2}\left[\left(\frac{\partial^2}{\partial\lambda^2} + \mu^2\frac{\partial^2}{\partial\phi^2} - \mu\sin\phi\frac{\partial}{\partial\phi} - 1\right)u - 2\sin\phi\frac{\partial}{\partial\lambda}v\right]$$

$$R_\phi = A_H(R\mu)^{-2}\left[\left(\frac{\partial^2}{\partial\lambda^2} + \mu^2\frac{\partial^2}{\partial\phi^2} - \mu\sin\phi\frac{\partial}{\partial\phi} - 1\right)v + 2\sin\phi\frac{\partial}{\partial\lambda}u\right] \ . \tag{4}$$

The author found additional terms by regarding the normal velocity profile (see Fig. 1) in the kinematic and dynamic boundary conditions:

$$T_\lambda = A_HR^{-2}(h+\zeta)^{-1}\left[u\left(\mu^{-2}\frac{\partial^2}{\partial\lambda^2} + \frac{\partial^2}{\partial\phi^2} - tg\phi\frac{\partial}{\partial\phi}\right) + 2\left(\mu^{-2}\left(\frac{\partial}{\partial\lambda}u\right)\frac{\partial}{\partial\lambda} + \left(\frac{\partial}{\partial\phi}u\right)\frac{\partial}{\partial\phi} - \mu^{-2}v\sin\phi\frac{\partial}{\partial\lambda}\right)\right]h - A_ru(h+\zeta)^{-1} + O(10^{-3}-10^{-4})$$

$$T_\phi = A_HR^{-2}(h+\zeta)^{-1}\left[v\left(\mu^{-2}\frac{\partial^2}{\partial\lambda^2} + \frac{\partial^2}{\partial\phi^2} - tg\phi\frac{\partial}{\partial\phi}\right) + 2\left(\mu^{-2}\left(\frac{\partial}{\partial\lambda}v\right)\frac{\partial}{\partial\lambda} + \left(\frac{\partial}{\partial\phi}v\right)\frac{\partial}{\partial\phi} + \mu^{-2}u\sin\phi\frac{\partial}{\partial\phi}\right)\right]h - A_rv(h+\zeta)^{-1} + O(10^{-3}-10^{-4}) \ . \tag{5}$$

The orders of magnitude of the (relative) errors lumped together into T from all the terms in Eqs. (2') originate from estimates of the neglected curvature in the radial integration (i.e., O(h/R)) and of the small-scale correlations in the kinematic boundary condition (i.e., probably very small terms like $\overline{v_1' \cdot \nabla_1 X'}$, $X = h,\zeta; 1 = \lambda,\phi$; see also Krauss 1973 and 1980).

A.2.2 The Balance Equation of Oceanic Angular Momentum

The balance equation of the (macroscalar relative) angular momentum in the direction of the Earth's rotation is gained by multiplying the "barotropic" equation of zonal motion [A.2.1: Eq. (2')] by $\rho R\mu(h+\zeta)$ and subsequently integrating over the ocean's surface:

$$\frac{d}{dt} P_{z,rel} = \int_{O_W} dO_R \left[\frac{\partial}{\partial t} \rho(h+\zeta)R\mu u + \tilde{\nabla}\cdot\tilde{v}\rho(h+\zeta)R\mu u\right] = \frac{\partial}{\partial t} P_{z,rel} \tag{1}$$

$$= \int_{O_W} dO_R \; \rho R\mu(h+\zeta)\left[-(R\mu)^{-1}\frac{\partial}{\partial\lambda}(\Phi_t+g\zeta) + uF_B + \eta_\lambda + 2\Omega\sin\phi v\right]$$

$$= -\langle P_W,P_M\rangle_F + \langle P_T,P_W\rangle_D + \langle P_T,P_W\rangle_B + \langle P_T,P_W\rangle_V + \dot{P}_{C,z} \quad .$$

Some remarks should be given:

$$\langle P_T,P_W\rangle_D = -\left[\int_{O_W} dO_R \rho\mu^{-1} h \frac{\partial}{\partial\lambda}\zeta + \frac{R^2}{2}\int_{\phi_1}^{\phi_2} d\phi(\zeta^2)\Big|_{\lambda_1(\phi)}^{\lambda_2(\phi)}\right]$$

(ϕ_1, ϕ_2; λ_1, λ_2 determined by the continents)

$\neq 0$ even in the period mean, because $\bar{\zeta} \neq 0$, namely even if $\zeta|_{\lambda_1} = 0 = \zeta|_{\lambda_2}$ is assumed and this already is a doubtful assumption.

$\langle P_T,P_W\rangle_V \neq 0$ in general, because it cannot be transformed in an integral depending on zero boundary values (as its three-dimensional analogon).

The last term of Eq. (1) can be transformed:

$$\dot{P}_{C,z} = -\int_{O_W} dO_R \frac{\partial}{\partial t}(\rho(h+\zeta)\Omega R^2\mu^2) = -\dot{P}_{R,z} \quad , \tag{2}$$

into the time change of the so-called Ω-momentum, i.e., $\overline{\dot{P}_{C,z}} = 0$. $\dot{P}_{R,z}$ is mediated by the variations of the polar moment of inertia. As the oceanic-lunar angular momentum transfer consists in the average during one period only of the contribution of the water elevation,

Table 3. Integrands of the individual angular momentum (a.m.) balance terms ($p_{z,rel}$) due to the undisturbed water depth (*) and the elevation (**). Coriolis contributions (causing $\dot{p}_{C,z}$) can be expressed by the rotational angular momentum quantities. In the first row the meaning of the resulting transfers is indicated, which are gained by $\int_{O_W} \cdot \, dO_R$

a.m. form	Storage	0	$\langle \cdot , P_M \rangle_F$ [a]	$\langle P_T , \cdot \rangle$ [b]	$\langle \cdot^{*} , \cdot^{**} \rangle$ [c]
$p^{*}_{z,rel}=R\rho\mu uh$	$\frac{\partial}{\partial t} p^{*}_{z,rel}$	$+\tilde{\nabla}\cdot\tilde{v}p^{*}_{z,rel}=\sigma[P^{*}_{z,rel}]$	$\sigma[P^{*}_{z,rel}]=$	$p^{*}_{z,rel}\left(u^{-1}\frac{d}{dt}u-vR^{-1}tg\phi\right.$	$\left.+h^{-1}\tilde{\nabla}\cdot\tilde{v}h\right)$
$p^{**}_{z,rel}=R\rho\mu u\zeta$	$\frac{\partial}{\partial t} p^{**}_{z,rel}$	$+\tilde{\nabla}\cdot\tilde{v}p^{**}_{z,rel}=\sigma[P^{**}_{z,rel}]$	$\sigma[P^{**}_{z,rel}]=$	$p^{*}_{z,rel}\left(\frac{\zeta}{h}\left(u^{-1}\frac{d}{dt}u-vR^{-1}tg\phi\right)\right.$	$\left.-h^{-1}\tilde{\nabla}\cdot\tilde{v}h\right)$
$p^{*}_{R,z}=\rho\Omega hR^2\mu^2$	$\frac{\partial}{\partial t} p^{*}_{R,z}$	$+\tilde{\nabla}\cdot\tilde{v}p^{*}_{R,z}=\sigma[P^{*}_{R,z}]$	$\sigma[P^{*}_{R,z}]$	$=-\dot{p}_{C,z}$	$-\sigma[P^{**}_{R,z}]$
$p^{**}_{R,z}=\rho\Omega\zeta R^2\mu^2$	$\frac{\partial}{\partial t} p^{**}_{R,z}$	$+\tilde{\nabla}\cdot\tilde{v}p^{**}_{R,z}=\sigma[P^{**}_{R,z}]$	$\sigma[P^{**}_{R,z}]$	$=$	$-\rho\zeta R\mu fv-p^{*}_{R,z}h^{-1}\tilde{\nabla}\cdot\tilde{v}h$

[a] Between ocean and Moon. [b] Between solid Earth and ocean. [c] Between the differently starred angular momentum forms; (d/dt)u: the total force equivalent to it has to be used, see A.2.1: Eq. (2').

it is useful to formulate individual balances depending on this part of the water depth and on the undisturbed one (Table 3).

A.2.3 The Balance Equation of the Oceanic Kinetic Energy

The balance equations of the (macroscalar relative) kinetic energy can be deduced from the velocity equations of a barotropic tide [multiplied by $\rho(h+\zeta)u$, respectively $\rho(h+\zeta)v$, summed, and after regarding the continuity equation, integrated over the ocean's surface; see A.2.1: Eqs. (1'), (2')]:

$$\frac{d}{dt} E_k = \int_{O_W} dO_R \left[\frac{\partial}{\partial t} \frac{\rho}{2} (h+\zeta)(u^2+v^2) + \tilde{\nabla}\cdot\tilde{v} \frac{\rho}{2} (h+\zeta)(u^2+v^2) \right] = \frac{\partial}{\partial t} E_k$$

$$= \int_{O_W} dO_R \rho (h+\zeta)[-\tilde{v}\cdot\tilde{\nabla}\Phi_t - g\tilde{v}\cdot\tilde{\nabla}\zeta - (u^2+v^2)^{3/2} r_B (h+\zeta)^{-1} + \tilde{v}\cdot\tilde{\eta}]$$

$$= \langle E_p, E_k \rangle_F + \langle E_p, E_k \rangle_D - \langle E_k, E_I \rangle_B - \langle E_k, E_I \rangle_V \quad .$$

In order to get a better insight into the imbalance, the individual balances of the energies associated with the zonal and meridional velocity components, respectively with the undisturbed water depth and the elevation (Table 4) should be evaluated.

A.2.4 On the Oceanic Potential Energies and Their Balances

1. Potential energy in the Earth's gravity field
   Usually (apart from integration $\int_{O_W} \cdot \, dO_R$; e.g., Zahel 1977, Hendershott 1972)

$$e_{pg} := \frac{1}{2} \rho g \zeta^2 \tag{1}$$

and: $$(\rho g)^{-1} \frac{\partial}{\partial t} e_{pg} = -\tilde{\nabla}\cdot\tilde{v}(h+\zeta)\zeta + (h+\zeta)\tilde{v}\cdot\tilde{\nabla}\zeta \quad . \tag{2}$$

This definition is not absolutely mandatory; another one is possible:

$$e_{pg} := \rho g (h+\zeta)\zeta \tag{3}$$

and: $$\frac{\partial}{\partial t} e_{pg} + \tilde{\nabla}\cdot\tilde{v}E_{pg} = \sigma[E_{pg}] = (h+\zeta)\rho g \frac{\partial}{\partial t}\zeta + (h+\zeta)\rho g\tilde{v}\cdot\tilde{\nabla}\zeta \tag{4}$$

or: $$Q[E_{pg}] = G(E_{pg}) - \langle E_{pg}, E_k \rangle_D \quad . \tag{5}$$

Table 4. Integrands of the individual kinetic energy balance terms due to the undisturbed water depth (*), the water elevation (**), the zonal ($_u$), and meridional ($_v$) currents. The first row shows the meaning of the conversions (gained by $\int_{O_W} \cdot\, dO_R$)

Energy form	Storage O	$\langle E_p,\cdot\rangle^a$; $\langle\cdot,E_I\rangle^b$	$\langle\cdot,X\rangle^c$ $X\in(E_{ku},\ldots,E_{kv})\ddagger$.
$e^*_{ku}=\frac{1}{2}\,h\rho u^2$	$\frac{\partial}{\partial t}\,e^*_{ku}+\tilde{\nabla}\cdot\tilde{v}e^*_{ku}=\sigma[E^*_{ku}]$	$\sigma[E^*_{ku}]=\rho hu\,\frac{d}{dt}\,u+e^*_{ku}h^{-1}\tilde{\nabla}\cdot\tilde{v}h$	
$e^*_{kv}=\frac{1}{2}\,h\rho v^2$	$\frac{\partial}{\partial t}\,e^*_{kv}+\tilde{\nabla}\cdot\tilde{v}e^*_{kv}=\sigma[E^*_{kv}]$	$\sigma[E^*_{kv}]=\rho hv\,\frac{d}{dt}\,v+e^*_{kv}h^{-1}\tilde{\nabla}\cdot\tilde{v}h$	
$e^{**}_{ku}=\frac{1}{2}\,\rho\zeta u^2$	$\frac{\partial}{\partial t}\,e^{**}_{ku}+\tilde{\nabla}\cdot\tilde{v}e^{**}_{ku}=\sigma[E^{**}_{ku}]$	$\sigma[E^{**}_{ku}]=\rho\zeta u\,\frac{d}{dt}\,u-e^{**}_{ku}\zeta^{-1}\tilde{\nabla}\cdot\tilde{v}h$	
$e^{**}_{kv}=\frac{1}{2}\,\rho\zeta v^2$	$\frac{\partial}{\partial t}\,e^{**}_{kv}+\tilde{\nabla}\cdot\tilde{v}e^{**}_{kn}=\sigma[E^{**}_{kv}]$	$\sigma[E^{**}_{kv}]=\rho\zeta v\,\frac{d}{dt}\,v-e^{**}_{kv}\zeta^{-1}\tilde{\nabla}\cdot\tilde{v}h$	
$e^*_k=e^*_{ku}+e^*_{kv}$	$\frac{\partial}{\partial t}\,e^*_k+\tilde{\nabla}\cdot\tilde{v}e^*_k=\sigma[E^*_k]$	$\sigma[E^*_k]=\rho h\left(u\,\frac{d}{dt}\,u+v\,\frac{d}{dt}\,v\right)+e^*_k h^{-1}\tilde{\nabla}\cdot\tilde{v}h$	
$e^{**}_k=e^{**}_{ku}+e^{**}_{kv}$	$\frac{\partial}{\partial t}\,e^{**}_k+\tilde{\nabla}\cdot\tilde{v}e^{**}_k=\sigma[E^{**}_k]$	$\sigma[E^{**}_k]=\rho\zeta\left(u\,\frac{d}{dt}u+v\,\frac{d}{dt}\,v\right)-e^*_k h^{-1}\tilde{\nabla}\cdot\tilde{v}h$	
$e_{ku}=e^*_{ku}+e^{**}_{ku}$	$\frac{\partial}{\partial t}\,e_{ku}+\tilde{\nabla}\cdot\tilde{v}e_{ku}=\sigma[E_{ku}]$	$\sigma[E_{ku}]=\rho(h+\zeta)u\,\frac{d}{dt}\,u$	
$e_{kv}=e^*_{kv}+e^{**}_{kv}$	$\frac{\partial}{\partial t}\,e_{kv}+\tilde{\nabla}\cdot\tilde{v}e_{kv}=\sigma[E_{kv}]$	$\sigma[E_{kv}]=\rho(h+\zeta)v\,\frac{d}{dt}\,v$	

[a]Conversion to and from potential energy. [b]Conversion to internal energy. [c]Conversion between the other kinetic energy forms; (d/dt)u, (d/dt)v: the tidal-, bottom-friction-, eddy-viscosity-, pressure gradient-, Coriolis forces and the metric terms of the "barotropic" equations of motion have to be used; see A.2.1: Eq. (2'); the Coriolis force is responsible for $\langle E_{ku},E_{kv}\rangle$.

The generation $G(E_{pg}) = \int_{O_W} (h+\zeta)\rho g(\partial/\partial t)\zeta dO = -\int_{O_W} \rho g(h+\zeta)\tilde{\nabla}\cdot\tilde{v}(h+\zeta)dO$ represents the energy taken from the Earth's gravity: $\overline{G(E_{pg})} = 0$.

2. Potential energy due to tidal forces

$$\text{Def.: } e_{p\Phi} := \rho(h+\zeta)\Phi \tag{6}$$

$$\text{and: } \frac{\partial}{\partial t}\,e_{p\Phi}+\tilde{\nabla}\cdot\tilde{v}e_{p\Phi} = \sigma[E_{p\Phi}] = (h+\zeta)\rho\left(\frac{\partial}{\partial t}\,\Phi+\tilde{v}\cdot\tilde{\nabla}\Phi\right) \tag{7}$$

or: $Q[E_{p\Phi}] = G(E_{p\Phi}) - \langle E_{p\Phi}, E_k \rangle_F$.

The generation

$$G(E_{p\Phi}) = \int_{O_W} (h+\zeta)\rho(\partial/\partial t)\Phi dO \overset{M_2}{=\!=} (\sigma/2)\langle P_W, P_M \rangle_F; \quad (\sigma/2=\Omega-n) \qquad (8)$$

directly relates to the angular momentum transfer to the Moon: this fact seems not to be noted until now (because of the slightly different definitions; e.g., Zahel 1977, Lambeck 1980: work integrals).

The individual balances depending on the undisturbed depth and the elevation of the water are also given (Table 5).

Table 5. Integrands of the individual potential energy balance terms due to the undisturbed water depth (*) and the water elevation (**). The first row shows the meaning of the "interaction integrals" (gained by $\int_{O_W} \cdot\, dO_R$)

Energy form	Storage	0	$\langle E_k, \cdot \rangle^a$ $\langle \cdot^*, \cdot^{**} \rangle^b$
$e^*_{pg}=\rho h\zeta$	$\frac{\partial}{\partial t} e^*_{pg} + \tilde{\nabla}\cdot\tilde{v}e^*_{pg} = \sigma[E^*_{pg}]$	$\sigma[E^*_{pg}]=\rho h\left(\frac{\partial}{\partial t} + \tilde{v}\cdot\tilde{\nabla}\right)\zeta + e^*_{pg}h^{-1}\tilde{\nabla}\cdot\tilde{v}h$	
$e^{**}_{pg}=\rho\zeta^2$	$\frac{\partial}{\partial t} e^{**}_{pg} + \tilde{\nabla}\cdot\tilde{v}e^{**}_{pg} = \sigma[E^{**}_{pg}]$	$\sigma[E^{**}_{pg}]=\rho\zeta\left(\frac{\partial}{\partial t} + \tilde{v}\cdot\tilde{\nabla}\right)\zeta - e^*_{pg}h^{-1}\tilde{\nabla}\cdot\tilde{v}h$	
$e^*_{p\Phi}=\rho h\Phi$	$\frac{\partial}{\partial t} e^*_{p\Phi} + \tilde{\nabla}\cdot\tilde{v}e^*_{p\Phi} = \sigma[E^*_{p\Phi}]$	$\sigma[E^*_{p\Phi}]=\rho h\left(\frac{\partial}{\partial t} + \tilde{v}\cdot\tilde{\nabla}\right)\Phi + e^*_{p\Phi}h^{-1}\tilde{\nabla}\cdot\tilde{v}h$	
$e^{**}_{p\Phi}=\rho\zeta\Phi$	$\frac{\partial}{\partial t} e^{**}_{p\Phi} + \tilde{\nabla}\cdot\tilde{v}e^{**}_{p\Phi} = \sigma[E^{**}_{p\Phi}]$	$\sigma[E^{**}_{p\Phi}]=\rho\zeta\left(\frac{\partial}{\partial t} + \tilde{v}\cdot\tilde{\nabla}\right)\Phi - e^*_{p\Phi}h^{-1}\tilde{\nabla}\cdot\tilde{v}h$	

[a]Conversion to kinetic energy. [b]Conversion between the two potential energy forms.

A.3 List of Symbols

Tensors (t.) are marked by _, (3-dim.) vectors (v.) by $\vec{}$ (2-dim. quantities by ~). In general, symbols formed by adding super- or subscripted symbols are not listed separately. As for a (rarer) second meaning of a symbol or form of an operator, the section of the most frequent appearance is then indicated. Used values or formulas are also given.

A.3.1 Normal Symbols and Some Dependent Quantities

Symbol	Meaning
$a;\ a_o = 3.844 \times 10^8 m$	Earth-Moon-distance
$_*a$	distance of barydenters of ocean and solid Earth
a_*	distance of Moon or Earth from system barycenter
$\vec{A}$	apparent force
$A_H = 10^6\ m^2\ s^{-1};\ A_r$	horizontal, radial eddy-viscosity coefficient
$\vec{d}$	distance v. in the absolute system
(e) E	(density of) energy
E_*	peak mechanical energy (see definition of Q-factor)
$f = 2\Omega \sin\phi$	Coriolis parameter
$\vec{F}$	absolute bodily force
$F_B = -r_B (u^2+v^2)(h+\zeta)^{-1}$	bottom-friction "parameter"
$\vec{g}$	effective gravity
h	(undisturbed) water depth
$\vec{j}[X]$	density of flux in the "X"-balance
$\vec{L}$	torque
$m = m_E m_M (m_E+m_M)^{-1}$ $= 1.215 \times 10^{-2}\ m_E$	reduced mass of Moon
$m_E = 5.98 \times 10^{24}$ kg	mass of Earth
$m_M = 0.0123\ m_E$	mass of Moon
$\vec{M}$	momentum v.
n	motion in mean longitude
O	surface
p	pressure
$(\vec{p})\vec{P}$	(density of) angular momentum
$Q = 2\pi E_* X^{-1}$; $X = \langle E_k, E_I \rangle_V + \langle E_k, E_I \rangle_B$	Q-factor ($Q_B: X = \langle E_k, E_I \rangle_B$; $Q_V: X = \langle E_k, E_I \rangle_V$; $Q_F: X = \langle E_p, E_k \rangle_F$)
Q[X]	production of X per time unit inside the volume
$\vec{r}$	"relative" distance v.

r	radial distance in spherical coordinates
$r_B = 3 \times 10^{-3}$	bottom-friction parameter
$R = 6.37 \times 10^6$ m	Earth's radius
t	time
$(u,v) = \tilde{v}$	longitudinal, meridional (radially mean) velocities
$\vec{v} = \dot{\vec{r}} = (v_\lambda, v_\phi, v_r)$	(3-dim.) velocity v.
$\vec{W}$	heat flux v.
X, Y	dummy quantities
z	coordinate in direction of Earth's rotation
γ	(general) gravitational constant
$\gamma_L = 0.69$	reduction factor (i.e., Love number) in tidal potential
Δ (e.g., $\Delta\lambda$, Δt)	step "length" (e.g., zonal angular grid-point distance, time step)
δ_{ij}	Kronecker symbol
$\underline{\varepsilon}$	stress t. (elastic contribution of $\underline{\Pi}$)
ζ	water elevation (disturbance in depth)
$\underline{\eta}$	viscosity t. (dissipative contribution of $\underline{\Pi}$)
$\tilde{\eta} = \tilde{R} + \tilde{T}$	lateral eddy-viscosity (parts according to Zahel and the author)
$\underline{\vartheta}$	t. of deformation
$\underline{\Theta}$	t. of inertia
λ	longitude
λ, μ	Lamé-coefficients: torsion modulus, rigidity (see A.1.2)
$\mu = \cos\phi$	
ν (η_1, η_2)	kinematic (dynamic) viscosity coefficients
$\underline{\Pi} = \underline{\varepsilon} + \underline{\eta}$	t. of pressures
ρ	density
$\rho_* = \rho(h + \zeta)$	mass of a surface unit element (see A.3.5)
σ	angular velocity of the tide (M_2: 1.405×10^{-4} s^{-1})

$\sigma[X]$	source of X
ϕ	latitude
$\Phi\left[= -\frac{c}{2} R\mu^2 \cos(\sigma t+2\lambda)\right]$	tidal potential [of the M_2 partial tide: $c = 0.761 \times 10^{-7}$]
$\Psi[X]$	flux of X through surface
$\vec{\Omega}$	rotation v.
$\Omega_E = 7.292 \times 10^{-5}\ s^{-1}$	Earth's angular speed of rotation

A.3.2 Mathematical Operators

Those operators are listed, which correspond to a fixed (mathematical) instruction [contrary to quantities like e.g. $\sigma[X]$].

$\times$	vectorial product
$\bullet$	scalar product
$\frac{d}{dt} = \left(\frac{\partial}{\partial t} + \vec{v}\cdot\nabla\right)$; $\dot{}$ (A.1.2)	total or substantial time derivative
$\frac{\partial}{\partial t}$; $_{,t}$ (A.1.2)	local time derivative
∇; $_{,j}$ (A.1.2)	Nabla operator
$\tilde{\nabla} = R^{-1}\left(\mu^{-1}\frac{\partial}{\partial\lambda}, \frac{\partial}{\partial\phi}\right)$	horizontal Nabla operator
$dO_R = R^2\mu d\lambda d\phi$	surface element of a sphere with radius R
$\overline{} = \frac{\sigma}{2\pi}\int_{t_1}^{t_1+(2\pi/\sigma)} \cdot\, dt$	period mean (t_1: initial moment of integration)
$\vert_X$	at the position "X"
$\vert_X^Y$	value at the position "Y" minus that at "X"

A.3.3 Subscripts

B	induced by bottom friction
C	originating from Coriolis force
D	induced by pressure gradient force
E	of (whole) Earth
F	induced by the forcing tidal potential

I	internal (e.g., at energy quantities)
M	lunar
R	rotational
T	of solid Earth
V	induced by the eddy-viscosity
W	oceanic
abs	in absolute reference system
c	due to contact (or reaction) forces
e	external (beyond the boundaries)
g	gravitational
i	(in the) interior (A.1.1)
i,j	of i,j component
k	kinetic
m	mechanical
n	along inside normal to surface
p	potential
rel	belonging to the relative coordinate system
t	tidal
u, (v)	originating from the zonal, (meridional) velocity components of the tidal currents
λ, ϕ, z	in λ-, ϕ-, z-direction
o	of the present value

A.3.4 Superscripts

*	due to the undisturbed water depth
**	due to the water elevation
'	small-scale quantities

A.3.5 Transfers, Conversions and Storages

As for the general "interaction symbolics" (in which final subscripts indicate the mechanisms operating) $\langle X,Y\rangle = -\langle Y,X\rangle$ means a transfer, respectively conversion from quantity X into quantity Y. (Abbreviations: a.m. = angular momentum; a.m.t. = angular momentum transfer;

e. = energy; e.c. = energy conversion). These symbols are used in a formal sense (A.1.4) and in their specified meaning for the modelled (macroscalar) interactions (the formulas of which thus are also given in the list). This yields in the individual special cases of ($\rho_* = \rho(h+\zeta) \neq$ const!):

- angular momentum quantities

$$\langle P_W, P_M \rangle_F = \int_{O_W} dO_R \rho_* \frac{\partial}{\partial\lambda} \Phi$$

oceanic-lunar a.m.t.

$$\overline{\langle P_W, P_M \rangle_F} = \int_{O_W} dO_R \overline{\rho\zeta \frac{\partial}{\partial\lambda} \Phi}$$

its period average

$$\langle P_T, P_W \rangle_D = -\int_{O_W} dO_R \rho_* g \frac{\partial}{\partial\lambda} \zeta$$

solid Earth-ocean a.m.t. via pressure gradient force

$$\langle P_T, P_W \rangle_B = \int_{O_W} dO_R \, R\rho_* \mu F_B u$$

solid Earth-ocean a.m.t. via bottom friction

$$\langle P_T, P_W \rangle_V = \int_{O_W} dO_R \, R\rho_* \mu \eta_\lambda$$

solid Earth-ocean a.m.t. via (lateral) eddy-viscosity

$$L_{C,z} = \int_{O_W} dO_R \, R\rho_* \mu f v = -\int_{O_W} dO_R \frac{\partial}{\partial t} (\rho_* \Omega R^2 \mu^2) = -\dot{P}_{R,z}$$

Coriolis contribution to the a.m. balance

$$P_{z,rel} = \int_{O_W} dO_R \rho_* R\mu u$$

relative a.m. in the direction of Earth's rotation

$$\frac{\partial}{\partial t} P_{z,rel} = \int_{O_W} dO_R \frac{\partial}{\partial t} (\rho_* R\mu u)$$

storage of relative a.m. in the direction of Earth's rotation

$$P_{R,z} = \int_{O_W} dO_R \rho_* \Omega R^2 \mu$$

absolute a.m. of "parcels at the Earth's surface" or Ω-momentum

$$\frac{\partial}{\partial t} P_{R,z} = \int_{O_W} dO_R \frac{\partial}{\partial t} \rho_* \Omega R^2 \mu$$

its storage (i.e., time change)

- energy quantities

$$\langle E_p, E_k \rangle_F = -\int_{O_W} dO_R \rho_* \tilde{v} \cdot \tilde{\nabla} \Phi$$

potential-kinetic e.c. due to tidal potential

$$\langle E_p, E_k \rangle_D = -\int_{O_W} dO_R \rho_* g \tilde{v} \cdot \tilde{\nabla} \zeta$$

potential-kinetic e.c. due to pressure gradient force

$$\langle E_k, E_I \rangle_B = \int_{O_W} dO_R \rho r_B (u^2+v^2)^{3/2}$$

kinetic-internal e.c. due to bottom friction

$$\langle E_k, E_I \rangle_V = -\int_{O_W} dO_R \rho_* \tilde{v} \cdot \tilde{\eta}$$

kinetic-internal e.c. due to (lateral) eddy-viscosity

$E_{k,rel} = \int_{O_W} dO_R \frac{\rho_*}{2} (u^2+v^2)$	relative kinetic e.
$\frac{\partial}{\partial t} E_{k,rel} = \int_{O_W} dO_R \frac{\partial}{\partial t} \left[\frac{\rho_*}{2} (u^2+v^2)\right]$	storage (= time change) of relative kinetic e.
$E_{pg} = \int_{O_W} dO_R \frac{1}{2} \rho g \zeta^2$	potential e. due to Earth's gravity
$\frac{\partial}{\partial t} E_{pg} = \int_{O_W} dO_R \frac{\partial}{\partial t} \frac{1}{2} \rho g \zeta^2$	its storage (i.e., time change)
$E_{p\Phi} = \int_{O_W} dO_R \rho_* \Phi$	potential e. due to tidal forces
$\frac{\partial}{\partial t} E_{p\Phi} = \int_{O_W} dO_R \frac{\partial}{\partial t} \rho_* \Phi$	its storage
$G(E_{p\Phi}) = \int_{O_W} dO_R \rho_* \frac{\partial}{\partial t} \Phi$	generation of (tidal) potential e.

As for the (analogous) individual balance terms due to the undisturbed water depth (marked by *) and the water elevation (**) respectively due to the zonal ($_u$) and meridional ($_v$) currents (in the energy budgets), see Tables 3, 4, and 5.

Acknowledgements. The computations were executed on the IBM 370-168 of the RHRZ, University Bonn. W. Hövel ran the HN-program generating the necessary velocity and elevation data.

This research was supported by the Deutsche Forschungsgemeinschaft (under Grant Su 69/2).

References

Baader H-R (1978) Eine Klimatologie des ersten Hauptsatzes. Dissertation, Meteorol Inst, Univ Bonn

Bartels J (1957) Gezeitenkräfte. In: Flügge S (ed) Handbuch der Physik, Geophysik II, Vol 48. Springer, Berlin Göttingen Heidelberg, p 734

Bronstein IN, Semendjajew KA (1966) Taschenbuch der Mathematik, 6th edn. Harry-Deutsch, Zürich, Frankfurt/M

Brosche P (1980) Gezeitenreibung im Erde-Mond-System. Mitt Astron Ges 51:81-100

Brosche P, Hövel W (1980) Restströme, Gezeitenstromellipsen und Bodenreibung. Dtsch Hydrogr Z 33:110-123

Brosche P, Sündermann J (1971) Die Gezeiten des Meeres und die Rotation der Erde. Pageophysics 86:95-117

Brosche P, Sündermann J (1979) Oceanic tidal friction: principles and new results. In: Mc Carthy DD, Pilkington JD (eds) Time and the Earth's rotation. IAU, Reidel, Dordrecht, p 317

Fortak H (1962) Concerning the vertically averaged hydrodynamic equations with respect to basic storm surge equations. In: Mathematical-hydrodynamical methods of physical oceanography. Mitt Inst Meeresk, Univ Hamburg, Vol 1

Fortak H (1968) Die Methode der Filteranalyse in der atmosphärischen Energetik. Contribution to 2nd SPAAZ-Seminar. Inst Theor Meterol, FU Berlin

Glansdorff P, Prigogine I (1971) Thermodynamic theory of structure, stability and fluctuations. Wiley Interscience, London New York Sydney Toronto

Groot De SR (1960) Thermodynamik irreversibler Prozesse. BI, Mannheim

Hendershott MC (1972) The effects of solid Earth deformation on global ocean tides. Geophys J R Astron Soc 29:389-402

Holton JR (1973) An introduction to dynamic meteorology, 2nd edn. Academic Press, New York

Krauss W (1973) Methods and results of theoretical oceanography I: Dynamics of the homogeneous and the quasi-homogeneous ocean. Gebrüder Bornträger, Berlin Stuttgart

Krauss W (1980) Basic hydrodynamics and thermodynamics. In: Sündermann J, Holz KP (eds) Mathematical modelling of estuarine physics, Vol I of Lecture notes on coastal and estuarine studies. Springer, Berlin Heidelberg New York, p 1

Lambeck K (1977) Tidal dissipation in the oceans: Astronomical, geophysical and oceanographic consequences. Philos Trans R Astron Soc Ser A 287:545-592

Lambeck K (1980) The Earth's variable rotation: Geophysical causes and consequences. Cambridge Univ Press, Cambridge

Landau LD, Lifschitz EM (1966a) Lehrbuch der theoretischen Physik I: Mechanik, 4th edn. Akademie, Berlin

Landau LD, Lifschitz EM (1966b) Lehrbuch der theoretischen Physik VI: Hydrodynamik. Akademie, Berlin

Monin AS, Kadenkovich VM, Kort VG (1977) Variability of the oceans. Wiley Interscience, New York London Sydney Toronto

Munk WH, MacDonald GJF (1960) The rotation of the Earth, a geophysical discussion. Cambridge Univ Press, London

Oort AH, Bowman II HD (1974) A study of the mountain torque and its innerannual variations in the northern hemisphere. J Atmos Sci 31: 1974-1982

Parke ME (1978) Open ocean tide modelling. In: Proc 9th GEOP-Conf. Ohio State Univ, Ohio

Pekeris CL, Accad Y (1969) Solution of Laplace's equations for the M_2-tide in the world oceans. Philos Trans R Astron Soc Ser A 265: 413-436

Schwiderski EW (1980) On charting global ocean tides. Rev Geophys Space Phys 18:243-268

Sündermann J, Brosche P (1978) The numerical computation of tidal friction for present and ancient oceans. In: Brosche P, Sündermann J (eds) Tidal friction and the Earth's rotation. Springer, Berlin Heidelberg New York, p 125

Wesson PS (1980) Does gravity change with time? Phys Today 33:32-37

Weizel W (1962) Physikalische Formelsammlung I. BI, Mannheim

Zahel W (1970) Die Reproduktion gezeitenbedingter Bewegungsvorgänge im Weltozean mittels des hydrodynamisch-numerischen Verfahrens. Mitt Inst Meeresk, Univ Hamburg, Vol 17

Zahel W (1977) A global hydrodynamic-numerical 1°-model of the ocean tides; the oscillation system of the M_2-tide and its distribution of energy dissipation. Ann Geophys 33:31-40

The Resonance Behaviour of the World Ocean

J. Sündermann[1]

1. Introduction

At present it seems to be unlike from the geological point of view as well as from celestial mechanics that the moon has approached the Earth during its history up to a few Earth radii (Gerstenkorn event). This dramatic situation will appear tracing back the Moon's orbit with the presently valid rate of momentum transfer from the Earth to the Moon (Gerstenkorn 1955).

It is commonly accepted today that this momentum transfer is produced mainly by oceanic tides (Lambeck 1980). Most important, therefore, is the question whether in ancient times these tides have been similar to the present ones and how the corresponding momentum transfer has been changed.

The essential influences on oceanic paleotides are given by a higher angular velocity of the Earth and by a different topography of the oceans (Webb this Vol.). The first factor is studied in Webb's paper; it results in a significant decrease of the momentum transfer. The second factor has been investigated for several periods of the Earth's history up to the Cambrian by means of numerical models (Sündermann and Brosche 1978, Krohn and Sündermann this Vol.). It turns out that the momentum transfer during the younger history of the Earth has varied at least by a factor 3. The present value, obviously, is relatively high, but similar magnitudes are obtained also for the Ordovician. Due to these changing values an exact backward integration of the Moon's orbit is not possible.

The question remains in which manner a modified topography (or topology) of the oceans influences the tides and hence the momentum transfer. Equation (1) describes the conservation of momentum in the oceanic water masses (Brosche and Sündermann, in press)

$$\left\langle \iint_{\text{ocean}} \left[\rho R^2 \cos^2\varphi \, DR_\lambda + \rho R g \cos\varphi \, D \frac{\partial \zeta}{\partial \lambda} - \frac{1}{2} \rho g C R^2 \cos^3\varphi \, \zeta(t) \sin(\sigma t + s\lambda) \right] d\lambda \, d\varphi = 0 \right. \quad . \qquad (1)$$

---

1 Institut für Meereskunde der Universität Hamburg, Heimhuder Str. 71, D-2000 Hamburg 13, Fed. Rep. of Germany

The following notations have been used:

- C tidal constant
- D total water depth
- R Earth radius
- R_λ East component of total friction
- g gravity acceleration
- s tidal type
- t time
- λ geographical longitude
- φ geographical latitude
- ρ density of the water
- ζ water surface elevation
- σ tidal angular velocity
- $\langle\ \rangle$ means the average over one tidal cycle.

It is seen from (1) that the momentum transfer between the ocean and the Moon (third term) is balanced by the momentum transfer between the solid Earth and the ocean. The latter is produced by an alteration of ocean currents due to bottom friction (first term) and pressure force of inclined bottom (second term).

An amplification of oceanic currents by tidal resonance will lead to an increased bottom friction. This could mean a higher rate of momentum transfer. The same effect is possible, however, by an alteration of the current field due to modified continental positions. The latter point is still to be investigated systematically. The first one is subject of the present paper which is restricted to the present ocean. The broad spectrum of obtained resonance periods, however, permits conclusions also for ancient oceans.

2. Tidal Resonance and Momentum Transfer

The tides of the world ocean are oscillations of the water masses forced by the sinusoidal tidal generating potentials. They must have, therefore, the same periods T_c as the generating celestial forces:

$$T_c = \frac{2\pi}{\sigma}$$

where σ is the corresponding frequency.

On the other side, the world ocean on the rotating Earth shows free oscillations which are classified in the longperiod part (periods >1 h) into vorticity modes and gravity modes (Platzman et al. 1981). The first group is governed by the Coriolis force, the second one by the gravity of the Earth. If the periods T_i respectively T_g of these free modes coincide with the periods of the tidal generating potential, resonance tides are achieved with increased elevations and currents.

The period of the inertial waves is given by

$$T_i = \frac{2\pi}{f}$$

where f is the Coriolis frequency.

Table 1. Period of inertial mode depending on geographical latitude

φ degree	T_i hours
0	∞
10	71.8
20	36.5
30	24.9
40	19.4
50	16.3
60	14.4
70	13.3
80	12.7
90	12.5

Table 1 shows this period depending on the geographical latitude. While a generation of vorticity modes by semidiurnal tides practically rules out, the diurnal tides could find resonance nearly at 30^0 North or South. But the amplitudes of the diurnal tidal generating potentials are relatively small. Thus, the resonance of vorticity waves are not further investigated. (In the real ocean the picture is complicated by the natural depth distribution resulting in topographically controlled vorticity modes. But they are still of minor importance in the tidal frequency band (Platzman et al. 1981)).

The period of the free gravity waves depends on the geometrical properties of an ocean basin. It is described approximately by

$$T_g = \frac{2L}{n\sqrt{gD}} \quad , \quad n = 1,2,3,... \tag{2}$$

where L is the length and D the depth of the sea.

Table 2 contains numerical values of this period for the basic mode (n = 1). Corresponding to the global topography of the oceans the

Table 2. Period of long gravity waves (h) depending on the length (L) and the depth (D) of an ocean basin (n = 1)

L \ D	1.000 m	2.000 m	3.000 m	4.000 m	5.000 m
100 km	0.6	0.4	0.3	0.3	0.3
500 km	2.8	2.0	1.6	1.4	1.3
1.000 km	5.6	4.0	3.2	2.8	2.5
3.000 km	16.8	11.9	9.7	8.4	7.5
5.000 km	28.0	19.8	16.2	14.0	12.5
10.000 km	56.1	39.7	32.4	28.0	25.1
20.000 km	112.2	79.3	64.8	56.1	50.2

marked three values are most frequent as well as the whole continuum limited by the quantities. This intervall contains the periods of semidiurnal and diurnal tides resulting in the principal possibility of tidal resonance. The situation is, however, considerably complicated by the complex geometry of the real oceans. It is extremely difficult to interprete the manifold oscillation phenomena of the world ocean as a superposition of the free modes given by Tables 1 and 2 (Platzman et al. 1981).

What is the effect of tidal resonance on the momentum transfer between Earth and Moon? The answer can not be given without a detailed consideration of the modified tidal amplitudes and phases. If resonance would only mean that the local water elevations $\zeta(\lambda,\varphi,t)$ are raised by a factor $\alpha(\alpha > 1)$

$$\zeta_r = \alpha\zeta$$

but the phases would remain unchanged, then the momentum transfer M given by the third term of (1) would be multiplied by the same factor

$$M_r = \alpha M \quad .$$

But this simple conclusion is not true. Resonance means also that the oscillation processes are changed with respect to time and location. This implies that the expression

$$\cos^3\varphi \, \zeta(\lambda,\varphi,t) \, \sin(\sigma t + s\lambda)$$

in the third term of (1) obtains a value which cannot be simply predicted, even not in tendency. Therefore, "resonance" does not automatically mean a higher deceleration of the Earth's rotation. The celestial effect can be determined only after a careful analysis of the oscillation systems in the ocean. This is possible only by means of numerical models (Brosche and Hövel, this Vol.).

3. The Eigen Modes of the Present Ocean

In the following a short presentation is given of the free ocean modes as known by recent model computations. It is based on the work of Gaviño (1982), Gotlib and Kagan (1982), Platzman et al. (1981) and Schulte (1979).

Starting from the Laplace tidal equations, Gaviño and Platzman are computing the eigen values and functions of the governing oscillation operator in a principally similar way. Gotlib and Schulte produce in their numerical models free waves by an initial disturbation and calculate the geographical distribution and the time depending run. The obtained time series are Fourier-analyzed giving the significant free modes.

A rough characteristics of three of the models is seen by Table 3. A comparison of the calculated oscillation systems in the frequency band of diurnal and semidiurnal tides may demonstrate that the results coincide fairly well, see Figs.1 and 2.

A detailed analysis of Figs.1 and 2 show, however, that there exist differences, especially in the Pacific. These can be explained most probably by the different discretizations.

Table 3. General characteristics of three models analyzing the free modes of the world ocean

Author	Period range [h]	Number of modes
Gaviño	65.7 - 10.7	85
Gotlib	51.1 - 8.4	34
Platzman	79.8 - 9.0	49

A compilation of the most important eigen modes is given in Table 4. Due to the different resolution of the models it is sometimes difficult to give a unique correspondence of modes. In the long period bands are only a few identical values. Only Gotlib notes the normalized amplitude of the eigen oscillations. Platzman tries to identify the phenomena as local modes in different parts of the world ocean.

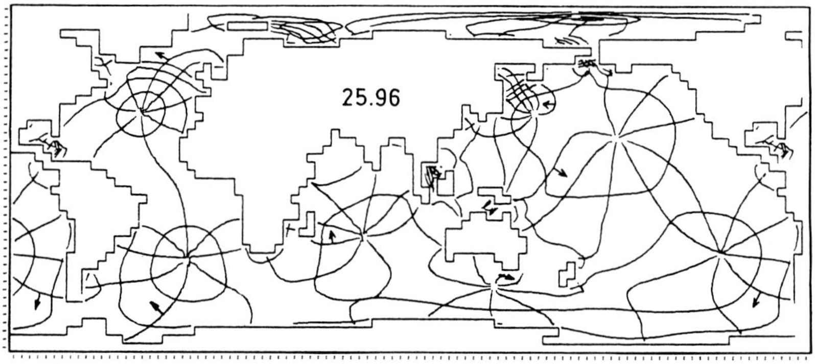

Gavino 1981

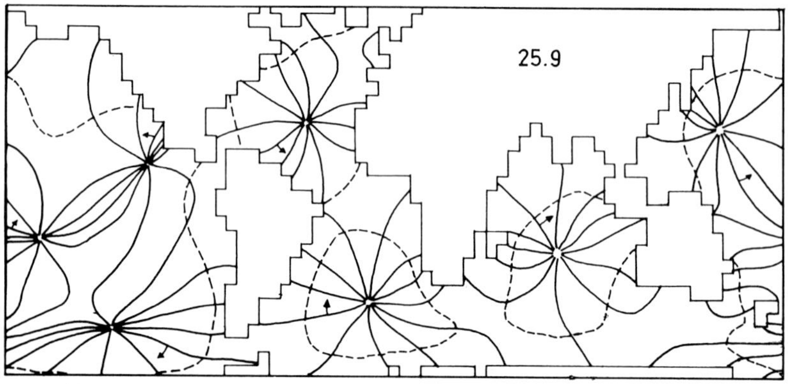

Gotlib and Kagan 1982

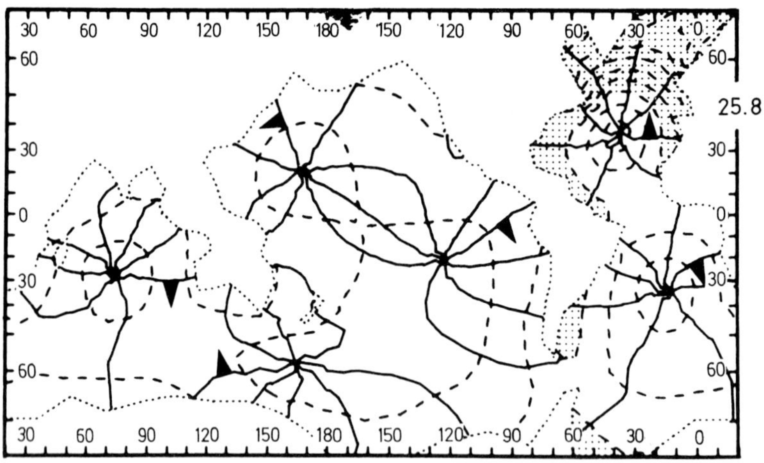

Platzman et al. 1981

Fig. 1. Cotidal and corange lines for a diurnal free mode in three numerical models. The rotational sense is indicated by an arrow. Gaviño: Cotidal and corange lines full. Gotlib: Cotidal lines full, corange lines dashed. Platzman: Cotidal lines full, corange lines dashed

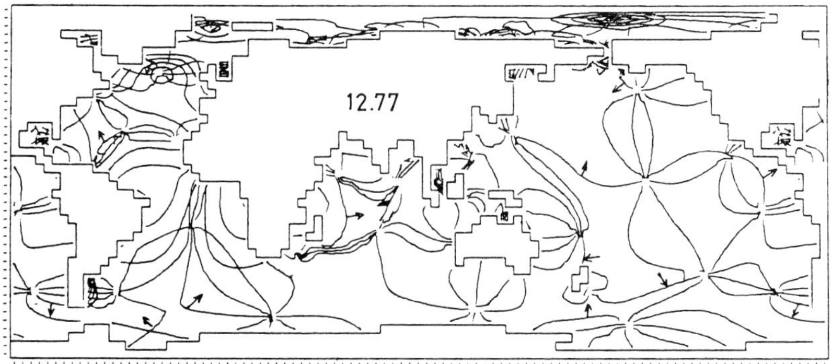

Gavino 1981

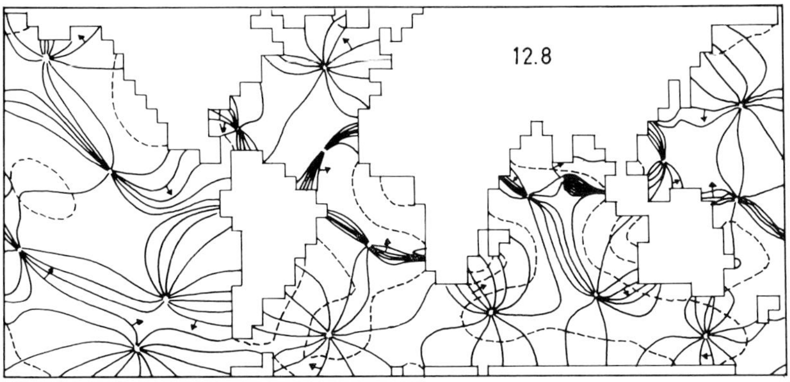

Gotlib 1982

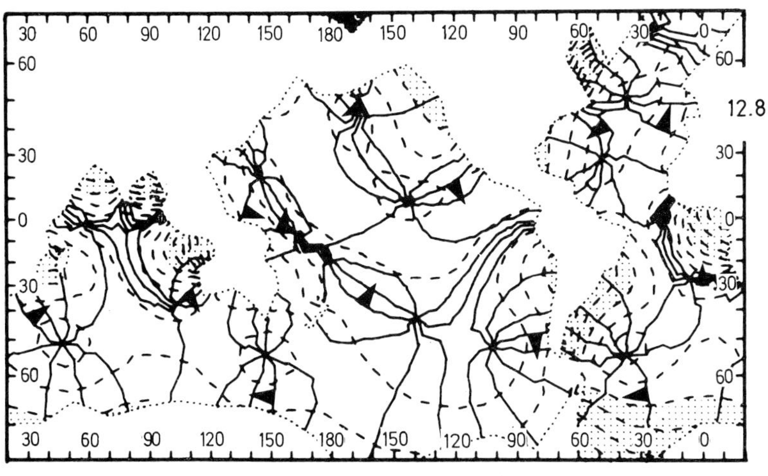

Platzman et al. 1981

Fig. 2. Cotidal and corange lines for a semidiurnal free mode in three numerical models. The rotational sense is indicated by an arrow

Table 4. Some important eigen modes of the world ocean after model results of Gaviño, Gotlib and Platzman

Author	Gaviño	Gotlib		Platzman	
Quantity	Period hours	Period hours	Relat. ampl. %	Period hours	Interpretation
	-	-	-	79.8	
	50.9	51.1	40	50.6	
	36.3	39.8	100	35.8	
	30.0	30.0	50	28.7	
Diurnal tides	26.0	25.9	47	25.8	North Atlantic longitudinal 1/2 wave
	23.5	23.9	32	23.7	North Atlantic longitudinal 3/4 wave
	21.9	22.5	49	21.2	Pacific 2/2 wave
	20.3	20.2	52	20.2	
	17.6	17.8	51	17.7	
	16.5	16.7	77	16.3	
	14.2	14.4	43	14.1	
Semidiurnal tides	12.8	12.8	32	12.8	Gulf of Guinea 1/4 wave
	12.7	12.7	58	-	-
	12.5	12.5	41	12.5	Arabian Sea, North Pacific
	12.0	12.0	31	11.9	Arabian Sea, Labrador Sea
	11.2	11.2	43	11.0	

It is seen by Table 4, and more detailed by the presentation in the mentioned papers, that in the real world ocean appear many eigen oscillations of different basins and bathymetric regions. They include periods in the range of diurnal and semidiurnal tides. But these are to be explained after Platzman, as seen from Table 4, by completely distinct local oscillation phenomena which are more random distributed than sources of a systematic rectification.

Moreover, it can be stated from the calculated amplitudes of various modes that there are differences of energy density by orders of magnitude. This is also demonstrated by a power spectrum derived by Schulte (1979 and pers. comm.), see Fig. 3.

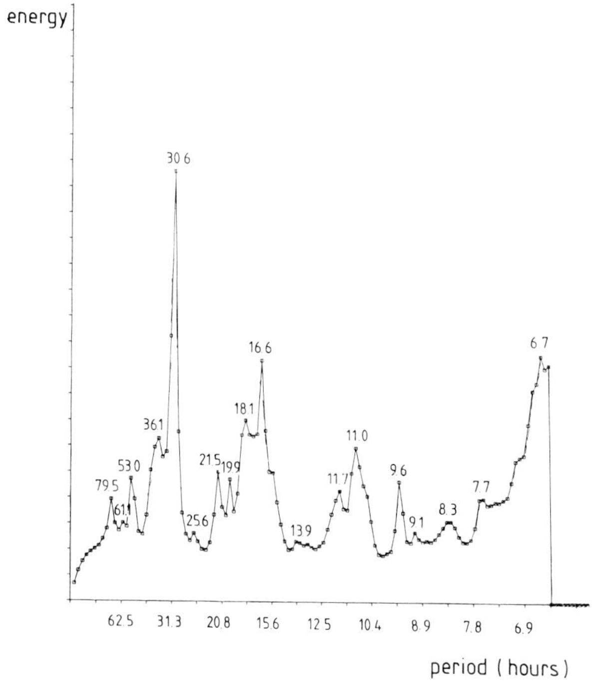

Fig. 3. Power spectrum of the free modes of the world ocean after Schulte (1979). The ordinate gives the integrated squares of amplitudes in arbitrary units on a linear scale. The numbers at the curve are periods in hours

The periods around 30 h which are close to the period of the antarctic Kelvin wave represent a certain exception. These modes show remarkably high energy densities in a broad belt around the antarctic continent.

The local resonance phenomena in small adjacent seas and bays (e.g., Newfoundland Bay) are not resolved by these models. But they hardly contribute to the global tides and are practically of no relevance for the momentum transfer between Earth and Moon.

4. Conclusions for the History of the Oceans

During the oceans' history especially the topographical situation has been changed significantly (Piper; Thiede; Ziegler, this Vol.). Even if the hypsographical curve remained without major alterations, and hence quantity D in formula (2), the horizontal dimensions L of ocean basins varied permanently. Consequently, the spectrum of the free modes in the ocean has been changed continuously during its history.

On the other side, it is not probable that this caused a completely different resonance behaviour. Plate tectonics and continental drift will steadily have deleted existing tidal resonances in parts of the world oceans and created new ones in other parts (Webb 1982, this Vol.). The broad spectrum of the present eigen oscillations gives rise to the assumption of similar conditions for ancient times. It is unlike that the present state of resonance forms an extreme exception during Earth's history.

The same is not necessarily true for the momentum transfer from the Earth to the Moon and hence for the deceleration of the Earth's rotation. It can be seen from formula (1) that this transfer is influenced also by the global distribution of depth gradients in East-West direction. Presently, we have a situation which is characterized by three large oceans and three meridionally positioned groups of continents. This was completely different, for instance, in the Permian when one super continent (Pangaea) and one ocean covered the Earth. At that time the momentum transfer was smaller (Sündermann and Brosche 1978). Compared with this the Silurian is characterized by a series of continents corresponding to zonal depth gradients especially in equatorial latitudes. This might be connected with the increased momentum transfer at that time (Krohn and Sündermann, this Vol.). This possible relation between the topology of continents and the deceleration of the Earth's rotation has still to be investigated.

References

Brosche P, Sündermann J (1982) Tidal friction and dynamics of the Earth-Moon system. In: Soffel H, Fuchs K (eds) Landolt-Börnstein-Geophysics. Springer, Berlin Heidelberg New York, in press

Gaviño JH (1981) Über die Bestimmung von reibungslosen barotropen Eigenschwingungen des Weltozeans mittels der Lanczos-Methode. Ph D thesis, Univ. Hamburg, 66 pp

Gerstenkorn H (1955) Über Gezeitenreibung beim Zweikörperproblem. Z Astrophys 26:245-274

Gotlib VYu, Kagan BA (1982) Numerical simulation of tides in the world ocean. Part III: A solution to the spectral problem. Dtsch Hydrogr Z 35: (in press)

Lambeck K (1980) The Earth's variable rotation: geophysical causes and consequences. Cambridge Univ Press, Cambridge, 449 pp

Platzman GW, Curtis GA, Hansen KS, Slater RD (1981) Normal modes of the world ocean. Part II: Description of modes in the period range 8 to 80 hours. J Phys Oc 11:579-603

Schulte D (1979) Die Ermittlung der Eigenschwingungsperioden des Weltozeans mittels des HN-Verfahrens und der Fast Fourier Transform auf der Grundlage eines 4°-Modell-Weltozeans. Thesis, Univ Hamburg, 53pp

Sündermann J, Brosche P (1978) Numerical computation of tidal friction for present and ancient oceans. In: Brosche P, Sündermann J (eds) Tidal friction and the Earth's rotation. Springer, Berlin Heidelberg New York 243pp

Tidal Friction for Times Around the Presence

P. Brosche and W. Hövel[1]

1. Introduction

In recent years a large number of investigations were devoted to the eigenfrequencies of the world ocean (see Sündermann, this Vol.). Besides other aspects, one motivation of our calculations was the suspicion that the present world ocean is very close to a resonance situation.

Comparing the period of 12.42^h of the dominating M_2-tide-generating potential with the eigenfrequency-computations of Platzman (1978), Platzman et al. (1981), Gotlib and Kagan according to Sündermann (1982, this Vol.), and Schulte (1979), who found eigenfrequencies with periods of 12.2^h and 12.5^h, such a situation is probable. The results of these authors show a very dense eigenfrequency-spectrum, and one might suppose that because of the limitations of the numerical scheme, this spectrum is not complete. The consequences of comparably short but intense variations of tidal parameters, which would result from an accumulation of resonance situations, are numerous and may not be discounted. As an example we mention Brosche's (1981) presumption that the geomagnetic dynamo could be affected by "short" time variations of the Moon's torque on the Earth.

The problem of Earth rotation and tidal friction, especially, poses some important questions:

a) Is there a monotonous correlation between the transfer of momentum (between Earth and Moon) and the distance from a resonance situation of the oceans? Does it always imply a relatively great (or small) torque?
b) What kind of variation can we expect for energy, energy dissipation, tidal elevations, velocities etc.? Which parameters change significantly? Furthermore: Does such a resonance make its entry abruptly or gradually?
c) Is the present ocean close to a resonance, as suspected, and if so, how close?
d) May we extrapolate data of the Earth-Moon system into the past (especially until the "Gerstenkorn-event") with a smooth inter-

1 Sternwarte der Universität Bonn, D-5300 Bonn, Fed. Rep. of Germany

Tidal Friction and the Earth's Rotation II
ed. by Brosche/Sündermann

action, or is the present ocean a very special ocean and are we consequently obliged to consider a "fine-structure" of the time dependence of tidal parameters (especially the torque between Earth and Moon)?

The available world-ocean models cannot provide answers to the above questions (which are certainly not a complete list). The information we obtain from these models belongs to specific instants quite separated from each other. This is because the ocean-continent configurations of these models are very different. Since the results differ by factors of 2 to 3, long time extrapolations are based on the uncertain assumption that the short time variations are negligible. But because of the dense structure of the world-ocean's spectrum this is not necessarily true.

The model we present here is a first example for "short" time variations. The construction of a realistic ocean model (based on Zahel's (1970) 4° model) with flexible ocean-continent configuration allows for the simulation of the drift of the whole American continent. In a time interval around the present we study the variations of energy, dissipation, momentum and torque until about 10 to 15 million years into the past and the future. Other information, like the movements of the center of mass of the Earth and the development of tidal elevations and velocities is also obtained.

2. Extension of the Hydrodynamic-Numerical Model

In time scales, which are long in comparison with one tidal cycle, one has to take into account (a) continental drift, which changes the ocean's geometry and thereby the eigenfrequencies of the oceans, and (b) the changes in both the strength and the frequencies of the tidal forces due to the dissipational changes of the lunar orbit. While the second might be discounted on time scales of several 10 million years, the first can cause rapid variations especially within short time scales.

In the above-mentioned works, Platzman (1978), Platzman et al. (1981), Schulte (1979), Gotlib and Kagan (according to Sündermann 1982, this Vol.) calculate the spectra of the *present* oceans. Furthermore, some of their results are based on linear models. Hence a comparison between the spectra of these models and the more realistic (nonlinear) hydrodynamic-numerical model of Zahel (1970) must be made with care. Moreover, the "power" attributed to the eigenfrequencies in those studies (if any, some contain lists of frequencies only) is not an a priori measure of the quantities we are interested in, namely energy and angular momentum transfer. On the other hand, we do not know how the spectrum reacts when the ocean's geometry is changed slightly. We therefore constructed a model which is independent of any eigenfrequency calculations.

This was done by an extension of Zahel's 4^0 world ocean model. Basic properties of this model were not changed. Thus the depth structure of the (present) ocean as digitalized by Zahel remained unchanged. First of all, the representation of the continental drift by an omission or addition of otherwise unchanged surface elements would have meant an introduction of discontinuity ab ovo. Presumably the results would have been masked by this source of "ruggedness". Moreover, the whole time span we had in mind corresponds almost to a one-element wide stripe in the continental margins, thereby prohibiting any finer time resolution. For these reasons, we took advantage of the fact that the present picture of continental drift can be represented — in a zeroth order approximation — by an east-west shift of the two Americas relative to the rest. So the simulation of the drift of the continental plates could be realized by a special adaption of the difference method, which makes it possible to calculate longitudinal movements. Whereas so far the east-west extent of the surface elements was a function of latitude only, the possibility of a continuous variation in this length has been introduced. In such a way the Atlantic and Pacific Oceans could be "squeezed" and "blown up" reciprocally, while the Indian Ocean remained unchanged.

This paper presents the consequences of such schematic movements of the two American continents. Expanding the Atlantic ocean and compressing the Pacific ocean corresponds to the future and the reverse to the past (see Fig. 1). According to modern values of the mean motion of continental plates from Thiede (1979), our longitude limits of America correspond to times of ~20 million years in the past and ~10 million years in the future.

The technical and numerical effort involved was great. To obtain a sufficient degree of accuracy, a large number of iterations had to be performed. The problems of the balances of energy and angular momentum and a description of special features, e.g., to reduce the number of iterations by external filtering processes, stability of the scheme etc. will be given in Hövel (1982). Here we should notice only that all data presented (integrals) show a precision of two significant digits at least. The total number of models we have computed is 34, so the time resolution is of the order of one million years.

All calculations were performed on the IBM 370/168 of the University Bonn with double precision.

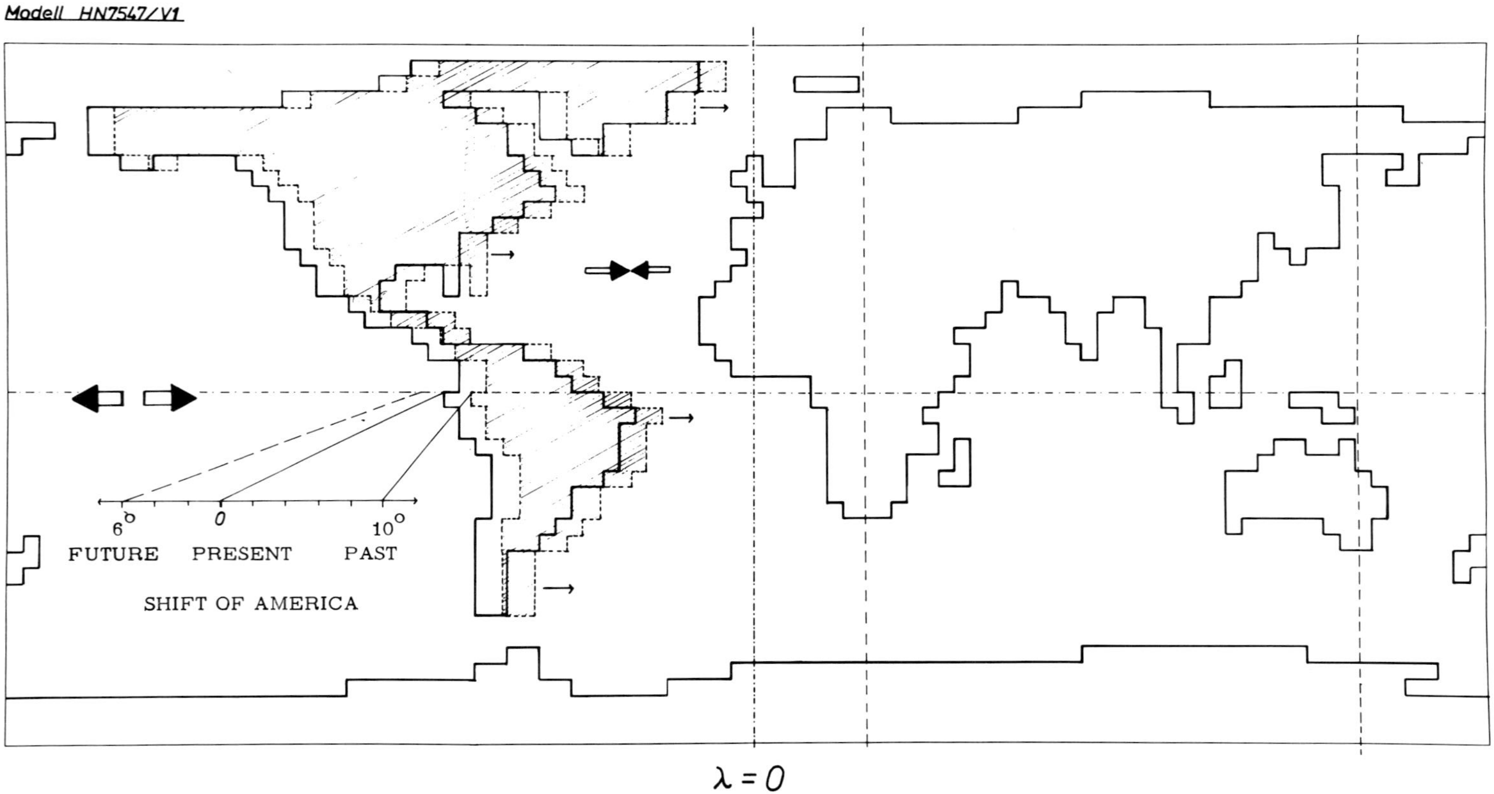

Fig. 1. Limits of a sequence of 4°-Ocean models with respect to movements (longitudinal) of the American continents. For calculations into the past, the Atlantic Ocean has been squeezed, and the Pacific Ocean has been expanded. A "tentative time scale" is represented here by the rate of translation of the two Americas in degrees. The limits correspond to times of ~20 million years in the past and to ~10 million years in the future

3. Results

Most of the parameters analysed (e.g., energy, angular momentum, center of mass, water elevation etc.) show strongly correlated behaviour: Over a relatively large range of the past, the influence of continental drift is small. About 10 million years ago some of the values (energy, torque Moon-Earth) reach a flat minimum. In the opposite direction (the more recent past and the future) there is a strong increase to an extreme. This extreme will be reached in about 5 million years. It is followed by a quick decrease (similar to the increase) up to the end of our computations.

Although the general behaviour does not seem to be like a catastrophic resonance, the extreme is very well and (in comparison with the above-mentioned minimum) sharply defined. We are not able to decide here whether the future ocean will meet one of its eigenfrequencies, because we do not know its spectrum, but our results are an indication that it will. Moreover, the determination of the spectrum of the world ocean as described in our nonlinear equation model is fraught with great mathematical difficulties and in most cases reduced to the determination of extremes of characteristic parameters (e.g., energy). In this sense, and taking into account the works of Platzman, Schulte and Gotlib and Kagan (see above) too, we have confirmed the presumption that the world ocean is close to a resonance. On the other hand, it was somewhat surprising to find only one such resonance in the whole time span calculated. Possibly we can interpret this one extreme as a blurred effect of several eigenfrequencies which the numerical scheme is not able to separate. This question must be postponed until a better knowledge of the spectrum of the world ocean (and its time dependence) is at hand. Nevertheless, the variation of the tidal parameters within our time span is astonishingly high. In the following sub-chapters the results are given in detail.

3.1 Tidal Elevations and Velocities

The mean amplitudes of the velocity components (r.m.s.-mean, averaged over the whole world ocean) descend to a minimum of ~2 cm/s in the past, while the resonance-values are ~3 cm/s (north-south component) and ~3.5 cm/s (east-west component), the present values are ~2.5 cm/s and ~3 cm/s, respectively. It seems remarkable that a difference in the values of the components can be seen near to the maximum only (see Fig. 2). Perhaps this can be interpreted as a preponderance of more excentric tidal ellipses and a better adaption of the currents to the tidal forces. The water elevations show an analogous behaviour. The total variation is ~0.3 m, with a maximum of $\bar{\zeta} \sim 43$ cm (Fig. 3).

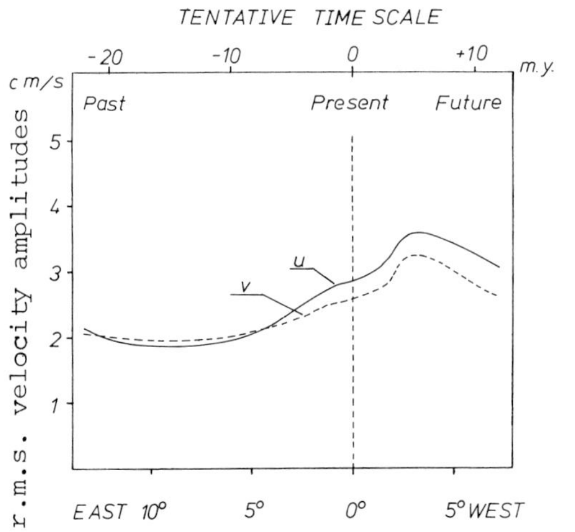

Fig. 2. r.m.s.-velocity amplitudes. East-West component (——) and North-South component (----)

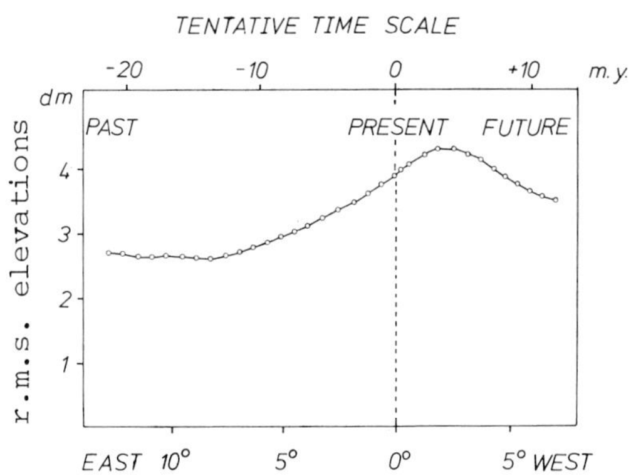

Fig. 3. The computed r.m.s.-elevations of the models. Each model is represented by an *open circle* in the figure. The mean distance between two succeeding models corresponds to about 1 million years on the time scale

3.2 Movements of the Center of Mass

In what follows we use a co-rotating coordinate system (see Brosche and Sündermann 1971) with x and y axes lying in the equatorial plane and directed towards longitudes 0° and $+90^{\circ}$ respectively and the z axis towards the north pole.

The center of mass of the Earth gives us information about the asymmetrical behaviour of ocean tides and the total amount of momentum in the oceans (see Hövel 1982). As usual, we computed only the deviations from a "mean" position, which show a harmonic time dependence (with M_2-tide frequency), although there are many disturbing effects and although the differential equations are nonlinear. The path which the barycenter describes in one tidal cycle is an ellipse, more or less excentric, and inclined in relation to the equatorial plane. For comparison with other authors (e.g., Yoder et al. 1981), we give here the deviations in the form of amplitudes and phases (counted from the lunar meridian transit at Greenwich) for each cartesian component of the corotating system.

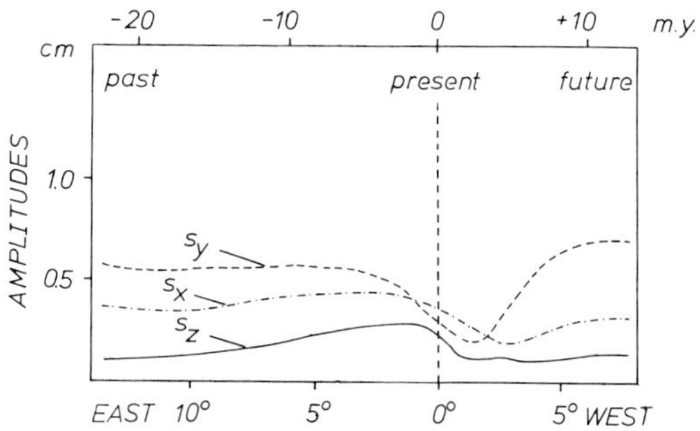

Fig. 4. The harmonic (main) part of the movement of the Earth's center of mass: Amplitudes of the dominating oscillations with M_2-tide frequency. S_x (-·-·-·-), S_y (----) and S_z (——) denoting the oscillations in the x,y,z directions in the co-rotating system. The movements are calculated as differential movements against a mean position of the center of mass, and based on the elevation of the oceans

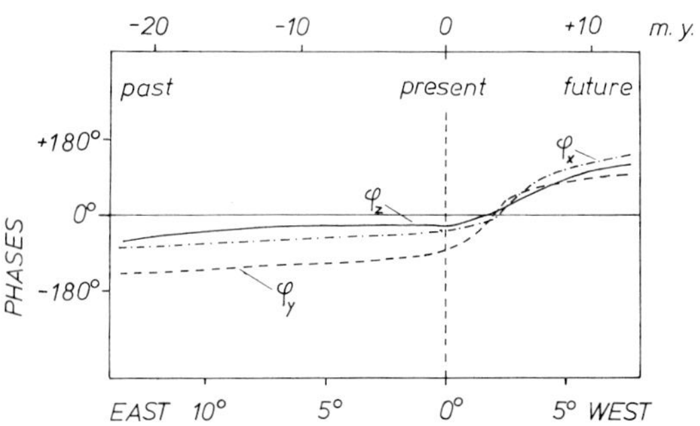

Fig. 5. The harmonic (main) part of the movement of the Earth's center of mass: Phases of the oscillation. These correspond to the amplitudes in Fig. 4

The time dependence of the amplitudes and the phases is shown in Figs. 4 and 5. It is remarkable that the amplitudes reduce to a minimum, when they approach the resonance. Outside this region the variation is relatively small.

The behaviour of the phases is also interesting. They jump at the resonance and reach values about 180° more than before. This seems to be a jump similar to that known for other oscillating systems at resonance.

The order of magnitude of the whole movement of the barycenter is about 0.5 to 1.5 cm (length of the main axis of the corresponding ellipse) and the main axis, regarded as a pointer in space, is directed into the Pacific ocean and its antipodal position, describing an open slope if we draw it for the full sequence of models (see the schematic Fig. 6).

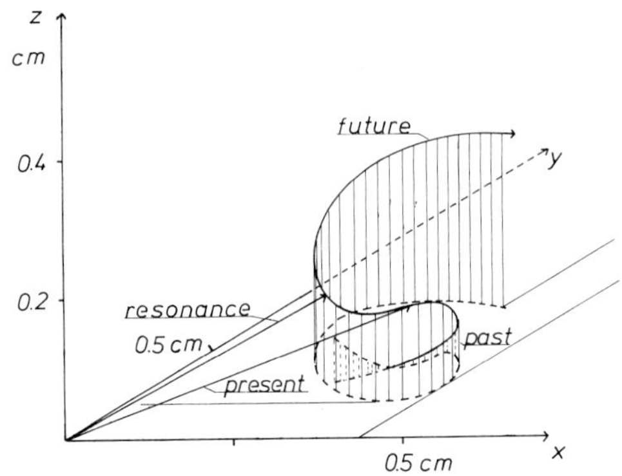

Fig. 6. The harmonic (main) part of the movement of the Earth's center of mass: Direction and length of the major axis of the ellipse of the movement (in the co-rotating system). The space curve is shown as full line above the equatorial plane and (-·-·-·-) below. Its projection into the plane is given by the *broken line*

3.3 Balance of Energy

The time-averaged values (where time-averaging is denoted by the brackets < >) of the total energy ($<E_{tot}>$), the potential energy ($<E_{pot}>$) and the kinetic energy ($<E_{kin}>$) of the M_2-tide which were calculated in the co-rotating coordinate system, show the same behaviour as described at the beginning of this chapter (see Fig. 7). Compared with the present value of the total energy, the minimum value in the past

$$<E_{tot}>_{min} = 4 \times 10^{24} \text{ erg}$$

is very low. At the maximum about two and a half times this value is reached:

$$<E_{tot}>_{max} = 9.5 \times 10^{24} \text{ erg} \quad .$$

The present value is

$$<E_{tot}>_{pres} = 7.3 \times 10^{24} \text{ erg}$$

which is somewhat higher than Zahel's (1977) value 6.9×10^{24} erg and in very good agreement with his calculations.

We notice that the quotient $q = <E_{pot}>/<E_{kin}> \sim 0.95$ is almost constant. Even at resonance this value remains nearly constant. Probably this is another fact which indicates the proximity of an eigenfrequency.

The total energy dissipation in the world ocean shows a considerable variation, too. The main part of the energy is dissipated by inner friction (see Fig. 8). Bottom friction is on a comparatively low level and shows only little dependence on continental drift.

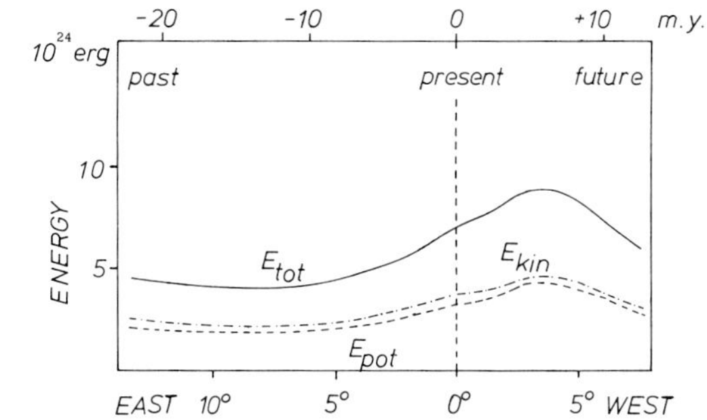

Fig. 7. The energy of the World Ocean. These curves belong to calculations in the co-rotating system. E_{kin} (-·-·-) E_{pot} (---) and E_{tot} (——) correspond to the time-averaged means of the kinetic, potential and total energy of the ocean respectively

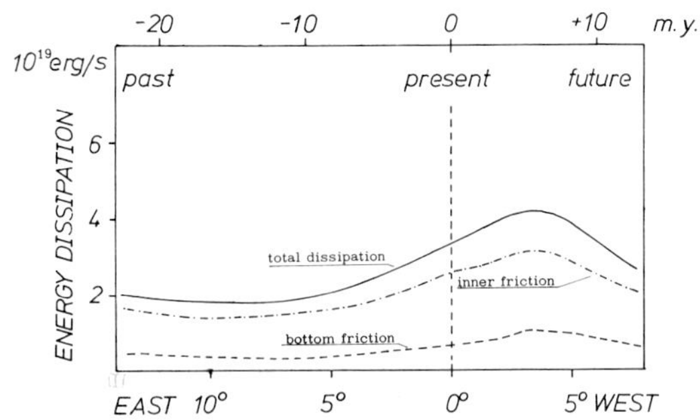

Fig. 8. Energy dissipation in the World Ocean. The curves show the dissipation rates (time-averaged) of bottom friction (---), turbulent friction (inner friction) (-·-·-) and the total dissipation rate (——)

The maximum value of the total energy dissipation is

$$<-\dot{E}_{diss}>_{max} = 4.2 \times 10^{19} \text{ erg/s}$$

while the minimum value is

$$<-\dot{E}_{diss}>_{min} = 1.7 \times 10^{19} \text{ erg/s} \quad ,$$

so the total variation is a factor of 2.5! The present ocean shows a total dissipation of

$$<-\dot{E}_{diss}>_{prst} = 3.2 \times 10^{19} \text{ erg/s} \quad ,$$

in good agreement with Zahel's (1977) value (3.3×10^{19} erg/s).

As shown in Fig. 9, there seems to be an almost linear correlation between the total energy of the M_2-tide and the total dissipation of energy. More precise information gives the specific dissipation factor Q, defined as usual by

$$Q^{-1} = \frac{1}{2\pi E^*} \oint \frac{dE}{dt} dt \quad .$$

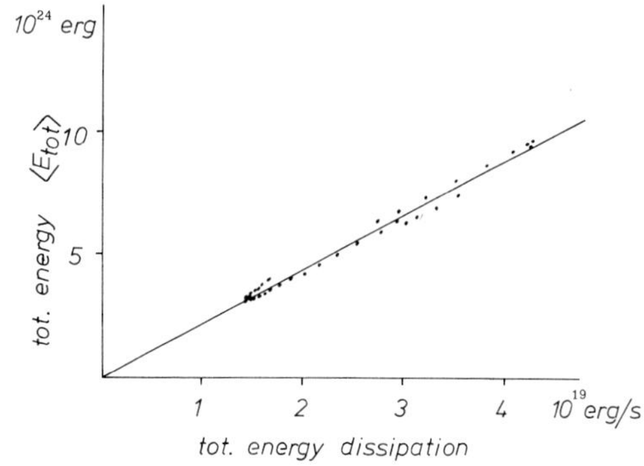

Fig. 9. Correlation between the mean dissipation rate and the total energy (time-average) stored in the oceans

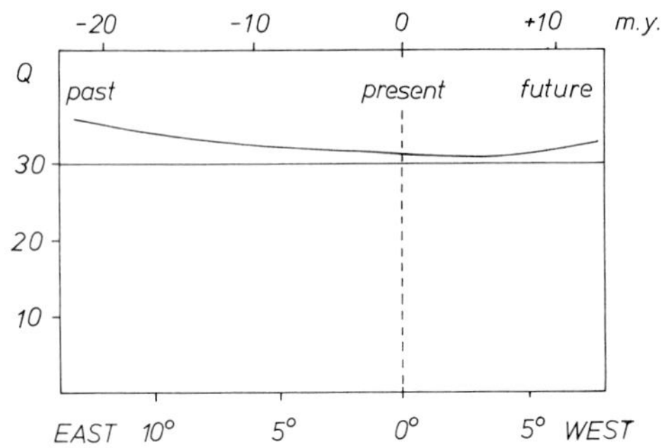

Fig. 10. Q factor (specific dissipation function) of the oceans (M_2-tide potential)

The line integral over dE/dt measures the dissipation of energy during one complete cycle. E^* is the peak energy stored in the oceans during that cycle.

Figure 10 shows the variation of Q for the time span in question. The variation is not large:

$$30 \leq Q \leq 36$$

and the minimum value of Q is taken at resonance, that is, at resonance the dissipation is most effective. The values of Q we found in our calculation lay within the range of Q's which had been otherwise determined. Zahel (1977) found a value $Q \sim 27$, Hendershott (1972) estimates a Q of 34. It is clear from Fig. 8 that this Q does not result from bottom friction in its main parts, but from the inner friction dissipation mechanism.

3.4 Balance of Angular Momentum and the Moon's Torque on the Earth

The component of the angular momentum we are most interested in is the component p_z, which is parallel to the Earth's rotation axis. The time-averaged quantity $\langle \dot{p}_z \rangle$ enables us to determine the rate of change of the angular velocity of the Earth. Several methods are

Table 1. Deformation of the Earth, analysed by calculation of the inertia tensor of the Earth (differential contribution)

Shift of America	C_2^2 (cm)	ε_2^2 (°)
9° East	3.01	84.0
6° East	2.64	87.2
3° East	2.92	81.6
0°	3.95	72.6
2° West	4.46	72.0
3° West	5.57	85.0
6° West	4.56	96.8

available to calculate this mean torque (see e.g., Brosche and Sündermann 1971), but all these methods encounter great problems, because the value $\langle\dot{p}_z\rangle$ is very small compared with the variations of $\dot{p}_z$ during one tidal cycle (a detailed discussion of these problems can be found in Baader 1982, this Vol. and Hövel 1982). The most reliable method is the direct calculation of the Moon's torque on the deformed ocean surface (see Parijskij et al. 1972).

The tidal deformations of the oceans show a very complex and complicated structure, as can be seen from the very well known tidal charts (for example see Zahel 1978), and an understanding and interpretation of the relations between the tidal deformations and $\langle\dot{p}_z\rangle$ is not easy. One way to simplify it is the elimination of all effects which do not affect $\langle\dot{p}_z\rangle$. This can be done by a decomposition of the computed tides into spherical harmonic waves. It is easy to see (Lambeck 1980) that only the wave corresponding to the tide-generating potential affects $\langle\dot{p}_z\rangle$ and that this reduces the problem at least formally.

This determination of the mean elliptic deformation of the Earth can also be achieved by a direct calculation of the differential oceanic contribution to the tensor of inertia of the Earth. Figure 11 shows the results of this analysis very schematically. The view is onto the equatorial plane with the projection of the tensor's ellipsoid.

The axis indicated by ☾ connects the center of the Earth with the Moon, the angle between the main axis of the projection ellipse and that axis corresponds to the phase delay of the tidal bulge. The amplitude of the mean deformation is denoted by ζ_1. Some values for this amplitude and the phase delay are given in Table 1. For a better comparison we transformed these values into the corresponding amplitudes C_2^2 of the spherical harmonics of second degree.

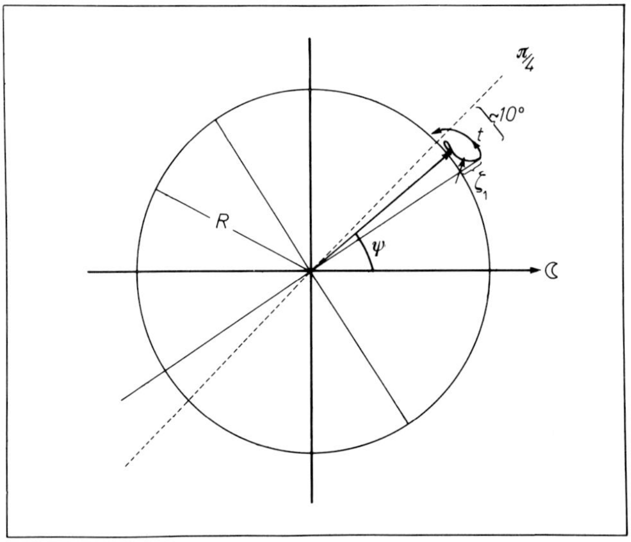

Fig. 11. Mean deformation of the Earth (schematically). ψ = phase lag, ζ_1 = amplitude of deformation, R = radius of the Earth

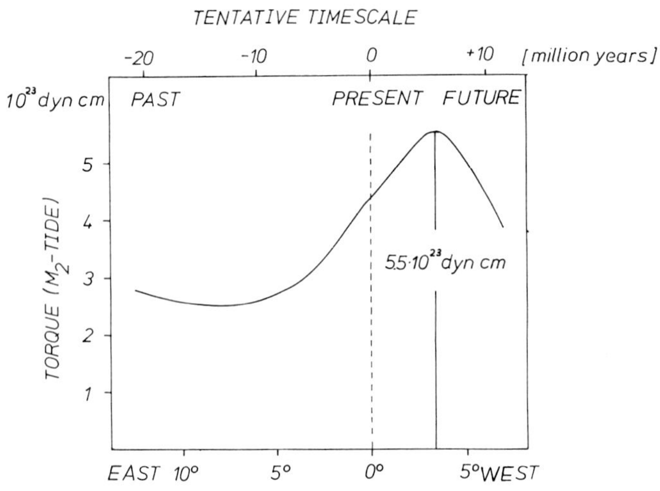

Fig. 12. Time-averaged torque $\langle\dot{p}_z\rangle$ of the Moon acting on the deformed Earth

The phase and the amplitude of the deformation do not change very much in the time span we discuss here. Analogous to the movements of the center of mass, the main axis of the spheroid regarded as a pointer in space describes a very small path (Fig. 11). The most interesting fact is that the mean position of this pointer is always around the $\pi/4$ situation which one would expect for a simplified ocean (without continents and without depth structure) at resonance, (see, e.g., Zahel 1978). This again is one indication of the near resonance situation.

Figure 12 shows a direct computation of $<\dot{p}_z>$ by integration of the locally acting torques over the whole globe. It is in good agreement with the previous results. Obviously the curve shows behaviour similar to most of the others. The total variation factor of the torque is about 2 to 2.5. The maximum value will be reached in the future (at resonance) and is

$$-<\dot{p}_z>_{max} = 5.5 \times 10^{23} \text{ dyn cm}$$

while the minimum value is only

$$-<\dot{p}_z>_{min} = 2.3 \times 10^{23} \text{ dyn cm} \quad .$$

The present value

$$-<\dot{p}_z>_{pres} = 4.3 \times 10^{23} \text{ dyn cm}$$

differs slightly from Zahel's (1977) 4.9×10^{23} dyn cm.

The variation obtained is very remarkable. The calculations into the past, especially, show that the transfer of angular momentum between the Earth and the Moon must have been smaller in the more recent past, that is, in the last million years. Consequently the slowing down of the Earth's rotation has been less intensive.

We have performed our calculations within a "short" time scale only and are therefore unable to extrapolate the behaviour towards larger scales, but our results indicate that the transfer of angular momentum (and the corresponding energy dissipation) shows a different time dependence from that which has been assumed until now.

4. Conclusions

One of the most important results of the sequence of models presented is the fact that besides the known long-scale variations of the tidal parameters, short-scale variations of the same order of magnitude may appear. It is quite surprising, that even small changes in the topographic structures of the oceans can cause such large variations.

Contrary to the usual point of view, our calculations show that a "fine structure" of the important parameters like energy dissipation and the torque Moon-Earth must be taken into account. A detailed investigation of this structure and its dependence on inner and outer factors is necessary, especially if we wish to extrapolate the history of the Earth-Moon System. Because the Earth's rotation is mainly determined by the variation of the torque between Earth and Moon this variation should be known for a large time span with a sufficient time resolution.

To sum up, all results seem to confirm the surmise that the present ocean is close to a resonance. Consequently the often implicit supposition that the present ocean is a representative ocean for long time scales must be critically examined. Further activities could be concentrated on similar computations of short time variations for other geological tide models (such as glaciation, pangaea, etc.). Perhaps it would be possible to produce an overlapping sequence of tide models representing a larger part of the total time scale.

Acknowledgment. We gratefully acknowledge the support of the Deutsche Forschungsgemeinschaft for this research.

References

Brosche P (1981) Geomagnetic reversals and tidal friction. Naturwissenschaften 68:139

Brosche P, Sündermann J (1971) Die Gezeiten des Meeres und die Rotation der Erde. Pure Appl Geophys 86:95-117

Hendershott MC (1972) The effects of solid earth deformations on global ocean tides. Geophys J 29:389-402

Hövel W (1982) Dissertation (in prep)

Lambeck K (1980) The Earth's variable rotation: Geophysical causes and consequences. Cambridge Monogr Mech Appl Math

Parijskij NN, Kuznetsov MV, Kusnetsova LV (1972) On the influence of ocean tides on the secular deceleration of the Earth's rotation. Fiz Zemli 2:3-12 (in Russian)

Platzmann GW (1978) Normal modes of the world ocean. J Phys Oceanogr 8:323-343

Platzmann GW, Curtis GA, Hansen KS, Slater RD (1981) Normal modes of the World Ocean II. J Phys Oceanogr 11:579-603

Schulte D (1979) Die Ermittlung der Eigenschwingungsperioden des Weltozeans mittels des hydrodynamisch-numerischen Verfahrens und der Fast Fourier Methode auf der Grundlage eines 4°-Modell-Ozeans. Diplomarbeit, Universität Hamburg

Thiede J (1979) Paleogeography and Paleobathymetry of the Mesozoic and Cenozoic North Atlantic Ocean. Geojournal 3.3:263

Yoder CH F, Williams JG, Parke ME (1981) Tidal Variations of Earth Rotation. J Geophys Res 86/B2, 881

Zahel W (1970) Die Reproduktion gezeitenbedingter Bewegungsvorgänge im Weltozean mittels des hydrodynamisch-numerischen Verfahrens. Mitt Inst Meereskd Univ 17:1-50

Zahel W (1977) A global hydrodynamical-numerical 1°-model of the ocean tides; the oscillation system of the M_2-tide and its distribution of energy dissipation. Ann Geophys 33:31-40

Zahel W (1978) The influence of solid earth deformations on semidiurnal and diurnal oceanic tides. In Brosche P, Sündermann J (eds) Tidal friction and the Earth's rotation I. Springer, Berlin Heidelberg New York, pp 98-124

Paleotides Before the Permian

J. Krohn and J. Sündermann[1]

1. Introduction

There is evidence from paleontological data that the number of days per year has not been constant at least for the last 600 m.y. As is commonly understood, this fact arises mainly from the tidal coupling of the Moon and the Earth. The dominating process is tidal friction between the Moon and the solid Earth via the oceans. Other effects as friction within the Earth's fluid core and mantle or between solid Earth and atmosphere or other external forces (e.g., solar wind) seem to be of less importance (Sasao et al. 1977, Kaula 1964, Zschau 1978, 1979, Lambeck 1980, Hirshberg 1972). There is considerably good agreement among different types of observations (pre-telescopic and telescopic), Earth satellite data and Lunar Laser Ranging data (e.g., Stephenson 1978, Morrison 1978, Cazenave and Daillet 1981, Calame and Mulholland 1978). They all yield similar results, the rate of dissipation of rotational energy being about 4×10^{19} erg s^{-1}. Results obtained from different ocean-tide models explain a large amount (at least 75%) of the observed rate as the result of tidal friction between the Moon and the Earth (Peceris and Accad 1969, Hendershott 1972, Kusnetzov 1972, Pariiskiy et al. 1972, Zahel 1977, 1980, Sündermann and Brosche 1978).

It is well known that if the present effect of tidal coupling is extrapolated back in time allowing for the increased tidal forces as the Moon approaches the Earth there follows an Earth-Moon distance of a few Earth radii only 1 or 2 billion years before the present (b.y.B.P.). As no evidence in the geological or paleontological data of the Earth or the Moon exists for such an event (huge tidal waves would have run around the globe), the question arises whether the present dissipation process is representative of geological history. Therefore, apart from all interesting problems involved in the determination of the lunar tidal torque itself for the present and for historic times, in this paper the change of the Earth's rotation rate over the period of the last 450 million years (m.y.) will be discussed, in the attempt to gain some insight into the underlying processes.

1 Institut für Meereskunde der Universität Hamburg, Heimhuder Str. 71, D-2000 Hamburg 13, Fed. Rep. of Germany

Tidal Friction and the Earth's Rotation II
ed. by Brosche/Sündermann

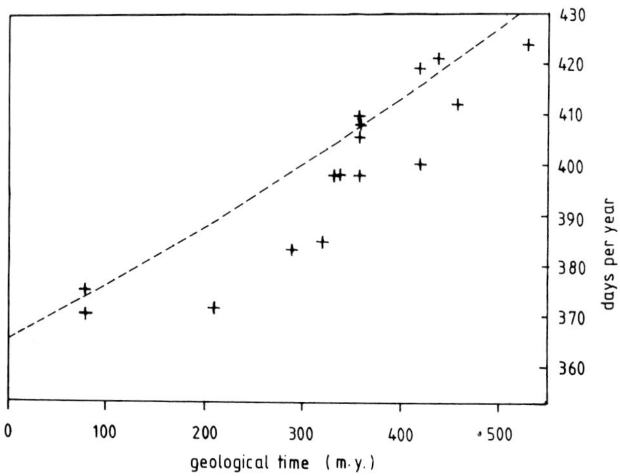

Fig. 1. Summary of paleontological data showing days per year. (After Scrutton 1978). The broken line extrapolates the present rate of deceleration

Figure 1 is taken from a review of Scrutton (1978) presented at the first Bielefeld meeting on tidal friction. It summarizes all paleontological data available and indicates that the length of the day (l.o.d.) has changed remarkably during the last 500 m.y. But the assumption of a constant lengthening of the day of 2 ms/century (about the present rate) can only roughly fit the data as shown by the dashed line. Instead, there is an indication of a relatively low dissipation rate during the last 300 m.y. (up to the Permian), whereas before that time a rate even somewhat higher seems reasonable.

As is commonly accepted and indicated by the concepts of continental drift and sea-floor spreading, the shape of the oceans has not been constant, neither the coastlines nor the bottom topography. Consequently the formation of the ocean tides will not have stayed constant according to the changed resonance properties of the ocean. The tide-generating forces have also varied throughout geological history, due to the changes in the motion of Moon and Earth. To extrapolate the present process of energy dissipation backwards in time is therefore a questionable procedure. In our opinion the only realistic way to estimate the tidal torque for ancient oceans is to use numerical ocean tide models. Only these models are able to consider the actual shape of the ocean basins available from paleogeographical maps back to the Mesozoic. First results for the Permian (240 m.y.B.P.) have already been presented (Sündermann and Brosche 1978).

After a brief outline of the methods calculating the tidal dissipation rate, results for some other Mesozoic and Paleozoic situations will be presented.

2. Method

2.1 Calculation of Tidal Torques

The determination of the present tidal torque between the Moon and the Earth is obtained from a balance of angular momentum or energy in the oceans. This requires detailed knowledge of the surface elevation and of the tidal currents. As the observed ocean data available cover only coastal areas and some very few stations in the open ocean, the results of numerical models have to be used instead.

Tidal dynamics are described by the Hydrodynamical Differential Equations (HDE) for a barotropic ocean of variable depths. Only a few models, however, consist of this complete set of nonlinear equations (Zahel 1970, 1977, 1980, Schwiderski 1978, Estes 1980); most authors use the linearized Laplace Tidal Equations (LTE), some of them introducing a linear bottom friction law (e.g., Parke and Hendershott 1978, Accad and Peceris 1978). As models involving grid resolutions up to 1° or 2° can resolve shallow water areas where nonlinear effects might appear, those using the nonlinear HDE should yield better results.

The transfer of angular momentum from the Earth to the Moon caused by tidal friction can be calculated by two different but equivalent ways: directly by a balance of angular momentum or indirectly by an energy balance, both in the ocean. The former has the advantage that there exist no inner sinks of angular momentum within the ocean that would have to be taken care of, whereas energy could be dissipated by turbulence (viscosity) and transferred into heat. The balance equation for the angular momentum is derived directly from the HDE after multiplication with the radius vector and subsequent integration over the total ocean surface and one tidal cycle (indicated by < >):

$$\left\langle \iint \Big[g\, R^2 \cos^2\varphi\, DR_\lambda + \rho R\, g \cos\varphi\, D \frac{\partial\zeta}{\partial\varphi} - \frac{1}{2} g^2 \rho C R^2 \cos^3\varphi\, \zeta \sin(\sigma t + s\lambda) \Big] d\lambda\, d\varphi \right\rangle = 0 \quad .$$

The following notations have been used:

ζ	surface elevation
D	total water depth
g	acceleration due to gravity
R	radius of the Earth
R_λ, R_φ	east and north component of total friction
ρ	density of water ($\rho = 1\ \mathrm{g\ cm^{-3}}$)

C	tidal constant (amplitude of tidal forces)
t	time
σ	frequency of the tide
s	index of the tide (s = 2 for semidiurnal tides)
λ,φ	east longitude, north latitude

The third term in the above equation denotes the transfer of angular momentum (tidal torque) between the water and the Moon we are interested in. This quantity has to be compared with the observed values. It is balanced by the sum of the transfer between the bottom and the ocean due to friction, i.e., the sum of bottom friction and inner friction (viscosity) (first term), and the transfer caused by the lateral pressure gradient (second term).

The energy balance is obtained by multiplying the HDE by the velocity vector (u,v) and integrating over the ocean surface and one tidal cycle:

$$\left\langle \iint \left[-\frac{g}{R\cos\varphi}\left\{\frac{\partial}{\partial\lambda}(Du\phi) + \frac{\partial}{\partial\varphi}(Dv\cos\varphi\phi)\right\} + \frac{g}{R\cos\varphi}\left\{\frac{\partial}{\partial\lambda}(Du\zeta) + \frac{\partial}{\partial\varphi}(Dv\cos\varphi\zeta)\right\} + D(R_\lambda u + R_\varphi v)\right] d\lambda\, d\varphi \right\rangle = 0$$

The first term represents the energy flux into the ocean due to the work done on the surface by the tide generating potential ϕ, the horizontal energy flux via tidal currents is specified by the second one. These two terms are balanced by the energy dissipation by bottom friction and horizontal viscosity (third term).

In principle, the dissipation rate can be calculated by both methods. The angular momentum balance, however, makes use of the geographical distribution of the surface elevations only, whereas the energy dissipation term involves tidal velocities. The latter are variables that can be obtained from ocean tide models with sufficient accuracy at relatively high computational costs only. In addition it has not always been noted that energy is dissipated by both bottom friction and inner friction (Jeffreys 1920, Heiskanen 1921).

For the above reasons the more direct method of the angular momentum balance has been applied. According to the dominating role of the present M_2-tide, we have restricted this study to paleo-M_2-tides only.

2.2 Calculation of Paleotides

Numerical ocean tide models need information about the shape of the ocean basins, the amplitude and phase of the tide-generating potential and the spin rate of the Earth. In the last few hundred million years, tidal parameters have only varied slightly and are sufficiently known. As there is rather poor information on global paleo-depths, a schematic bottom topography is applied with two different depths only: one representative for the deep ocean, the other for the shallow shelf areas (Sündermann and Brosche 1978). It is expected that Deep Sea Drilling results will yield more detailed information up to the Mesozoic; charts of the history of the Atlantic ocean (Thiede 1979) already exist and new global paleogeographical maps will be published soon (Ziegler 1981, pers. comm.).

3. Results

Table 1 summarizes the geological situations tides have been calculated for. The selection is characterized by great differences in the land-sea distribution and differing extents of shallow water areas. This paper concentrates on the Silurian (about 420 m.y.B.P.) and Ordovician (about 450 m.y.B.P.) as examples for pre-Permian situations. Paleotides and the corresponding tidal torques for the other geological situations (Upper Permian, Upper Cretaceous, Pleistocene glaciation) are given by Sündermann and Brosche (1978) and Krohn et al. (1981).

Table 1. Summary of geological epochs tides have been calculated for

Epoch	m.y.B.P.	Remarks
Present	0	
Pleistocene	0.2	Worldwide glaciation, lowered sea-level
U. Cretaceous	70	Worlwide transgression of the oceans, large extent of continental shelves
U. Permian	240	Super continent Pangaea
M. Silurian	420	Large southern continent Gondwana, more shallow seas than at any other time in Phanerozoic history
M. Ordovician	450	Most of the continents concentrated in southern hemisphere, ocean-covered poles

3.1 Silurian Paleotides

For Paleozoic times quite detailed paleogeographical global reconstructions exist (Ziegler et al. 1979, Scotese et al. 1979), which could be used for our purposes directly. Figure 2 shows a map for the Middle Silurian (420 m.y.B.P.), slightly modified to lay emphasis on the topography and shape of the coastlines of the ocean at that particular time. The Silurian is characterized by the large southern continent Gondwana and a ring of smaller continents and islands forming a nearly enclosed ocean in lower and middle latitudes. But the outstanding feature is the large area of shallow shelves covering more of the continental plates than at any other time in the Earth's history.

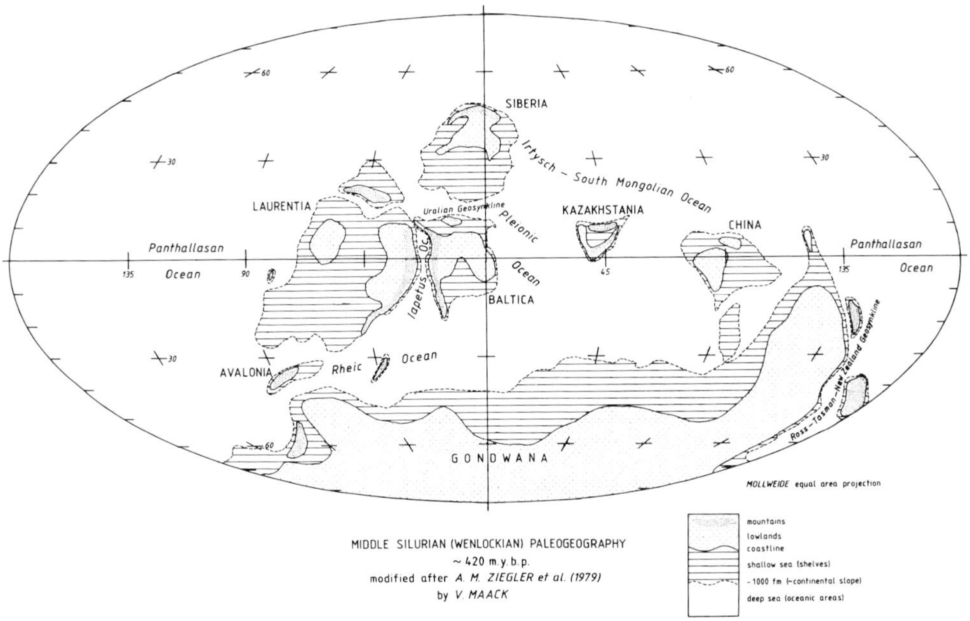

Fig. 2. Paleogeographical map of the Middle Silurian (420 m.y.B.P.) (Maack 1981)

In Fig. 3 the geographical distribution of land, shelves and deep sea areas of the ocean tide model is indicated. As the grid distance is 4°, only the main properties of the paleogeographical map are resolved by the model.

The co-tidal (Fig. 4) and co-range lines (Fig. 5) as computed by the numerical model reveal the complicated oscillation system of the tide. Amplitudes are within the range of present values, the highest located at coastal regions of the Australian plate. Because the resolution of the grid is rather coarse, the calculated values are

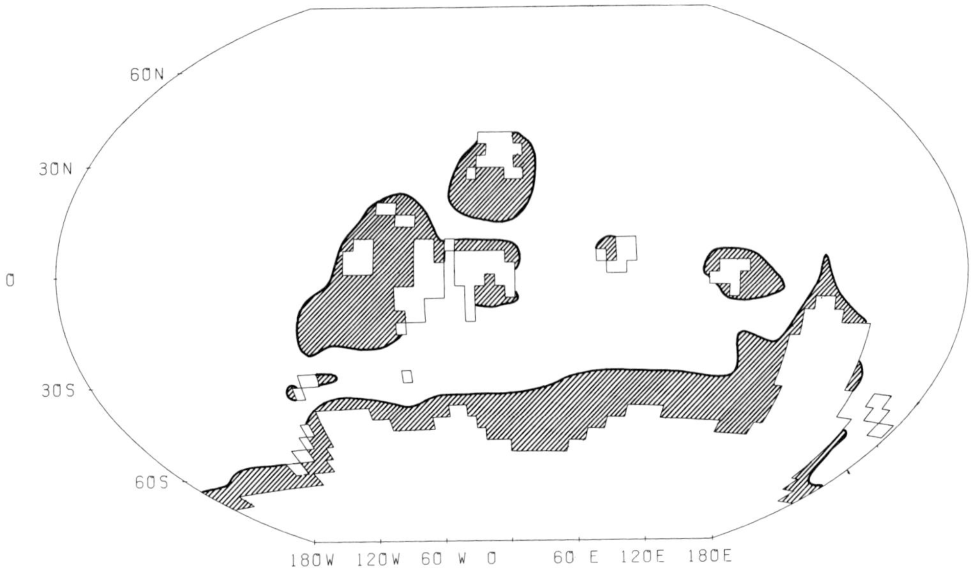

Fig. 3. Model bathymetry for the Middle Silurian ocean; shaded areas indicate shallow seas

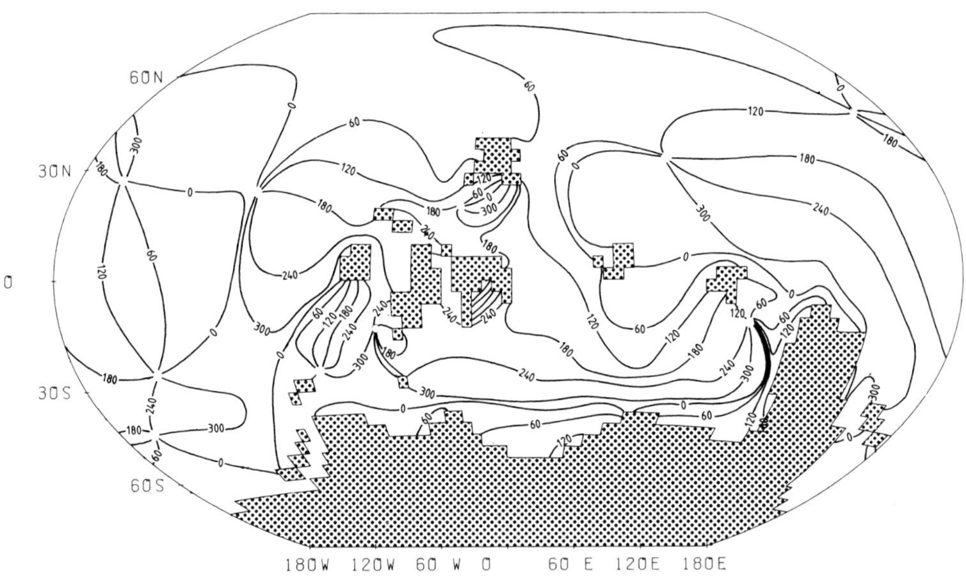

Fig. 4. Cotidal lines for the M_2-tide of the Middle Silurian ocean; in degrees referred to the Moon's transit at zero-meridian

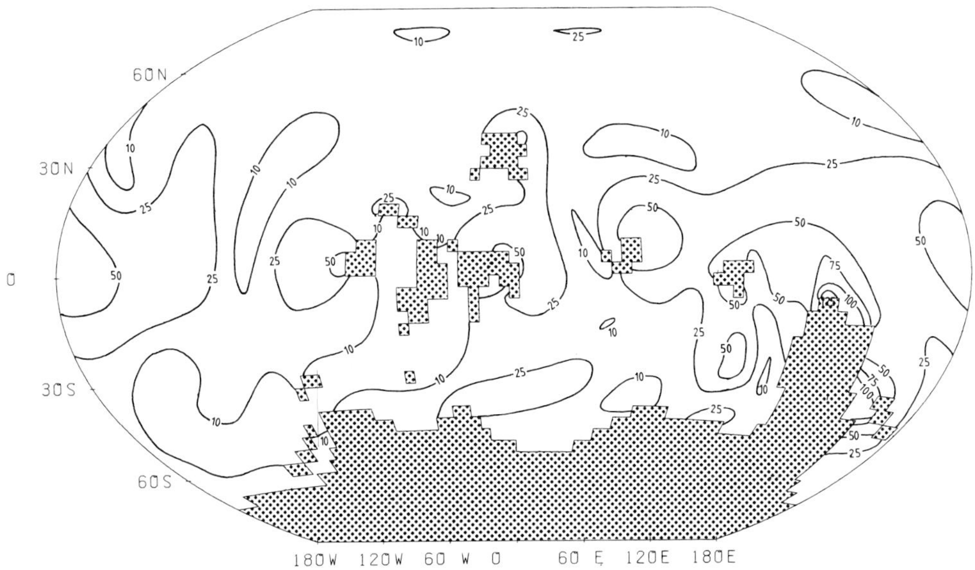

Fig. 5. Corange lines of the M_2-tide for the Silurian ocean; in centimeters

representative for a large area and cannot resolve local phenomena. In particular it is likely that amplitudes higher than the maximum of 120 cm as calculated by the model may have occurred. Besides its undisturbed behavior in the Panthallassan ocean, the tidal wave is characterized by progressing southward onto the large Gondwana shelf where it is dissipated.

3.2 Ordovician Paleotides

During the Middle Ordovician (450 m.y.B.P.) Gondwanaland on its southern movement has not yet reached the South Pole, so both poles are covered by the ocean. Most of the other continents are placed at low latitudes, only Baltica is located at high latitudes close to the South Pole. In spite of the relatively coarse model all essential characteristics of the paleogeographical map (Fig. 6) can be found in Fig. 7, showing the bottom topography of the numerical model.

The main features of the modelled tide are the wave entering the Iapetus ocean (the proto-Atlantic) at its western part and leaving it through its eastern gaps (Fig. 8). The chart of the co-range lines (Fig. 9) is marked by a broad band of high amplitudes (50-100 cm) at low latitudes.

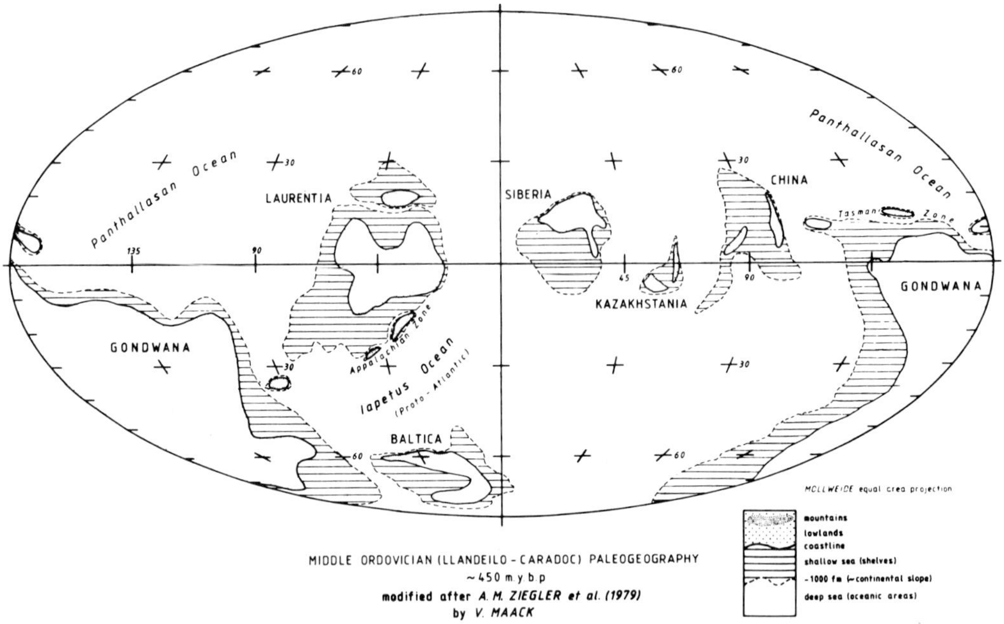

Fig. 6. Paleogeographical map of the Middle Ordovician (450 m.y.B.P.)

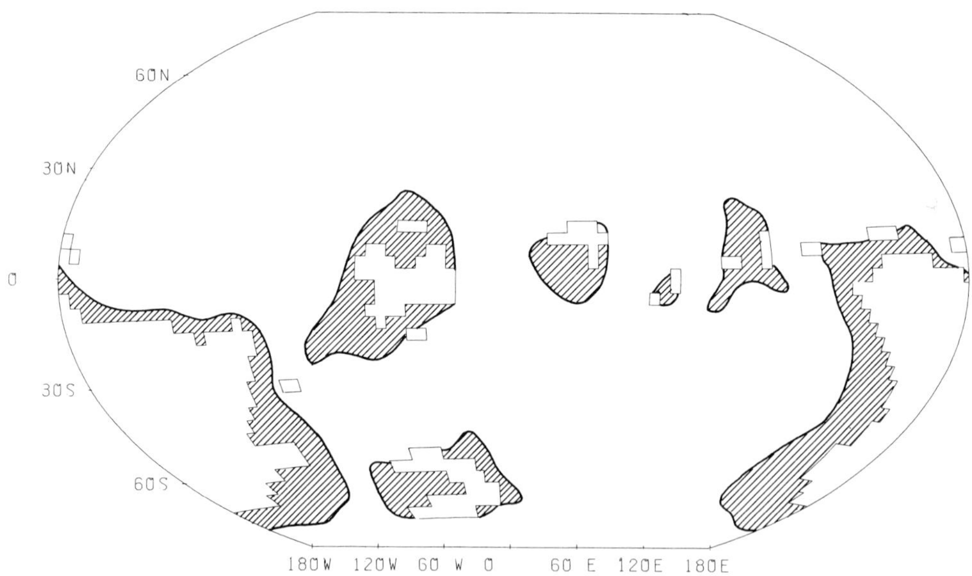

Fig. 7. Model bathymetry for the Middle Ordovician ocean; shaded areas indicate shallow seas

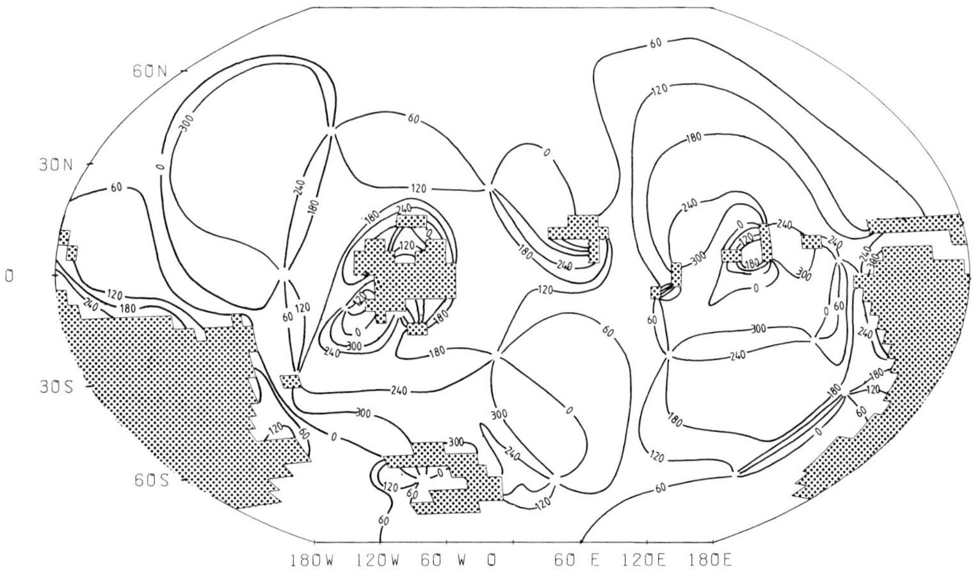

Fig. 8. Cotidal lines of the M_2-tide for the Middle Ordovician ocean; in degrees referred to the Moon's transit at zero-meridian

Fig. 9. Co-range lines of the M_2-tide for the Middle Ordovician; in centimeters

3.3 Tidal Torques for Ancient Oceans

The calculation described in Sect. 2 results in a global chart of local torques. In Figs. 10 and 11 charts for the Cretaceous and Permian (for the sake of completeness and comparison) are shown, while the results for the Silurian and Ordovician are displayed in Figs. 12 and 13. An arrow to the right indicates a positive, i.e., accelerating contribution, while a negative (decelerating) torque is represented by an arrow to the left. In all figures (Figs. 10 to 13) both areas of positive and negative signs exist, meaning that in certain parts of the world ocean a tidal torque could be generated that would even accelerate the Earth's rotation. It could well be observed that according to the more zonal arrangement of continents during the pre-Permian stages, a more zonal and somewhat smoother distribution of torques exists than during those stages with a more meridional arrangement such as during the Permian and Cretaceous, where the picture is similar to that of the present world ocean as presented by Sündermann and Brosche (1978).

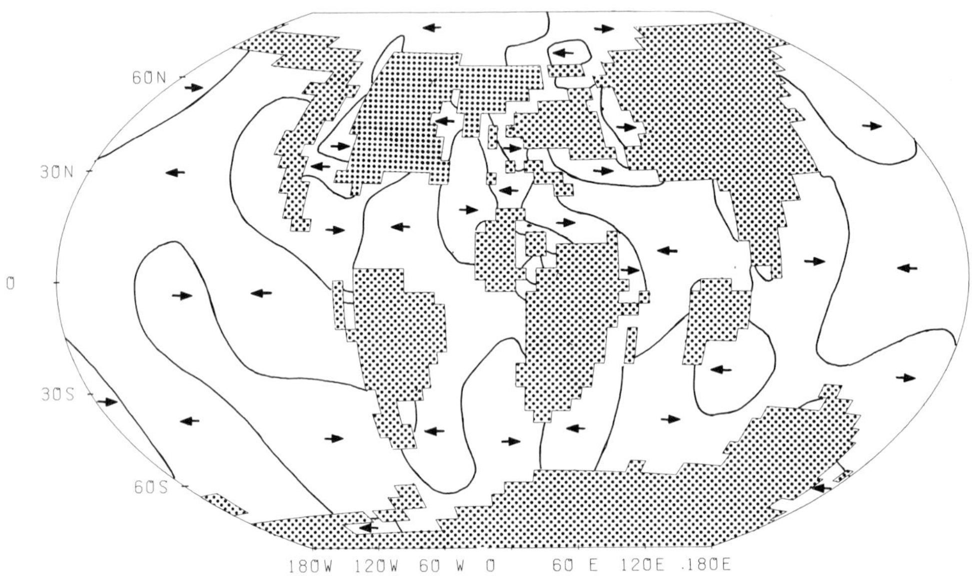

Fig. 10. Upper Cretaceous M_2-tide; direction of tidal torques

Integrating these local values over circles of equal latitude yields information about the variation with latitude (Fig. 14). Even after this integration intervals of positive torques still remain, especially in middle and higher latitudes. Apart from all other curves in this figure where the maximum is centered around the equator, the peak of the Silurian torque appears at subtropical northern latitudes. The reason is the large southern continent Gondwana covering the South Pole and extending 30° to 80° northward, leaving space for the generation of a tidal torque at northern and lower southern latitudes only.

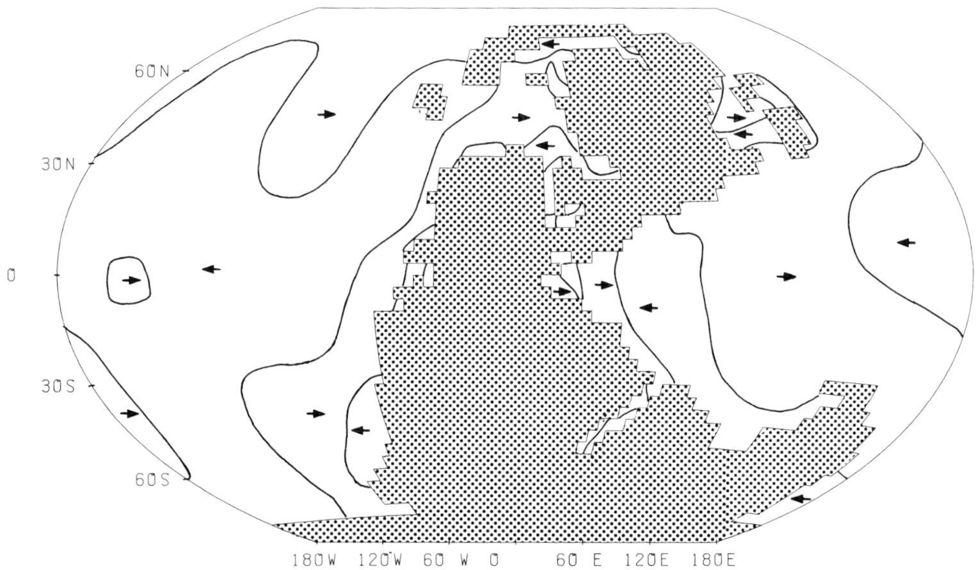

Fig. 11. Upper Permian M_2-tide; direction of tidal torques

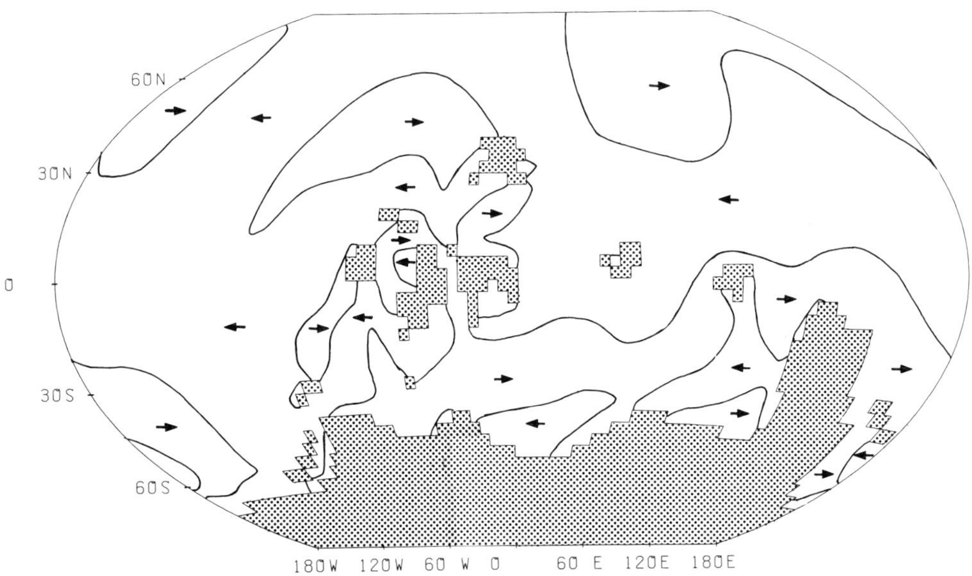

Fig. 12. Middle Silurian M_2-tide; direction of tidal torques

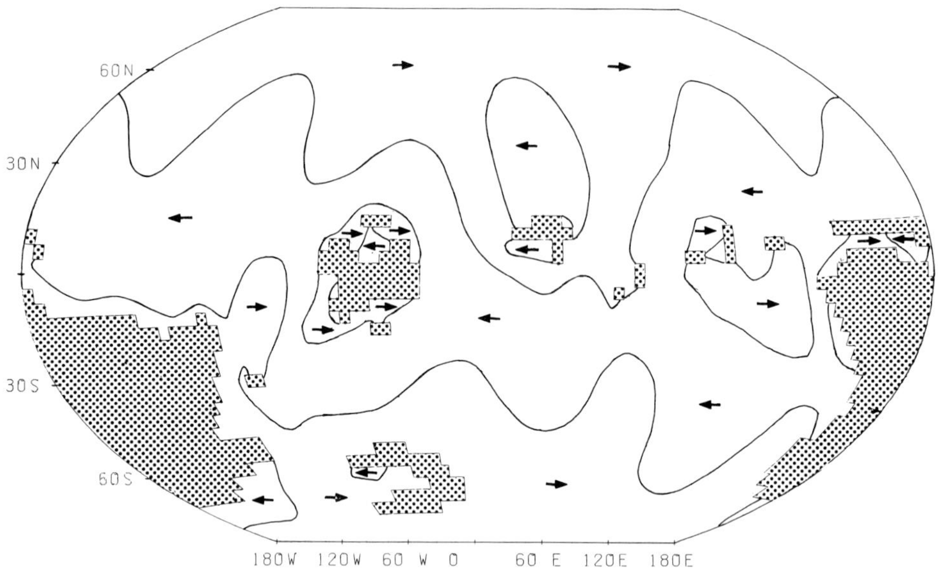

Fig. 13. Middle Ordovician M_2-tide; direction of tidal torques

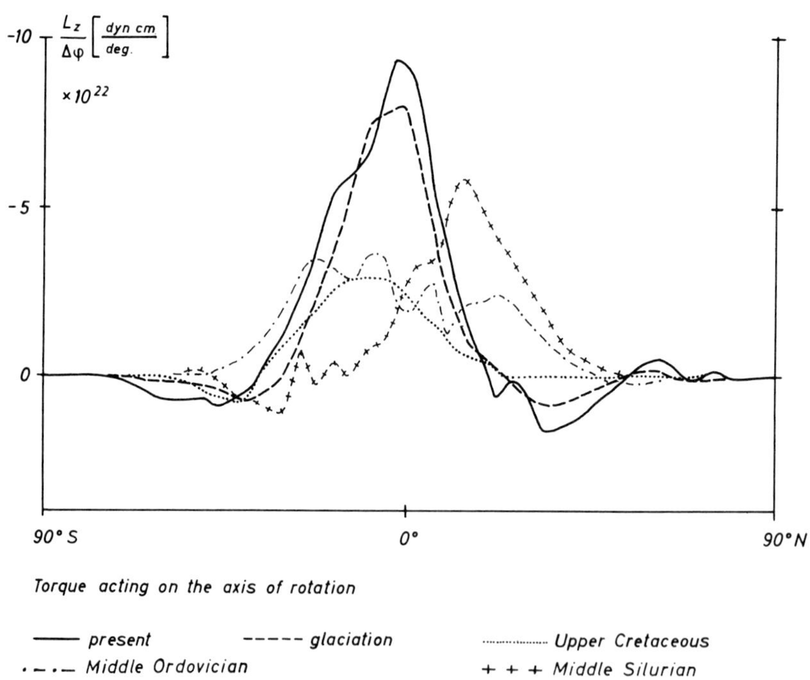

Fig. 14. Zonally integrated tidal torques for ancient oceans

Table 2. Summary of results

Epoch	m.y.B.P.	Topography	Earth's angular velocity normalized	Amplitude of tidal forces normalized	$-\langle \dot{E}_{rot} \rangle$ 10^{19} erg s^{-1}	$-\langle L_z \rangle$ 10^{23} dyn·cm
Present	0	Detailed	1	1	3.6	4.9
Present	0	Schematic	1	1	2.3	3.1
Glaciation	0.2	Detailed (−100 m)	1	1	3.3	4.5
U.Cretaceous	70	Schematic	1.01	1	1.6	2.2
U.Permian	240	Schematic	1.04	1.002	1.4	2.0
M.Silurian	420	Schematic	1.08	1.003	2.9	3.6
M.Ordovician	450	Schematic	1.08	1.003	3.5	4.3

The total tidal torque L_z around the Earth's axis of rotation (or the corresponding change in the Earth's rotational energy $\dot{E}_{rot}$) is obtained after integration over one tidal cycle and the entire ocean surface. Table 2 summarizes the results for all considered paleo-oceans. The astronomical parameters (normalized by the present value) and information on the ocean depths are indicated for each paleo-ocean. The term "schematic topography" indicates that the topography has been approximated by two representative depth values only as described above in Sect. 2.2. The depths of the glaciation stage model have been lowered by 100 m with respect to the present data to account for the large amount of inland ice and glaciers at that time. In the last two rows the change of the Earth's rotational energy and the tidal torque respectively have been listed, both calculated using the balance equation of angular momentum.

The present value of $-\dot{E}_{rot} = 3.6 \times 10^{19}$ erg s^{-1} is in good agreement with other authors. Their results derived from numerical models range from 3.3×10^{19} erg s^{-1} to 4.2×10^{19} erg s^{-1} (Peceris and Accad 1969, Hendershott 1972, Kusnetzov 1972, Pariiskiy et al. 1972, Sündermann and Brosche 1978, Zahel 1977, 1980) if the originally given values are partly modified to account for the elastic tidal response of the Earth's crust in all models (Lambeck 1980). [However, it should be mentioned that two recently obtained results yield significantly lower dissipation rates: 2.2×10^{19} erg s^{-1} by Parke and Hendershott (1978) and 2.6×10^{19} erg s^{-1} by Accad and Peceris (1978)]. The results mentioned so far are all obtained considering the M_2-tide only. If the effects of other tidal constituents are added, the present total rate of energy dissipation by the tidal coupling between the Earth, Moon and Sun as computed by numerical models can explain the observed rate of about 4×10^{19} erg s^{-1} (Lambeck 1980).

4. Discussion

The dissipation rates for ancient oceans indicate variations of about the factor of 2.5 in the past 450 m.y. The rates of the Cretaceous and Permian suggest a long period of smaller rates than today, whereas during the Silurian and Ordovician tidal torques are of the present order of magnitude or even higher (see also Lambeck and Pullan 1980).

Assuming the calculated torques to be representative for large time intervals so that the whole time axis is covered, it is even possible to "explain" the period of relatively constant l.o.d. from 100 to 300 m.y.B.P. and the more rapid change before. In Fig. 15 the data of Table 2 have been plotted together with the paleontological data given by Scrutton (1978). Curve "A" has been obtained relating the ancient torques to the result of the model for the present ocean

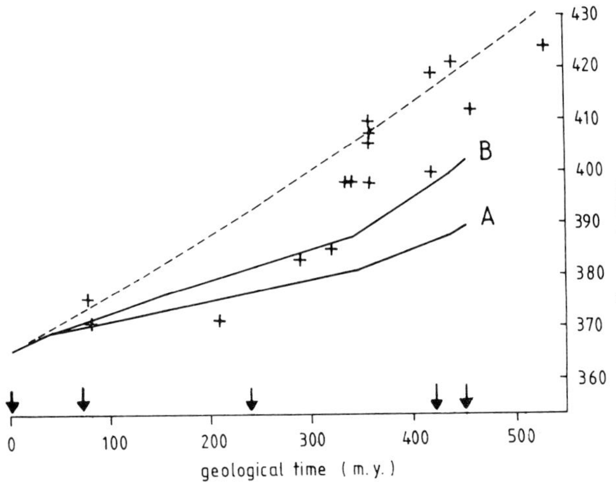

Fig. 15. Estimated evolution of the number of days per year derived from the data of Table 2 plotted into Fig. 1. Vertical arrows indicate the periods tides have been calculated for. Curve A: assuming ancient torques relative to the present one considering the detailed ocean bottom topography; curve B: same exept for considering the present schematic topography

where the actual detailed topography is included. As the models for the geological past involve a schematic depth distribution only (see Sect. 2), a comparison between the ancient torques and that resulting from the present model also considering a schematic topography should yield more reliable results (curve B). The curves have been constructed as follows: the calculated torques have been assumed to be representative (constant) not only for the specific geological situation considered (as indicated by the small vertical arrows) but also for a certain time interval. The boundaries of those intervals have been defined by the (temporal) mid-distance between two adjacent arrows. By this means one gets a rather rough estimate of the time dependence of L_z during the last 450 m.y. and as the change of the l.o.d. is proportional to L_z the corresponding number of days per year (assuming a constant length of the year). It should be mentioned, however, that very few data points exist on the time axis only and that nothing is known about the tidal torques belonging to the geological situations between them. Results that do not fit the interpolated parts of the curves in this figure can of course distort the lines A and B considerably. Brosche and Hövel (this Vol.) have pointed out the sensitivity of the calculated torque regarding small variations of the present ocean's geometry.

On the other hand, there are restrictions in reconstructing very old oceans. Global paleogeographical maps are available back to the Paleozoic (Ziegler et al. 1979, Scotese et al. 1979). However, this period covers only a short time compared with the time of the existence of shallow seas (3.8 b.y.B.P.) or a hydrosphere of the approximate present volume (2.5 to 3.2 b.y.B.P.) with tidal indications. Because the Precambrian oceanic crust has been nearly entirely destroyed, detailed global reconstructions of Precambrian oceans seem not to be possible. Therefore we will have to turn over to more or less schematical representations of the Proterozoic ocean.

Very recently two attempts in this sense have been made (Hansen 1981, Webb 1980). Whereas Webb's ideas are described in this volume, Hansen's approach should be touched upon briefly. He considered two extremely different land-sea arrangements: a single continent of present area centered at the North Pole and in the other case at the equator. The LTE are solved in this domain by a finite element model regarding two different depths, representing the deep ocean and the shelf areas respectively. Calculations were made for weak and strong bottom friction differing by the factor 10, resulting in four different tidal torques. They were used as upper and lower limits together with other astronomical parameters in order to integrate the orbital equations of the two-body problem backward in time. The resulting Earth-Moon distance for the last 4.5 b.y. showed a minimum distance of about 35 Earth radii. This result (that the tidal coupling has been much weaker throughout nearly the entire geological past) is not supported by our results of even higher dissipation rates than today between 400 and 450 m.y.B.P. One reason may be that Hansen's ocean approximations are by no means representative for any specific geological situation. Especially the different longitudinal configurations of continents which are possibly crucial for the momentum transfer are not considered in Hansen's paper, see also Sündermann (this Vol.).

In a recent study Kagan (1981) related the tidal torque to parameters of the second spherical harmonic of the ocean tide. These parameters (amplitude and phase lag) were used together with the time depending motions of Sun and Moon to estimate ancient torques. He concluded that during the entire phanerozoic history of the Earth (back to 540 m.y.B.P.) the dissipation rate due to tidal friction between ocean and solid Earth has been higher than at present (up to factor 1.40) and changes nearly linearly. His approach is based however on the doubtful assumption that the amplitudes and phase lags of the ocean tide expansions did not vary throughout the geological time considered.

5. Summary

The method of calculating paleotides and the corresponding tidal torques for ancient oceans of Brosche and Sündermann (1977a,b) has been applied to some more geological epochs. The tidal oscillation systems for the Middle Silurian and Middle Ordovician are displayed, whereas the tidal torques are additionally given for a Pleistocene glaciation and the Upper Cretaceous. The calculated torques indicate a period of relatively low dissipation rates (the last 300 m.y.) when the l.o.d. changed less than today and times of somewhat higher rates prior to 300 m.y.B.P.

Acknowledgements. The geological part, namely the compilation of the paleogeographical maps, was done by V. Maack, most of the line drawings are by Miss R. Krautwald. This research was funded by the Deutsche Forschungsgemeinschaft under contract No. Su 69.

References

Accad Y, Peceris CL (1978) Solution of the tidal equations for the M_2 and S_2 tides in the world ocean from a knowledge of the tidal potential alone. Philos Trans R Soc London Ser A 290:235-266

Brosche P, Sündermann J (1977a) Tides around Pangaea. Naturwissenschaften 64:89

Brosche P, Sündermann J (1977b) Effects of oceanic tides on the rotation of the Earth. In: Mulholland JD (ed) Scientific applications of lunar laser ranging. Reidel Dordrecht Boston, p 133

Calame O, Mulholland JD (1978) Effect of tidal friction on the lunar orbit and the rotation of the Earth and its determination by laser ranging. In: Brosche P, Sündermann J (eds) Tidal friction and the Earth's rotation. Springer, Berlin Heidelberg New York, p 43

Cazenave A, Daillet S (1981) Lunar tidal acceleration from Earth satellite orbit analysis. J Geophys Res 86 (B3):1659-1663

Estes RH (1980) A simulation of global ocean tides using altimeter data with systematic orbit error. Mar Geod 3:75-139

Hansen KS (1981) Secular effects of oceanic tidal dissipation on the Moon's orbit and the Earth's rotation. Dissertation, Univ Chicago

Heiskanen W (1921) Über den Einfluss der Gezeiten auf die säkulare Acceleration des Mondes. Ann Acad Sci Fenn A18:1-84

Hendershott MC (1972) The effects of solid Earth deformations on global ocean tides. Geophys J 29:389-402

Hirshberg J (1972) Upper limit of the torque of the solar wind on the Earth. J Geophys Res 77(25):4855-4857

Jeffreys H (1920) Tidal friction in shallow seas. Philos Trans R Soc London Ser A 221:239-264

Kagan BA (1981) Tidal torque and dissipation of tidal energy in the paleoocean. Isv Acad Nauk SSSR (Fiz Atmos Okeana) 17(8):849-857

Kaula WM (1964) Tidal disspation by solid friction and the resulting orbital evolution. Rev Geophys Space Phys 2:661-685

Krohn J, Brosche P, Sündermann J (1981) Paläogezeiten und Erdrotation. Geol Rundsch 70:64-77

Kusnetzov MV (1972) Calculation of the secular retardation of the Earth's rotation from up to date cotidal charts. Isv Acad Sci USSR Phys Solid Earth 2:3-11

Lambeck K (1980) The Earth's variable rotation: Geophysical causes and consequences. Cambridge Univ Press, Cambridge

Lambeck K, Pullan S (1980) The lunar fossil bulge hypothesis revisited. Phys Earth Planet Inter 22:29-35

Maack V (1981) Zur paläogeographischen Rekonstruktion erdgeschichtlicher Ozeane. Geol Rundsch 70:78-90

Morrison LV (1978) Tidal accelerations of the Earth's rotation from astronomical observations in the period A.D. 1600 to the present. In: Brosche P, Sündermann J (eds) Tidal friction and the Earth's rotation. Springer, Berlin Heidelberg New York, p 22

Pariiskiy NN, Kusnetzov MV, Kusnetzova LV (1972) The effect of oceanic tides on the secular deceleration of the Earth's rotation. Isv Acad Sci USSR Phys Solid Earth 2:3-12

Parke ME, Hendershott MC (1978) M_2, S_2, K_1 models of the global ocean tide on an elastic Earth. Mar Geod 3:379-408

Peceris CL, Accad Y (1969) Solution of Laplace's equation for M_2 tide in the world ocean. Philos Trans R Soc London Ser A 265: 413-436

Sasao T, Okamoto I, Sakai S (1977) Dissipative core-mantle coupling and nutational motion of the Earth. Publ Astron Soc Jpn 20:83-105

Schwiderski EW (1978) Global ocean tides, part I: A detailed hydrodynamical interpolation model. Naval Surface Weapons Center, NSWC/DL TR-3866, Dahlgren VA

Scotese CR, Bambach RK, Barton C, Voo R van der, Ziegler AM (1979) Paleozoic base maps. J Geol 87(3):217-277

Scrutton CT (1978) Periodic growth rhythms in fossil organisms and the length of the day and month. In: Brosche P, Sündermann J (eds) Tidal friction and the Earth's rotation. Springer, Berlin Heidelberg New York, p 154

Stephenson F (1978) Pre telescopic astronomical observations. In: Brosche P, Sündermann J (eds) Tidal friction and the Earth's rotation. Springer, Berlin Heidelberg New York, p 5

Sündermann J, Brosche P (1978) The numerical computation of tidal friction for present and ancient oceans. In: Brosche P, Sündermann J (eds) Tidal friction and the Earth's rotation. Springer, Berlin Heidelberg New York, p 125

Thiede J (1979) Paleogeography and paleobathymetry of the mesozoic and cenozoic north Atlantic ocean. Geo J 3(3)

Webb DJ (1980) Tides and tidal friction in a hemispherical ocean centred at the equator. Geophys J R Astron Soc 61:573-600

Zahel W (1970) Die Reproduktion gezeitenbedingter Bewegungsvorgänge im Weltozean mittels des hydrodynamisch-numerischen Verfahrens. Mitt Inst Meereskd Univ Hamburg 17:1-50

Zahel W (1977) A global hydrodynamic-numerical 1°-model of the ocean tides; the oscillation system of the M_2-tide and its distribution of energy dissipation. Ann Geophys 33:31-40

Zahel W (1980) Mathematical modelling of global interaction between ocean tides and Earth tides. Phys Earth Planet Inter 21:202-217

Ziegler AM, Scotese CR, McKerrow WS, Johnson ME, Bambach RK (1979) Paleozoic paleogeography. Annu Rev Earth Planet Sci 7:473-502

Zschau J (1978) Tidal friction in the solid Earth: Loading tides versus body tides. In: Brosche P, Sündermann J (eds) Tidal friction and the Earth's rotation. Springer, Berlin Heidelberg New York, p 62

Zschau J (1979) Auflastgezeiten. Habilitationsschrift, Univ Kiel

On the Reduction in Tidal Dissipation Produced by Increases in the Earth's Rotation Rate and Its Effect on the Long-Term History of the Moon's Orbit

D. J. Webb[1]

1. Introduction

When extrapolating the orbits of the Earth and Moon back into the distant past, it has been usual to assume that the tidal torque acting on the Moon depends only on the magnitude of the forces exciting the tides and on no other factor such as the shape of the ocean or the rotation rate of the Earth. With this assumption it is then found that the Moon was within a few Earth radii of the Earth between 1.5×10^9 and 2.0×10^9 years ago. At this separation the tidal effects should have left some mark in the geological record (Munk 1968), but surprisingly there is no record that such an event occurred either then or at any other time within the last 3.8×10^9 years (Piper 1978, Burnett 1975).

This conflict has led to a number of studies of the underlying assumptions used to extrapolate the Moon's orbit, looking in particular at the effect of changing the shape of the ocean or changing the rotation rate of the Earth (Sündermann and Brosche 1978, Hansen 1981, Webb 1982a,b). These studies have shown that in the long term it is the changing rotation rate of the Earth which is most important and which has given rise to the apparent conflict.

In the past, the increased rotation rate of the Earth resulted in higher tidal frequencies and, as the average speed of tidal waves in the ocean is to first order independent of the rotation rate, this led to a reduction in the average wavelength of the tides. The shorter wavelength tidal waves were less efficiently excited by the long wavelength tidal forces and so the magnitude of the tides, the power they dissipated in the oceans and the torque they exerted on the Moon were all reduced.

In the calculations reported by Webb (1982b), a model of the tides in a hemispherical ocean was used with the ocean being averaged over all possible orientations relative to the Earth's axis. This was done in order to average out the effect of individual resonances of

---

1 Institute of Oceanographic Sciences, Wormley, Godalming, United Kingdom

Tidal Friction and the Earth's Rotation II
ed. by Brosche/Sündermann

the model ocean in much the same way that continental drift does for the real ocean. When the Earth's rotation rate was doubled from its present rate, the tidal dissipation was found to be reduced to a third of its present value and the torque exerted by the tides was reduced to a sixth of its present value.

Calculations with the averaged model also showed that the past reduction in the tidal torque, due to the faster rotation of the Earth, moved the time of close approach of the Moon and Earth to 5.3×10^9 years ago. As this is before the date of information of the Solar system it implies that the Moon was formed in orbit around the Earth.

This first calculation did not include the energy dissipated by the Earth tides, because at the present time the amount of energy involved is small and is usually neglected. However, early in the history of the Earth-Moon system, the reduced oceanic tides means that the energy dissipated within the Earth becomes important and should be included. The calculation was therefore repeated using a Q-factor of 150 for the energy dissipated within the solid Earth. This moved the time of the close approach of the Earth and Moon to 3.9×10^9 years ago.

The present paper develops the work in two directions. First it has been suggested (Zschau, pers. comm.) that a Q value of 150 is too low and that a value of 250 or 350 would be more realistic. The calculations are therefore repeated using these larger values.

Secondly Piper (this Vol.) proposes that between 580×10^6 and 2.8×10^9 years ago the continents were joined together forming a single massive continent which was centred somewhere near the south pole. The importance of this proposal, as far as tidal friction is concerned, arises because if it is correct, the averaging procedure that has been used is invalid and instead one has to allow for a systematic change in the tidal dissipation. This arises because the tidal waves then have a path encircling the globe near to the equator. As discussed by Webb (1982a), an equatorial path allows the existence of an ocean resonance which is a good match to the tidal forces. This strong resonance produces enhanced dissipation near to its resonance frequency and reduced dissipation at other frequencies. To investigate how such a continent distribution would affect the history of the Earth-Moon system, the calculations are repeated using a model southern continent during the period concerned.

Sections 2 and 3 of this paper describe the oceanic and astronomical models used in the calculations. The results from the work are presented in Sect. 4 and discussed in Sect. 5.

2. The Ocean Model

The ocean model is based on the use of Laplace's tidal equations. The equations are:

$$\frac{\partial \underline{u}}{\partial t} + \underline{f} \times \underline{u} + \frac{\kappa}{h}\underline{u} = -g\nabla(\zeta - \bar{\zeta}) \quad ,$$

$$\frac{\partial \zeta}{\partial t} + \underline{\nabla} \cdot (h\underline{u}) = 0 \quad . \tag{1}$$

$\underline{u}$ is the tidal velocity, ζ the tidal height, $\bar{\zeta}$ the height of the equilibrium tide, $\underline{f}$ the Coriolis vector, κ the friction coefficient, h the depth of the ocean and g the acceleration due to gravity. The model is solved for a hemispherical ocean of constant depth using expansions for ζ and $\underline{u}$ in terms of spherical harmonics. Further details about the method used are given in Webb (1982a). For the calculations reported here, the depth of the ocean is taken as 4400 m and the friction coefficient is chosen so that the decay time for energy in the model ocean is 30 h. In the actual ocean the decay time is believed to lie between 20 and 60 h.

The orientation of the model ocean to the Earth's axis may be varied and this property is used to define an averaged model in which the ocean's response is averaged over all orientations of the ocean relative to the Earth's axis. This process averages out the resonances of the model ocean producing a relatively smooth curve which it is hoped approximates to the average behaviour of the real ocean in the past. Continental drift has a time scale of 10^8 years which is much shorter than the age of the oceans and it might be expected to give a similar averaging effect.

An example of the results obtained with the averaged model is shown in Fig. 1. Curve (a) is the averaged power dissipated by ocean, at different frequencies, when excited by the forces corresponding to the Y_2^{-2} equilibrium tide of unit r.m.s. amplitude. These forces are the ones which drive the main semi-diurnal tides in the ocean (Munk and Cartwright 1966), the M2 equilibrium tide having a r.m.s. amplitude of 0.126 m and the S2 tide a r.m.s. amplitude of 0.058 m.

Before comparing the model results with observations and also before using them with the astronomical model, corrections must be made for the r.m.s. amplitude of the relevant equilibrium tide, for the effect of Earth tides on the ocean (Zahel 1980) and for the fact that the model ocean has only 80% of the area of the actual deep ocean. When these corrections are made, the power dissipated by the M2 tide in the averaged ocean is 3.07 TW (1 terrawatt = 10^{12} W). This compares with a recent estimate of 3.06 TW based on astronomical observations (Lambeck 1977). The close agreement is not significant,

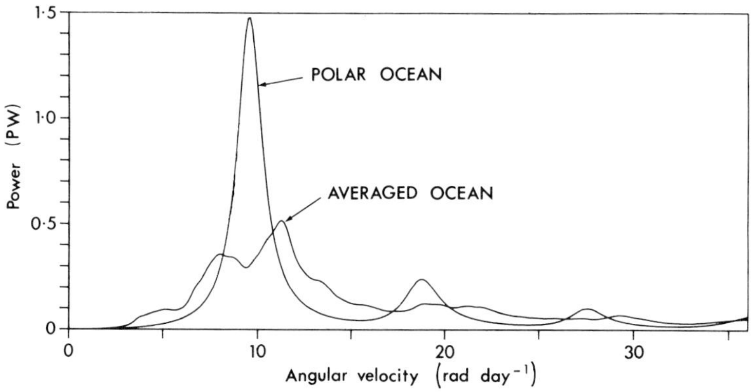

Fig. 1. The power dissipated in the averaged ocean and a hemispherical ocean centred at the pole, for the case when ω the angular velocity of the tide is 1.93 times the angular velocity of the Earth. The power is in pettawatts (1 PW = 10^{15} W). The angular velocity is measured in radians per present Earth day. The tidal forcing used corresponds to a Y_2^{-2} equilibrium tide of unit r.m.s. amplitude. Calculations were made at an angular velocity separation of $2\pi/100$ rad day^{-1}

especially as the present ocean is unlikely to be an average one, but it does indicate that the averaged model is reasonably realistic and should be suitable for the present qualitative study.

3. The Astronomical Model

The full solution of the evolution of the orbits of the Moon and Earth, including the changes in inclination and eccentricity of the orbits, requires a very lengthy calculation and the inclusion of the diurnal and long-period tides as well as the main semi-diurnal tides. Therefore, in the present work, which is concerned with the qualitative importance of different physical effects, a simplified model of the Earth-Moon system is used. The Earth is taken to be in a circular orbit about the Sun, the Moon is in a circular orbit about the centre of gravity of the Earth-Moon system, and the orbital planes are taken to coincide with the Earth's equatorial plane.

The angular momentum of the Moon and Earth about each other is m^*nr^2, where m^* is the reduced mass of the Earth-Moon system, n is the angular velocity of the Moon in its orbit and r is its distance from the Earth. If T_m is the tidal torque acting on the orbit, then the rate of change of the angular momentum is

$$\frac{\partial}{\partial t}(m^*nr^2) = T_m \quad . \tag{2}$$

For a circular orbit, n and r are related by,

$$n^2 r^3 = \text{constant} \quad . \tag{3}$$

Substituting into (2) this gives,

$$\frac{\partial n}{\partial t} = -3T_m/(m^* r^2) \quad . \tag{4}$$

The angular momentum of the Earth is $I\Omega$ where I is the Earth's moment of inertia and Ω its rate of rotation. Goldreich (1966) gives an expression for the change of I with Ω. The rate of change of the Earth's angular momentum, due to the tidal torques, is

$$\frac{\partial}{\partial t}(I\Omega) = -(T_m + T_s) \quad . \tag{5}$$

T_s is the torque acting on the Earth's orbit about the Sun due to the solar tides. This has a significant effect on the Earth's own rotation rate, but the effect on its orbit is small and will be neglected.

The torques T_m and T_s can be calculated from the power dissipated by the tides. The power lost from the Earth's rotational energy due to the torque T_m is $T_m\Omega$, and the power transferred to the Moon's orbit is $T_m n$. The difference between these quantities gives W_m, the power dissipated by the lunar tides in the ocean. Thus,

$$W_m = T_m(\Omega - n) \quad ,$$

or,

$$T_m = W_m/(\Omega - n) \quad . \tag{6}$$

Similarly,

$$T_s = W_s/(\Omega - s) \quad , \tag{7}$$

where W_s is the power dissipated by the solar tides in the ocean and s is the angular velocity of the Earth about the Sun.

If $W(\omega,\Omega)$ denotes the power dissipated by the tides in the model ocean, when the angular velocity of the tide is ω, the angular velocity of the Earth is Ω and the tidal forces correspond to the $\bar{Y}_2^2$ equilibrium tide of unit amplitude; then making the corrections referred to earlier, the power dissipated by the lunar tide when the Earth-Moon separation is r (Webb 1982b), is,

$$W_m = 11.4 \times 10^{-3}(r_0/r)^6 \; W(\omega_m,\Omega) \quad . \tag{8}$$

r_0 is the present Earth-Moon separation and ω_m the angular velocity of the M2 tide. The power dissipated by the solar tides is,

$$W_s = 2.40 \times 10^{-3} W(\omega_s, \Omega) \quad , \tag{9}$$

where ω_s is the angular velocity of the S2 tide.

To calculate the past behaviour of the Earth-Moon system, (4) and (5) are numerically integrated back in time, starting from the present orbital separations of the Earth, Moon and Sun, and using a Runge-Kutta-Merson subroutine to do the integration. Torques are calculated from (6), (7), (8), and (9), with the dissipation function $W(\omega,\Omega)$ being calculated by the ocean model.

In the astronomical model it is found that ω_m is always approximately equal to 1.93Ω and ω_s is approximately equal to 2.00Ω. Also $W(\omega,\Omega)$ varies only slowly with Ω. Therefore the method used to interface the two models is to run the ocean model at closely spaced values of ω, first with ω equal to 1.93Ω and then with ω equal to 2.00Ω. This gives two tables which are transferred to the astronomical model. Linear interpolation is used within each table to reach the required value of ω and between the tables to reach the required values of both Ω and ω.

Dissipation due to Earth tides can be included in the astronomical model. Equation (10.2.4a) of Lambeck (1980) is used with $\sin\varepsilon$ equal to Q^{-1}, where Q is the quality factor for the solid Earth. For the present orbit of the Moon, a Q of 250 results in 0.12 PW being dissipated within the solid Earth by the M2 tide. For extrapolating into the past, if Q remains constant, the power dissipated by the M2 Earth tide can be written as;

$$W_{me} = W_{m0}(r/r_0)^6(\omega/\omega_{m0}) \quad ,$$

where W_{m0} is the dissipation rate and ω_{m0} is the angular velocity of the tide at the present time. A similar expression is obtained for the solar S2 tide.

With the Moon and Sun constrained to lie in the equatorial plane the total power dissipated by the M2 and S2 ocean tides in the averaged model is 4.3 PW. When Earth tides are included this increases to 4.6 PW and, using the astronomical model, results in a secular acceleration of the Moon of -32.5" century^2. This may be compared with Lambeck's (1977) estimate of -28" century^2 based on astronomical observations and -24" century^2 from satellite observations.

4. Results

Curve (a) of Fig. 2 shows the Earth-Moon separation as a function of time when the averaged ocean model is used but Earth tides are not included. When Earth tides are included, with a Q value for the solid Earth of 150, then curve (b) is obtained. It is seen that including Earth tides, moves the time of closest approach from 5.3×10^9 years ago to 3.9×10^9 years ago. Both these calculations were reported in Webb (1982b).

Following the comments of Zschau mentioned previously the model has now been re-run using Q values for the Earth of 250 and 350. The results are shown as curves (c) and (d) respectively of Fig. 2. A Q value of 250 gives the closest approach 4.3×10^9 years ago and a Q value of 350 gives a time of 4.6×10^9 years ago.

The above calculations all assume the continued existence of the Earth, Moon and oceans as one goes back in time. In reality radioactive dating indicates that the Solar System was formed 4.6×10^9 years ago. Less is known about the age of the oceans but the Isua supercrustal sequence of W. Greenland, estimated as being 3.8×10^9 years old, are marine sediments and so imply that the oceans must be older. We shall assume that they are 3.9×10^9 years old.

If the dissipation due to ocean tides is set to zero in the model prior to 3.9×10^9 years ago, then this has no effect on the case when Q equals 150, but for Q equal to 250 the time of closest approach is moved to 5.1×10^9 years ago, and for Q equal to 350, the time of closest approach is moved to 6.2×10^9 years ago. If the extrapolation is halted 4.6×10^9 years ago when the Solar System was formed, the initial values for the Earth-Moon separation are found to be 31.3 Earth radii and 36.0 Earth radii respectively (1 Earth radii = 6.378×10^6 m). If there is no power dissipated by Earth tides then the initial separation is 41.6 Earth radii.

Piper (this Vol.) suggests that a large southern continent may have been in existence between 580×10^6 and 2.8×10^9 years ago. As discussed in the introduction, this may have had a large systematic effect on the power dissipated in the ocean. To illustrate the effect, Fig. 1 shows both the energy dissipated in the averaged ocean and in a polar ocean which has its boundary along the equator.

The polar ocean is seen to be dominated by one strong resonance. When its spatial structure is studied (see Webb 1982a), it is found to be a westward travelling equatorial wave which closely matches the tidal forces. The other resonances, which are orthogonal to the strong resonance, are thus also approximately orthogonal to the tidal forces and so are only weakly excited by the forces. This results in an ocean with a response which is enhanced near to the resonant frequency and reduced elsewhere.

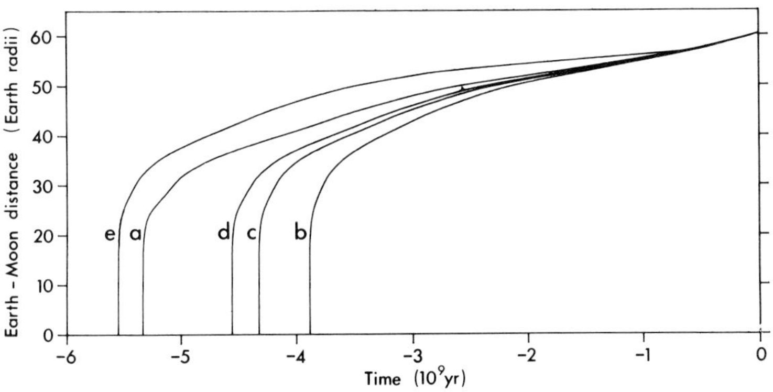

Fig. 2. The Earth-Moon separation in Earth radii (1 Earth radii $= 6.378 \times 10^6$ m) for the cases (a) no Earth tides; (b) Earth tides with Q = 150; (c) Q =250; (d) Q = 350; (e) Q = 250 and with a southern continent between 580×10^6 and 2.8×10^9 years ago

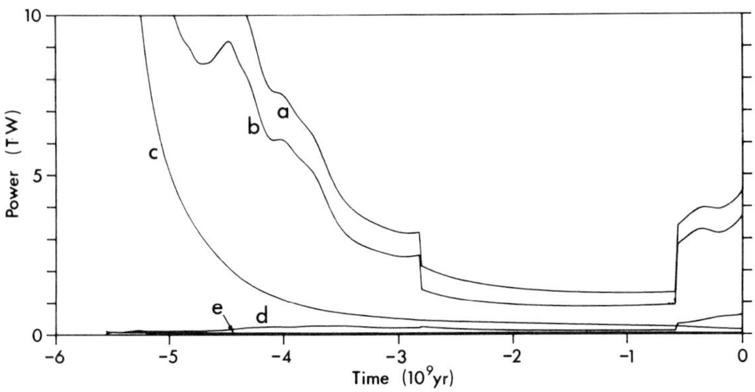

Fig. 3. The power dissipated within the oceans and the solid Earth for case (e) of Fig. 2. The power is in units of terrawatts (1 TW $= 10^{12}$ W) (a) Total power dissipated; (b) by the M2 tide in the ocean; (c) by the M2 tide in the solid Earth; (d) by the S2 tide in the ocean; (e) by the S2 tide in the solid Earth

Between 580×10^6 and 2.80×10^9 years ago, the calculations show that the Earth's angular velocity lay between 7.3 and 8.5 radians per present Earth day. This corresponds to a M2 tide of between 14.1 and 16.4 rad day^{-1}, and as shown by Fig. 1 the energy dissipated in a polar ocean would have been reduced by a factor of about two compared with the averaged ocean. To see how this would have affected the history of the Earth-Moon system the astronomical model was mofified so that between these dates it calculated the tidal torques from the energy dissipated in the polar ocean.

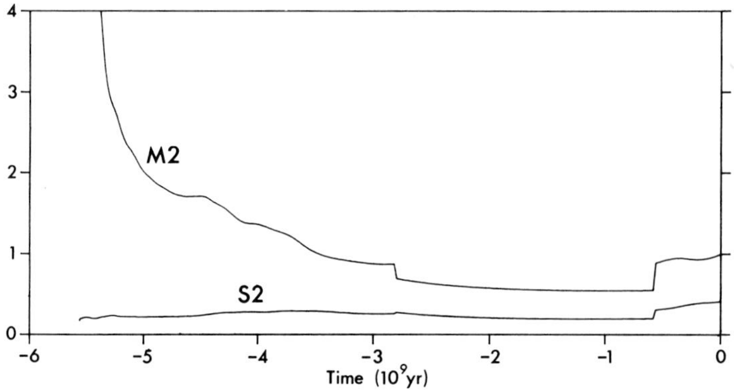

Fig. 4. The r.m.s. amplitude of the M2 and S2 tides relative to the present r.m.s. amplitude of the M2 tide for case (e) of Fig. 2

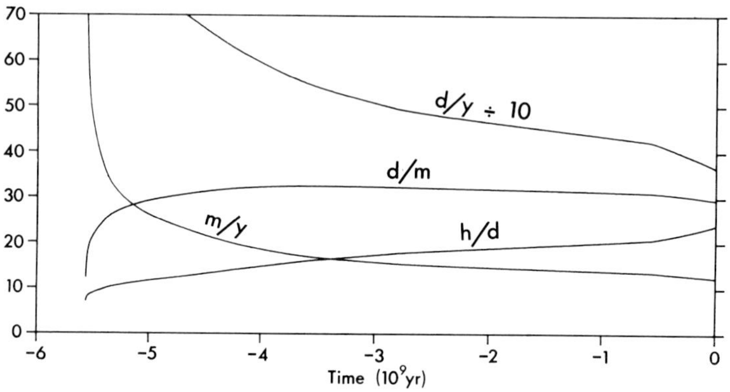

Fig. 5. The number of hours per mean solar day (h/d), the number of days per synodic lunar month (d/m), the number of months per year (m/y) and the number of days per year divided by 10 (d/y ÷ 10), for case (e) of Fig. 2

For this calculation, the Q of the Earth is taken to be 250 and the Earth, Moon and oceans are assumed to be in existence during the complete period of integration. The result is shown as curve (e) of Fig. 2. 4.6×10^9 years ago the Earth-Moon separation is seen to be 41.6 Earth radii and if the oceanic dissipation is set to zero prior to 3.9×10^9 years ago, this increases to 46.9 Earth radii.

The power dissipated in the model is shown in Fig. 3. Between 580×10^6 and 2.8×10^9 years ago the power dissipated is reduced because of the southern continent. The effect is smallest near 2.8×10^9 years ago because of the resonance seen in Fig. 1 near 19 rad day^{-1}. This affects the higher frequency S2 tide more than it does the M2 tide.

The r.m.s. amplitude of the tides is shown in Fig. 4. An interesting results is that except for the reduction due to the southern continent, the r.m.s. amplitude of the M2 tide has remained relatively constant over the history of the oceans; the increase in tidal amplitude resulting from the slowing down of the Earth's rotation rate being roughly balanced by the increasing Earth-Moon separation.

Figure 5 gives the number of present day hours per day, the number of days per lunar month, the number of months per year and the number of days per years. The number of days per lunar month is seen to be almost constant over the history of the Earth-Moon system, a result which might be useful in paleontological studies. Another important variable is the number of days per year which might also be measurable from biological records. In the recent past the number of days per year has been changing quite rapidly, but prior to the break up of the southern continent the model indicates that it changed much more slowly. Finally, the model predicts that 4.6×10^9 years ago the initial rotation period of the Earth was 12.7 h. This increases to 14.7 h if tidal dissipation is set to zero 3.9×10^9 years ago. These results are comparable to the range of 10 to 13 h suggested by MacDonald (1966).

5. Discussion

The results reported in this paper further confirm that if the effect of the Earth's rotation on the tides is allowed for, then either the time of close approach of the Moon and Earth is found to occur near to the time of formation of the Earth-Moon system or else the Moon is formed in orbit about the Earth.

Webb (1982b) showed that if the solid Earth had an effective Q of 150 then the models predicted that the time of close approach occurred 3.9×10^9 years ago. If the Q is 250 or 350, and these have been suggested to be more reliable values, then assuming the continued existence of the ocean the present calculations show that the time of close approach becomes 4.3×10^9 and 4.6×10^9 years ago respectively. If the oceanic dissipation is set to zero prior to the formation of the oceans 3.9×10^9 years ago then the models predict that no close approach occurred after the formation of the solar system 4.6×10^9 years ago. Instead the Moon is found to be initially in orbit about the Earth at a distance of 31.3 Earth radii if the Q is 250 or 36.0 Earth radii if the Q is 350.

Piper has suggested that the continents coalesced to form a massive southern continent between 580×10^6 and 2.8×10^9 years ago. Such a continent is found to reduce the tidal dissipation during the period to about only half of that obtained with the averaged ocean. It results in an Earth-Moon separation at the time of formation of the solar system of 41.6 Earth radii. If the oceanic dissipation is reduced to zero prior to 3.9×10^9 years ago this increases to 46.9 Earth radii.

On balance the present results favour the hypothesis that the Moon was formed by accretion in orbit about the Earth. However, they do not conclusively rule out the possibility that the Moon was captured by the Earth soon after the formation of the solar system. The main reason for rejecting this hypothesis is still that the initial rotation period of the Earth would have had to be less than six hours and this seems unlikely (MacDonald 1966).

The highly simplified nature of the models means that the results should be treated with a little caution. Further calculations need to be carried out, to check the importance of all the model parameters and using, where feasible, more realistic models. Possibly the most important parameter that needs further study is the decay time of energy in the ocean. In the present work this is assumed to be 30 h and it would be useful to repeat the calculations with a decay time of 60 h. It is also important to know how this parameter has actually changed in the past and how it has been affected by systematic changes in the depth, width and area of the continental shelves on which most of the dissipation presently occurs. Any historical information concerning the variation of these quantities would be valuable, as is information on any systematic changes in the depth of the ocean as this affects the average wavelength of the tides. The astronomical model could be improved by including the eccentricity and inclination of the orbits of the Earth and Moon, and by including the effect of the diurnal and long period tides. Hansen (1981) has done some important work along these lines.

Finally, although the results of this paper are theoretically interesting they also need experimental verification. Queries have been raised about the accuracy of paleontological measurements, but this remains the most promising source of data about rotation periods in the distant past. The calculations of Lambeck (1980), Webb (1982b) and the present paper all show the number of days per month to be roughly constant over the history of the Earth. This, therefore, might be used as a test of the reliability of a particular source of data, the number of days per year or the number of months per year then being used to verify the theoretical results.

Acknowledgements. I am indebted to Profs. Brosche and Zschau for discussions and for their suggestions concerning this study.

References

Burnett DS (1975) Lunar Science: The Apollo Legacy. Rev Geophys Space Phys 13:13-34

Goldreich P (1966) History of the lunar orbit. Rev Geophys Space Phys 4:411-439

Hansen KS (1981) Secular effects of oceanic tidal dissipation on the Moon's orbit and the Earth's rotation. Thesis, Univ Chicago

Lambeck K (1977) Tidal dissipation in the ocean: Astronomical, Geophysical and Oceanographic consequences. Philos Trans R Soc London Ser A 287:545-594

Lambeck K (1980) The Earth's variable rotation: geophysical causes and consequences. Cambridge Univ Press, Cambridge

MacDonald GJF (1966) Origin of the Moon: Dynamical Considerations. IN: Marsden BG, Cameron AGW (eds) The Earth-Moon system. Plenum Press, New York, p 288

Munk WH (1968) Once again-tidal friction. Q J R Astron Soc 9:352-375

Munk WH, Cartwright DE (1966) Tidal spectroscopy and prediction. Philos Trans R Soc London A 259:533-581

Piper JDA (1978) Geological and geophysical evidence relating to continental growth and dynamics and the hydrosphere in precambrian times. In: Brosche P, Sündermann J (eds) Tidal friction and the Earth's rotation. Springer, Berlin Heidelberg New York, p 243

Sündermann J, Brosche P (1978) The numerical computation of tidal friction for present and ancient oceans. In: Brosche P, Sündermann J (eds) Tidal friction and the Earth's rotation. Springer, Berlin Heidelberg New York, p 243

Webb DJ (1982a) The effect of polar wander on the tides of a hemispherical ocean. Geophys J R Astron Soc 68:689-707

Webb DJ (1982b) Tides and the evolution of the Earth-Moon system. Geophys J R Astron Soc 70:261-271

Zahel W (1980) Mathematical modelling of global interaction between ocean tides and Earth tides. Phys Earth Planet Inter 21:202-217

A Note on the Variability of Growth Increment Formation in the Shell of the Common Cockle *Cerastoderma edule*

T. Ohno[1]

1. Introduction

It is well known that the records of incremental growth of hard parts of invertebrates are preserved as sequences of growth increments (Clark 1974, Neville 1967). Wells (1963) suggested, taking a Devonian coral as an example, that growth increments and their cyclic patterns in fossils could be used to determine the length of days per year in the geological past. Since then analyses of such patterns are used in reconstructing the history of the Earth's rotation; a comprehensive review is given by Scrutton (1978). Among them are number of studies based on fossil bivalve shells by Berry and Barker (1968, 1975), Pannella (1972, 1975), Pannella and MacClintock (1968), Pannella et al. (1968).

The basic assumptions for such studies are:

1. existence of growth increment, which is formed with a constant basic rhythm, such as solar-daily, tidal rhythms, etc.
2. existence of long-term rhythms, recorded in growth patterns, for example, fortnightly, synodic monthly, annual rhythms, etc.

Informations about the growth rhythms of recent bivalves should provide a solid basis for the interpretation of fossil materials.

During work on a Ph.D. thesis at the Institute of Paleontology of the University of Bonn, I have had the chance to examine basic as well as superimposed long-term growth rhythms in the shells of living bivalve molluscs under the kind supervision of Prof. H.K. Erben, director of the institute. The results will be published in a separate paper (Ohno 1982).

Here, independent of my thesis, I would like to present a note on the variability of growth increment formation in the common cockle *Cerastoderma edule*. Even in a single tidal flat, the growth pattern of the cockles varies according to the level in which they live.

1 Institut für Paläontologie, Universität Bonn, Nußallee 8, D-5300 Bonn 1, Fed. Rep. of Germany

Tidal Friction and the Earth's Rotation II
ed. by Brosche/Sündermann

2. Different Types of Growth Patterns

The common cockle *Cerastoderma edule* is an inhabitant of European coasts and lives infaunally in intertidal as well as subtidal shallow water. In the radial section of shells of the cockles we can see two components of incremental growth. One is a thin growth line; the other is a thick increment, which is bordered by a pair of growth lines (Fig. 2,a2). In earlier works (House and Farrow 1968, Farrow 1971, 1972, Whyte 1975) it was suggested that the cockles form their increments with daily rhythm. But recent studies, which include intensive experiments with the marked, planted, and recovered materials, could not verify their proposals (Richardson et al. 1979, Ohno 1982).

Near the middle of the intertidal zone, where the cockles are constantly exposed and submerged with the 12.4-h cycle of semidiurnal tide, they build their increments with 12.4-h rhythm (Richardson et al. 1979, Ohno 1982). The pattern of increments in the shells of this level is quite regular (type I). The formation of shell is interrupted during the exposure at low tides, which results in the formation of well defined growth lines. The thickness of the increments is more or less similar to each other.

In subtidal zone – formed under the lack of rhythmic exposure – the growth pattern does not have a regular appearance (type II). Most increments are bordered with faintly defined growth lines. The thickness of the increments varies irregularly.

Richardson et al. (1980) suggested that the cockles of subtidal zone build their increments with endogenous tidal rhythm. However Ohno (1982) shows that cockles of subtidal zone form their increments without constant rhythm. Therefore the growth pattern type II is indicative of a non-rhythmic growth increment formation.

The aim of this note is to know the mode of increment formation of the cockle at the border of intertidal/subtidal zone, i.e., of levels between the Highest Low Water (HLW) and the Lowest Low Water (LLW).

3. Material and Method

Samples were collected on the tidal flat of Eyrac in the Bay of Arcachon along the French Atlantic coast. The tide of the locality is a semidiurnal one. The average range of the tide is about 3 m. There is a difference of more than 0.9 m between the water levels of the HLW and the LLW. Therefore the locality is suitable to study the variation of increment formation at the border of intertidal/

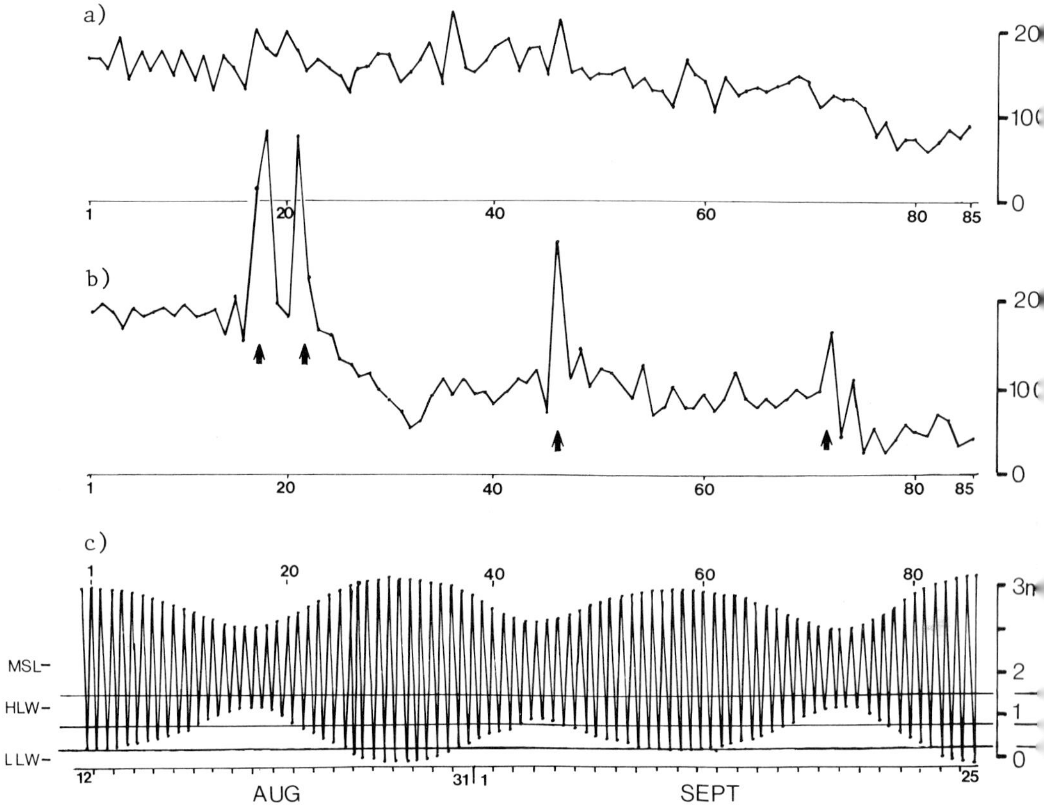

Fig. 1. (a) Growth curve of a cockle from the level A (sample no. P432). Thickness of the 85 most recently formed growth increments is plotted. (b) Growth curve of a cockle from the level B (sample no. P377). Thickness of the 85 most recently formed growth increments is plotted. Arrows indicate anormally thick increments. (c) Predicted tidal range of the last 85 tides before the sampling of the specimens at Eyrac, Bay of Arcachon (the diurnal inequality is neglected). A,B,C = three levels of sampling of the cockles; MSL = Mean Sea Level; HLW = Highest Low Water; LLW = Lowest Low Water

subtidal zone. On September 25, 1980 about 10 living specimens were collected in each of three different levels at Eyrac (Fig. 1c). The level A lies between the mean sea level (MSL) and the HLW. The samples from this level are exposed at every tide throughout one fortnight and their growth pattern serves as a control in examining the growth patterns of the samples from two lower levels. The levels B and C are situated between the HLW and the LLW. The cockles from the level B experience an exposure at each low tide during spring tides. They stay continuously under the water for several tidal cycles during neap tides. The cockles of the lowest level C are exposed semidiurnally only for several tidal cycles during spring tides. The rest of the fortnight they stay continuously under water.

The collected samples were killed immediately. In laboratory radial sections of the shells were made, polished and etched with 0.1 mol HCl for about 30 s. After that they were spattered with gold and observed under a SEM of the Institute of Paleontology, Bonn.

4. Observations

On the polished and etched radial sections of the shells one can see growth increments, which are bordered by thin, ridgelike growth lines (Fig. 2,a2).

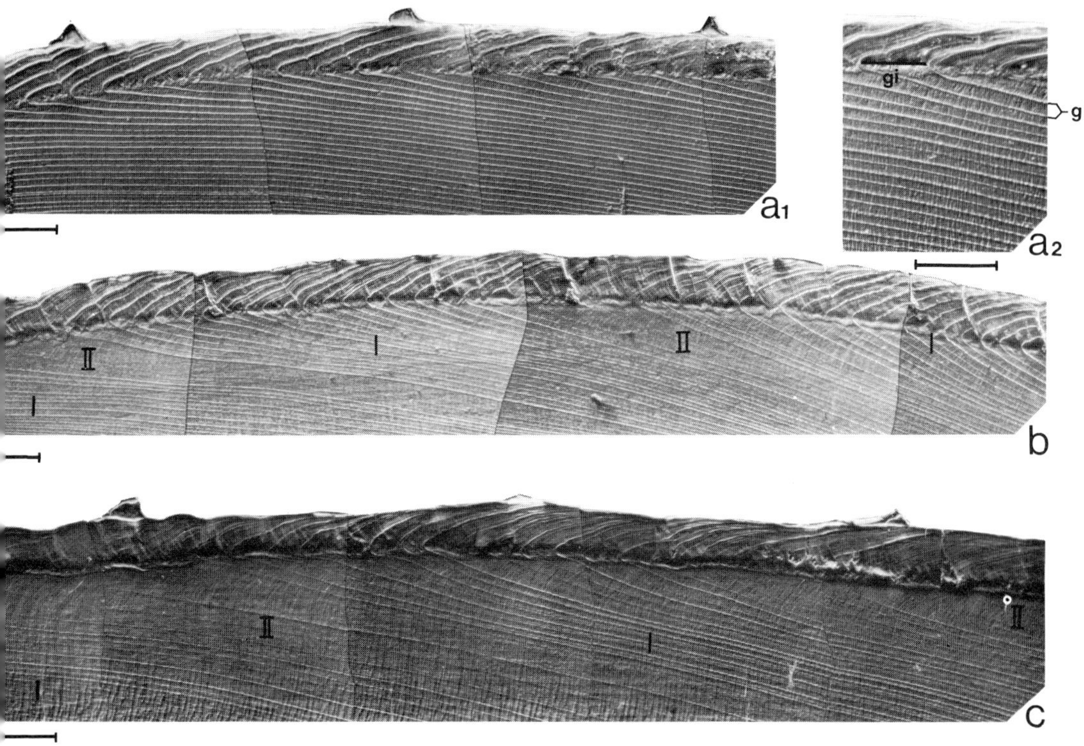

Fig. 2a-c. (Bars under each figure represent 200 μ). (a1) A radial section of a shell from the level A, showing exclusively the growth pattern type I (sample no. P432). (a2) Enlargement of a part of the shell shown in Fig. 2,a1. gi = a growth increment; gl = two growth lines bordering the increment (gi). (b) A radial section of a shell from the level B, showing the alternation of the type I (I) and type II (II) growth patterns. In this specimen the type I is dominant (sample no. P378). (c) A radial section of a shell from the level C, showing the alternation of the type I (I) and type II (II) growth patterns. In this specimen the type II is dominant (sample no. P426)

In the shells from the level A, one finds exclusively the type I growth pattern, which is typical for the cockles living near the mean sea level (Fig. 2,a1). Growth lines are well defined with strong relief. The thickness of increments is quite regular throughout the section. The shells from the border of intertidal/subtidal zone (the shells from the level B and C) have alternations of the growth patterns type I and II.

In the shells of the level B the type I is dominant. There are thick parts of type I growth pattern with about 20 well-defined growth lines. Between them there are thin parts of type II growth pattern with a few, faintly defined growth lines. The thickness of increments is irregular and there are sometimes very thick increments (Fig. 2b).

In the shells of the level C, type II is dominant (Fig. 2c). There are thin parts of type I with a few, more or less well-defined growth lines, alternating with thick parts of the type II with faintly defined, irregularly spaced growth lines.

The thickness of the 85 most recently deposited growth increments was measured in the shells of the level A and B. Two typical individual measurements are shown as growth curves in Fig. 1a (for a specimen of the level A) and in Fig. 1b (for a specimen of the level B). The growth curve of the level A does not show any anormally thick increments. On the other hand, the curve of the level B shows several anormally thick increments (arrows in Fig. 1b), which occur in the type II growth pattern. The occurrance of these thick increments corresponds to the times of neap tides.

5. Discussion

Even in one tidal flat, individual cockles from different levels build different growth patterns. There is no doubt, that the type I pattern of shells from the level A is a result of rhythmic exposure with 12.4-h period during low tides. Because shells examined here are not marked, planted and recovered ones, the interpretation of the growth pattern type II in the cockles of level B and C in regard of its rhythmicity should be done carefully. However, the occurrence of abnormally thick increments in the shells of the cockles of the level B and its correspondence to the total submergence during neap tides show convincingly, that the growth pattern type II represent a non-rhythmic increment formation. These thick increments must represent a growth of at least more than one tidal cycle. This observation suggests complications in the applicability of growth increments in the fossil records to the reconstruction of the history of the Earth's rotation.

Besides daily rhythm, there is tidal rhythm, with which growth increments are built. One should determine which kind of rhythms is recorded in the fossil materials. In fossils from the level between the HLW and the LLW it might be easy to find a fortnightly pattern as the alternation of the growth pattern type I and II or their analogues. The existence of the type II makes it possible to recognize the alternation. However, because of its non-rhythmic nature, the existence of the type II makes it impossible to determine the exact number of the tidal cycles in one fortnight. A fortnight pattern is recognizable, but its period is not determinable. Therefore this kind of alternation might have little application to the reconstruction of the history of the Earth's rotation. In this regard the published figures on the length of the fortnight in fossil materials should be treated with reservation.

As shown in the cockles from the level A, cockles of the middle of the intertidal zone do not show any remarkable pattern at first sight, which suggests the existence of fortnight rhythm. However efforts must be done to find a method to determine long-term rhythms in the shells of the middle of intertidal zones, because in this level the cockles build their growth increments almost exactly with 12.4 h tidal rhythm i.e. with a constant basic rhythm (Richardson et al. 1979, Ohno 1982). An attempt in this direction is done in Ohno (1982).

Before we can apply growth pattern analysis of fossil materials successfully to the reconstruction of the history of the Earth's rotation, we should collect more informations about the variable mode of growth increment formations in the living invertebrates.

Acknowledgement. I am deeply grateful to Prof. H.K. Erben, Director of the Institute of Paleontology, University of Bonn, who first made me acquainted with this interesting theme. I also wish to thank him for making working facilities of the Institute available to me. I am also thankful to Dr. C. Cazaux of the Station Biologique d'Arcachon, Institut Universitaire de Biologie Marine, Université de Bordeaux I, for his kind support during my research in Arcachon. Thanks are given to Mr. D. Waloßek, Institute of Zoology, University of Hamburg, for his valuable discussions. The research is supported by a grant from the Deutsche Forschungsgemeinschaft (Forschungsthema Er 4/39).

References

Berry WBN, Barker RM (1968) Fossil bivalve shells indicate longer month and year in Cretaceous than present. Nature (London) 217: 938-939

Berry WBN, Barker RM (1975) Growth increments in fossil and modern bivalves. In: Rosenberg GD, Runcorn SK (eds) Growth rhythms and the history of the Earth's rotation. Wiley, London, pp 9-25

Clark GR II (1974) Growth lines in invertebrate skeletons. Annu Rev Earth Planet Sci 2:77-99

Farrow GE (1971) Periodicity structures in the bivalve shell: experiments to establish growth controls in *Cerastoderma edule* from the Thames Estuary. Palaeontology 14:571-588

Farrow GE (1972) Periodicity structures in the bivalve shell: analysis of stunting in *Cerastoderma edule* from the Burry Inlet (South Wales). Palaeontology 15:61-72

House MR, Farrow GE (1968) Daily growth banding in the shells of the cockle, *Cardium edule*. Nature (London) 219:1384-1386

Neville AC (1967) Daily growth layers in animals and plants. Biol Rev 42:421-439

Ohno T (1982) Ph D thesis, Univ Bonn (in prep.)

Pannella G (1972) Paleontological evidence on the Earth's rotational history since early Precambrian. Astrophys Space Sci 16:212-237

Pannella G (1975) Paleontological clocks and the history of the Earth's rotation. In: Rosenberg GD, Runcorn SK (eds) Growth rhythms and the history of the Earth's rotation. Wiley, London, pp 253-284

Pannella G, MacClintock C (1968) Biological and environmental rhythms reflected in molluscan shell growth. J Paleontol 42 (Mem 2):64-80

Pannella G, MacClintock C, Thompson MN (1968) Paleontological evidence of variations in length of synodic month since late Cambrian. Science 162:792-796

Richardson CA, Crisp DJ, Runham NW (1979) Tidally deposited growth bands in the shell of the common cockle, *Cerastoderma edule* (L.). Malacologia 18:277-290

Richardson CA, Crisp DJ, Runham NW (1980) An endogenous rhythm in shell deposition in *Cerastoderma edule*. J Mar Biol Assoc UK 60:991-1004

Scrutton CT (1978) Periodic growth features in fossil organisms and the length of the day and month. In: Brosche P, Sündermann J (eds) Tidal friction and the Earth's rotation. Springer, Berlin Heidelberg New York, pp 154-196

Wells JW (1963) Coral growth and geochronometry. Nature (London) 197:948-950

Whyte MA (1975) Time, tide and cockle. In: Rosenberg GD, Runcorn SK (eds) Growth rhythms and the history of the Earth's rotation. Wiley, London, pp 177-189

Paleogeography and Paleobathymetry: Quantitative Reconstructions of Ocean Basins

J. Thiede[1]

1. Introduction

The possibility of reconstructing quantitatively the evolution of the physiography of ocean basins through time is one of the prerequisites for the understanding of the current regimes controlling distributions of their major water masses and changes of the water mass distributions through time. Our ability to carry out such reconstructions has increased remarkably since the age structure of the Jurassic to Recent crust under the modern oceans was resolved in considerable detail (Sclater et al. 1981) over the past decade. Sea-floor spreading and the subsidence of the surface of the ocean floor as a result of the thermal cooling of the underlying ocean crust after its formation at the mid-ocean ridge (Sclater et al. 1971) explain major trends of the morphology of the individual subbasins of the world ocean. The age structure of the ocean crust and its subsidence through time provide earth scientists with quantitative measures to describe the evolution of ocean basins through time as three-dimensional physiographic provinces of the earth's surface. The advance of quantitative three-dimensional reconstructions of major geological structures has not only improved our understanding of the paleogeographic and paleoenvironmental evolution of the ocean basins, but has also provided us with a base for attempts to model quantitatively specific paleoceanographic situations (Krohn et al. 1981).

This paper describes some of the basic principles of reconstruction of the physiographic evolution of former ocean basins. The Atlantic Ocean will be used as an example for the type of results which can be achieved because it is the best known and most easily understood major subbasin of the world ocean.

1 Department of Geology, University of Oslo, P.O. Box 1047, Blindern, Oslo 3, Norway

Tidal Friction and the Earth's Rotation II
ed. by Brosche/Sündermann

2. Paleogeographic Reconstructions

Paleogeographic reconstructions of the coast lines around ocean basins have to be based on two different types of data. The evolution of the deep ocean basins themselves can be traced using the age structure of the underlying oceanic crust, as far as this is known in the necessary detail. The shape of the surrounding epicontinental seas has to be deduced from a completely different type of data, namely the extent of marine sediments on and along the adjacent ocean margins and continents. It is often rather difficult to reconcile discrepancies arising from the different nature of these two data sources.

2.1 Ocean Basins

The age structure of the ocean crust under the individual subbasins of the world ocean can today be assumed to be reasonably well understood (Sclater et al. 1981). However, due to the consumption of large areas of oceanic crust along subduction zones, it is presently impossible to estimate how some of these basins have evolved through time. This applies especially to the Pacific Ocean which to a large degree is surrounded by subduction zones along active continental margins where large segments of ocean crust have been recycled into the earth's interior continuously during the past 150 m.y. In contradiction, ocean basins largely surrounded by passive continental margins such as the Atlantic, Indian and Polar oceans have preserved a record of most of their history.

The map of the distribution of magnetic anomalies in the North Atlantic Ocean (Fig. 1) is a good example for the detail and refinement of the data base which is needed to trace the various stages of horizontal growth of an ocean basin. The age of some of the major magnetic anomalies in the North Atlantic, their distance from the Mid-Atlantic Ridge and the resulting spreading rates (Fig. 2) give a measure for the speed of opening of this ocean which is now approx. 6000 km wide.

Following these data, the history of the North Atlantic Ocean can be subdivided into three major stages. It probably started to open during early to middle Jurassic times when it became part of the circumequatorial Tethys Ocean. The second stage comprises growth to the North and a break through to the South Atlantic Ocean in Cretaceous times. The last stage, which developed during Tertiary and Quaternary times when the Norwegian-Greenland Sea and a pathway to the Polar Ocean opened, but when the seaways to the tropical Pacific and Indian Ocean were disrupted, probably had the most important impact on paleoceanography because a wide and deep connection developed between the cool to cold water masses of the northern and southern hemispheres (Thiede 1979).

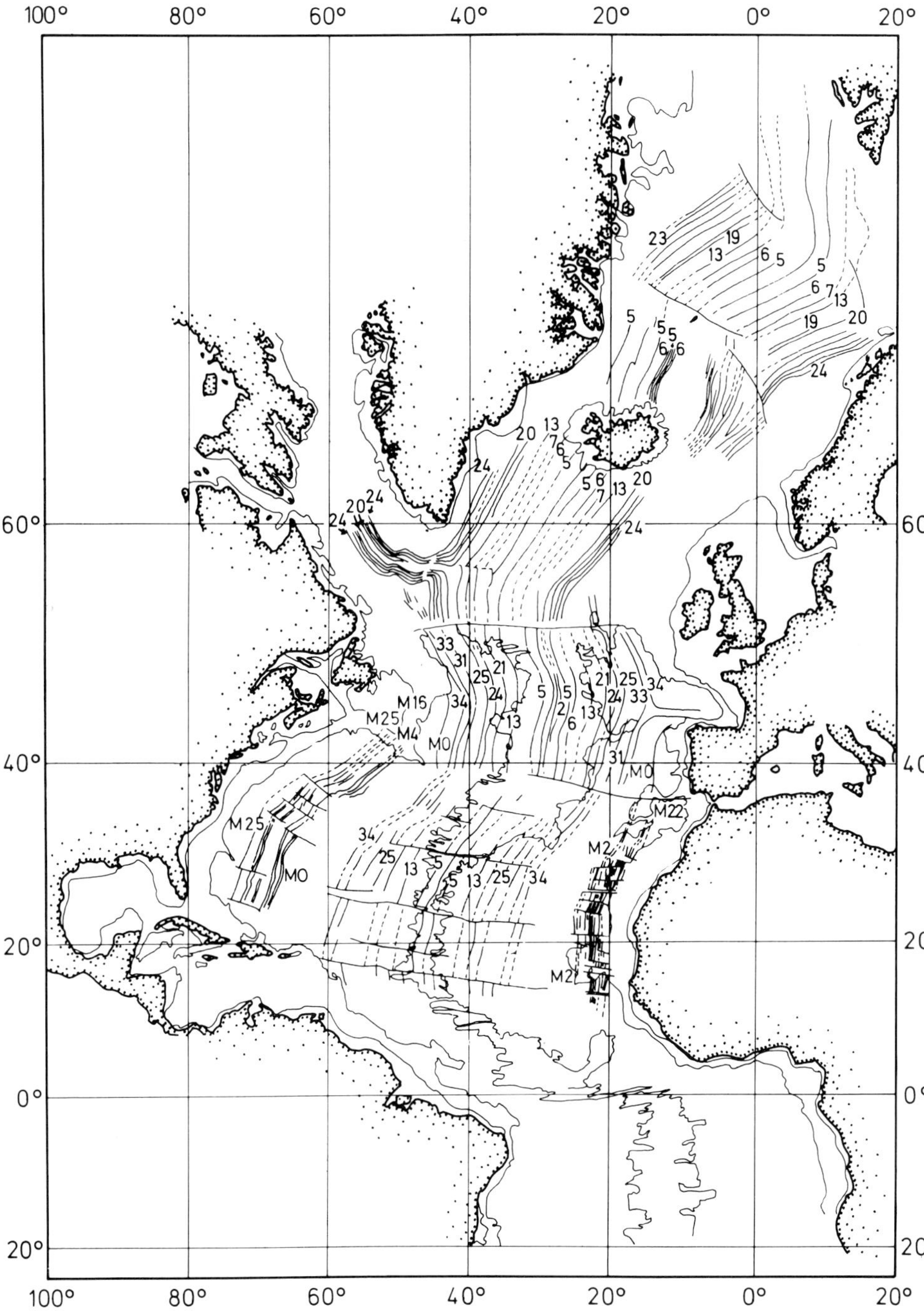

Fig. 1. Distribution of magnetic anomalies in the North Atlantic Ocean (Thiede 1979)

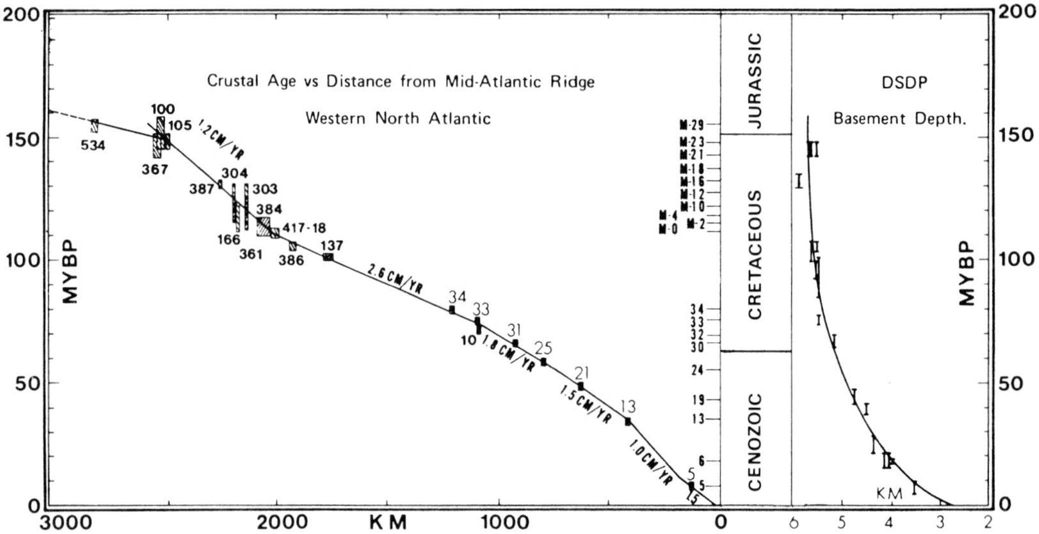

Fig. 2. North Atlantic spreading rates vs. time and the age-depth relationship of the oceanic basement. (After Perry et al. 1980.) The number along the ordinate in the center of the figure are magnetic anomalies; they are repeated along the younger part of the spreading rate curve

2.2 Epicontinental Seas

The reconstruction of the coast lines around former ocean basins requires detailed studies of the distribution of marine sediments along their continental margins and in areas which were once covered by shallow epicontinental seas. Trying to do so, one faces three principal problems which have resulted in a considerable reduction of the temporal and spatial resolution of the extent of fossil epicontinental seas. Firstly, it is difficult to correlate the very detailed stratigraphic records obtained in the oceans to epicontinental sea deposits because they are much more difficult to date. Many of the zone fossils of pelagic sediments are produced by oceanic planktonic organisms which do not enter the shallow seas. If they do their remains are often much less well preserved than in the oceanic sediments. Secondly, epicontinental seas have been subjected to sea level changes over several hundred meters (Vail et al. 1980) which resulted in the development of important hiatuses and in large areas from where no geological record has been preserved. Thirdly, the record of former coast lines along active continental margins is severely perturbed and completely destroyed due to tectonic processes. Therefore, the stratigraphic data available from shallow sea areas will never permit the same detail in paleogeographic reconstructions as can be obtained for the Mesozoic and Cenozoic ocean basins.

3. Paleobathymetric Reconstructions

The data for establishing the paleodepth of reconstructions of the paleogeography of shelf seas and ocean basins must also be collected from widely different sources. Shelf seas are usually located on continental crust and henceforth in general only a few hundred meters deep at maximum. Information from deep wells into the sediment cover providing data on sediment thicknesses as well as sediment distribution and on the paleodepth requirements of benthic fossils allows a further refinement of this general statement in many areas (van Hinte 1978). However, the incompleteness of the sedimentary columns deposited along continental margins and in epicontinental seas often impedes the accuracy which is required for quantitative oceanographic modelling experiments.

The situation is quite different in the ocean basins per se because of the age-depth constancy of oceanic crust which has resulted in very systematic changes of the hypsometry of the floor of ocean basins through time (Sclater et al. 1971). Ocean crust is subsiding (Fig. 3) after being generated along the mid-ocean ridge, approx. 1 km during the first 10 m.y. after its formation, another 1 km during the next 20-30 m.y., and an additional 1 km during the ensuing 60-70 m.y. (Berger and Winterer 1974). This regular subsidence is modified by the isostatic response to the increasing load of the overlying sediments and water column, resulting in a further depression of the surface of the oceanic basaltic crust (Fig. 2). It can further be disturbed when plate tectonic horizontal movements carry segments of this crust through regions of intraplate volcanism, over hot spots or over bulges generated by irregularities of deeper situated layers of the earth's interior.

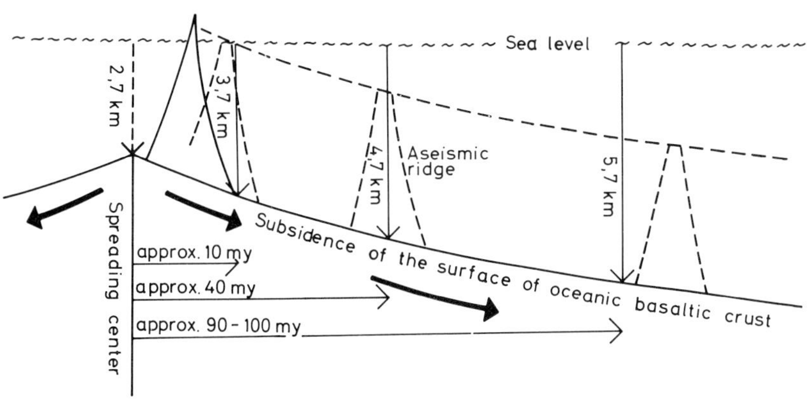

Fig. 3. Subsidence of the surface of oceanic crust generated along a mid-ocean ridge spreading center and of an aseismic ridge (After Detrick et al. 1977)

Although the crest of mid-ocean ridge system generally is located at 2.7 km water depth (Sclater et al. 1971), there are many regions with prominent depth anomalies (for example the shallow region between Scotland and Greenland in the North Atlantic Ocean). Aseismic rises which represent one example of anomalously shallow ocean floor and which often develop close to the mid-ocean ridge as a result of volcanism, seem to trail the subsidence of the underlying oceanic crust throughout most of the time of their existence albeit at much shallower depth levels than the surrounding ocean floor (Detrick et al. 1977). The subaerial parts of the volcanic cones on top of these rises are often subject to intensive erosion until they have subsided below wave base. However, through their sediment cover they represent monitors of the depositional environments on littoral to bathyal ocean floors. Regions of anomalous water depth in the oceans are often very important for the distribution of water masses because they channelize their movement along the ocean floor. In general, the depth of former ocean basins whose crust is still preserved under the modern oceans can today be estimated and reconstructed quantitatively with great precision (Sclater et al. 1971), for normal ocean floor within the range of ±100-200 m, along aseismic rises of ±300-500 m.

4. An Example: The Tertiary Atlantic Ocean

Following the above outlined techniques, reconstructions of the Atlantic Ocean during three Tertiary time slices are presented in Fig. 4. The plate rotations which have been used follow Sclater et al. 1977, the extent of the shallow seas has been compiled from a multitude of sources, mainly summarized in van Andel et al. 1977 and Thiede 1979. These Atlantic maps are examples for the type of data and their precision which are presently available from paleogeographic and paleobathymetric reconstructions of ocean basins surrounded by passive continental margins and of their adjacent epicontinental shallow seas. The maps show that the major basins were all approaching shapes and depths as we know them from today's Atlantic Ocean. Major events changing the pathways of the surface water circulation occurred in the northern part of the Norwegian-Greenland Sea when the passage between Greenland and Svalbard broke through, and in the tropical-subtropical areas when the connections to the Pacific Ocean and to the

---

Fig. 4a-c. Paleobathymetry and paleogeography of the Atlantic Ocean during (a) late Eocene/early Oligocene time (approx. 38 m.y.B.P.), (b) early Miocene time (approx. 21 m.y.B.P.), and (c) late Miocene time (approx. 10 m.y.B.P.). Isobaths are in kilometers (Figs. 4a-c see pages 235-237)

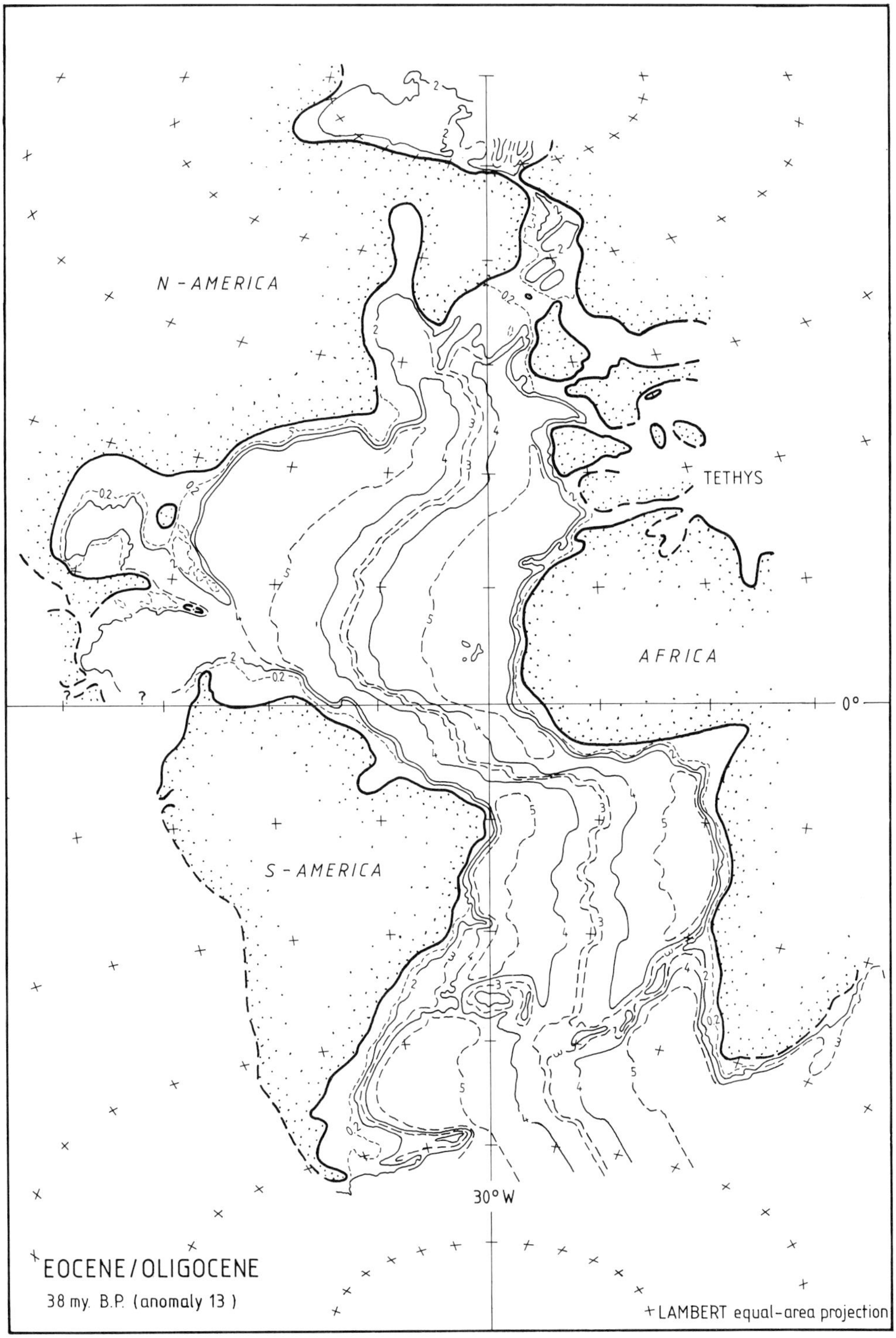

Fig. 4a (Fig. 4b,c see page 236 and 237)

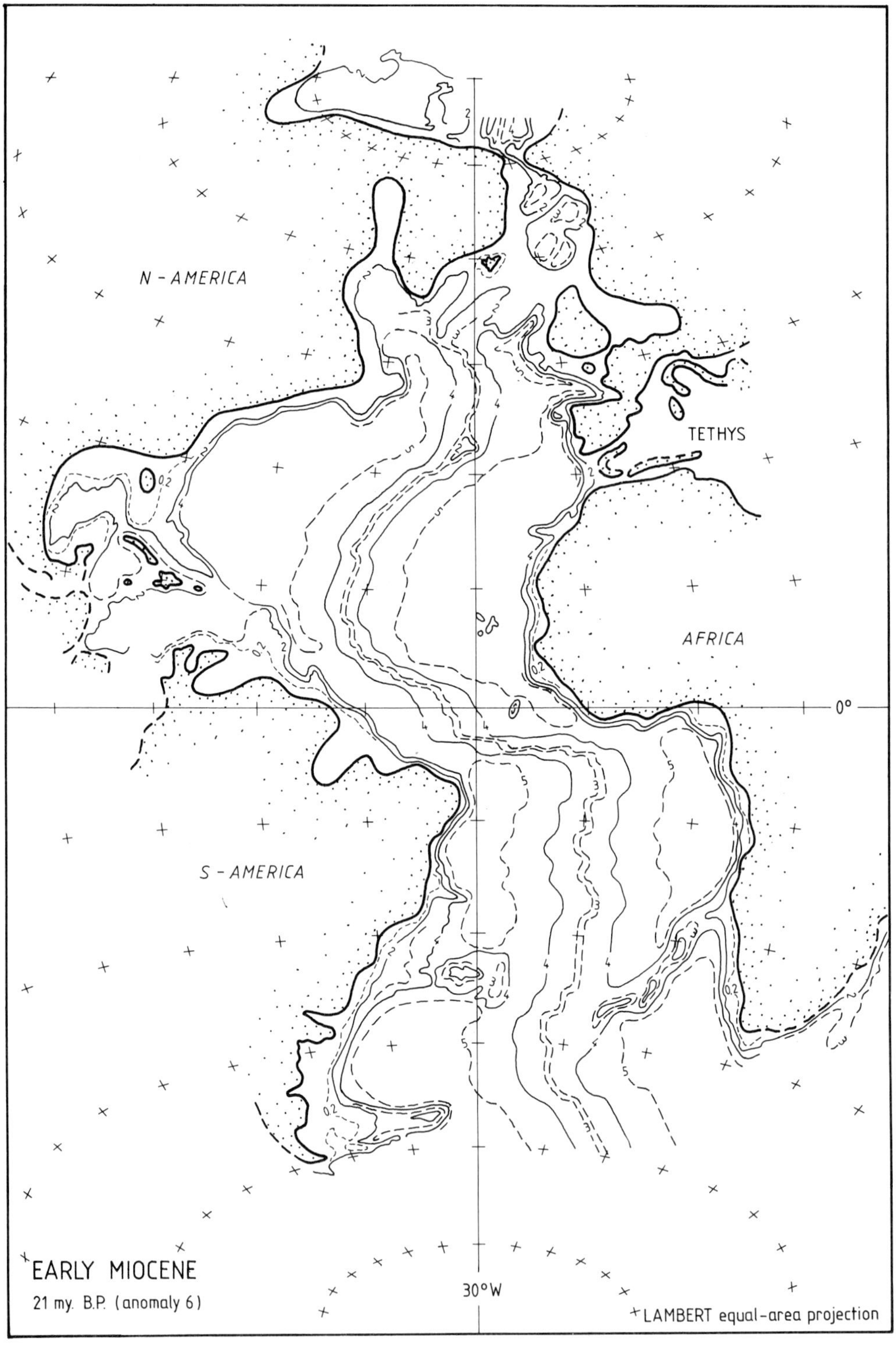
N - AMERICA
S - AMERICA
AFRICA
TETHYS
0°
30°W
EARLY MIOCENE
21 my. B.P. (anomaly 6)
LAMBERT equal-area projection

Fig. 4b

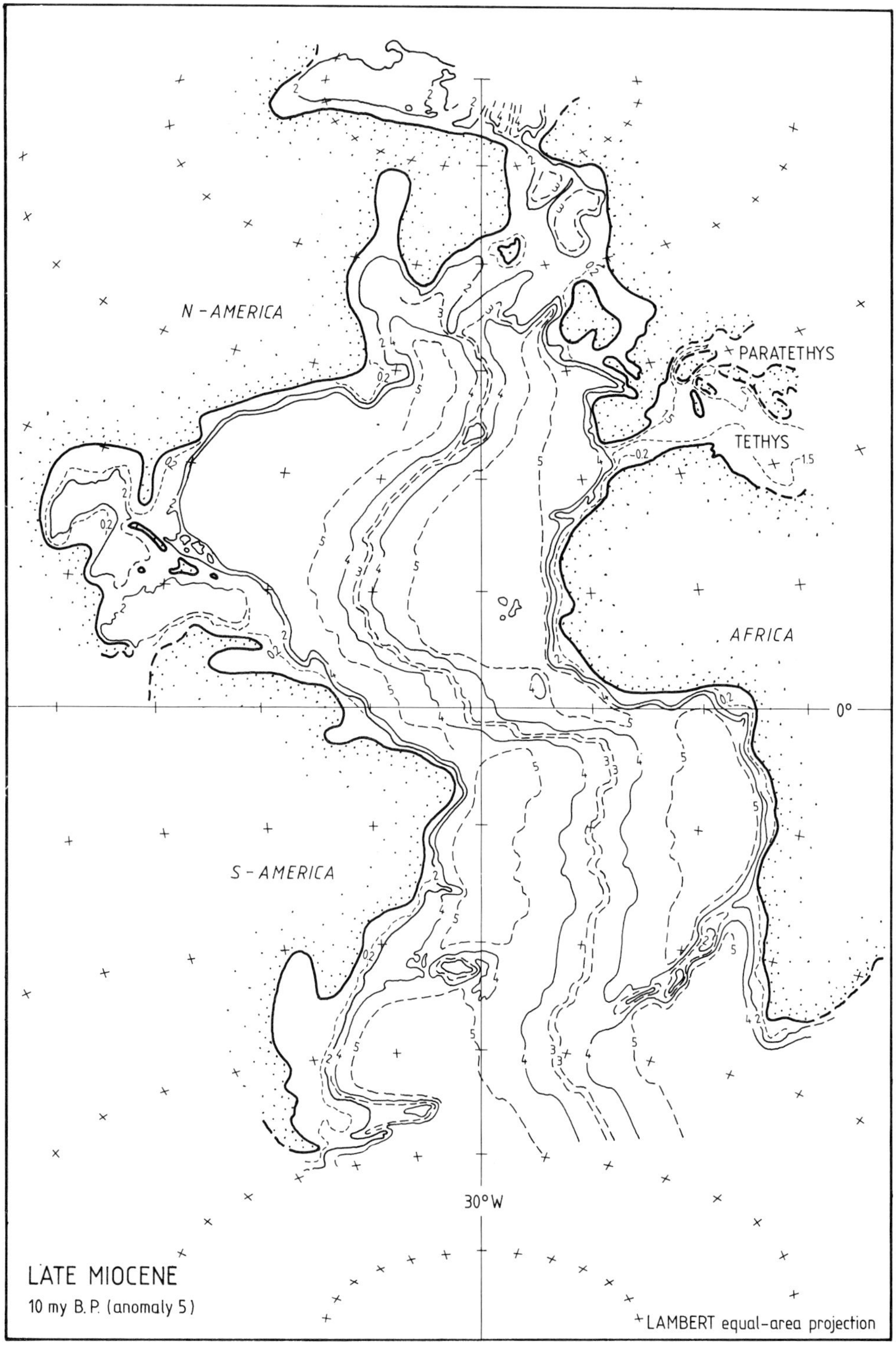
N-AMERICA
PARATETHYS
TETHYS
AFRICA
S-AMERICA
0°
30°W
LATE MIOCENE
10 my B.P. (anomaly 5)
LAMBERT equal-area projection

Fig. 4c

Tethys narrowed. The plate tectonic history of the latter two regions is extraordinarily complicated and will probably never be resolved accurately and in the same detail as that of the Atlantic Ocean proper.

The paleogeographies and paleobathymetries of the Greenland-Scotland Ridge and of the Rio Grande Rise-Walvis Ridge complex which traverse the North and South Atlantic oceans as shallow regions, are still debated (Thiede and Eldholm in press). They constrain the exchange of water masses between the different subbasins of the Atlantic Ocean today, and they are assumed to have done so more in the geologic past because of their shallower location. The Greenland-Scotland Ridge, for example, might well have supported large subaerial sections during mid-Tertiary times. The evolution of deep-sea channels connecting the eastern and western Atlantic basins poses another problem which cannot be solved adequately at the present time.

However, the new means of quantitative paleogeographic and paleobathymetric reconstructions of the ocean basins and their adjacent shallow epicontinental seas represent a major scientific advancement. Such reconstructions are today only available for the ocean basins surrounded by passive continental margins, but they hopefully will be further developed to also include the Pacific Ocean in the near future. In particular, this type of reconstruction will advance possibilities of applying quantitative oceanographic models to types of oceans which only existed in the geological past.

Acknowledgments. This study is part of a group effort to describe the history of the oceans throughout the past 150 million years. J. Sclater (Cambridge, Mass.) has kindly given the author access to part of the geophysical base data at a very early stage. V. Maack (Hamburg) has produced the Atlantic reconstructions used in Fig. 4. The help of the colleagues and the opportunity to present part of these data in Bielefeld is gratefully acknowledged.

References

Andel van TH, Thiede J, Sclater JG, Hay WW (1977) Depositional history of the South Atlantic Ocean during the past 125 million years. J Geol 85:651-698

Berger WH, Winterer EL (1974) Plate stratigraphy and the fluctuating carbonate line. Spec Publ Int Assoc Sediment 1:11-48

Detrick RS, Sclater JG, Thiede J (1977) Subsidence of aseismic ridges. Earth Planet Sci Lett 34:185-196

Hinte van JE (1978) Geohistory analysis — Application of micropaleontology in exploration geology. Am Assoc Petrol Geol Bull 62:201-222

Krohn J, Brosche P, Sündermann J (1981) Paläogezeiten und Erdrotation. Geol Rundsch 70:64-77

Perry RK, Flemming HS, Vogt PR, Cherkis NZ, Feden RH, Thiede J, Strand JE (1980) North Atlantic Ocean: Bathymetry and plate tectonic evolution. Map, Naval Res Lab, Washington DC

Sclater JG, Anderson RN, Bell ML (1971) Elevation of ridges and evolution of the central eastern Pacific. J Geophys Res 76:7888-7915

Sclater JG, Hellinger S, Tappscott T (1977) The paleobathymetry of the Atlantic Ocean from the Jurassic to present. J Geol 85:509-552

Sclater JG, Parsons B, Jaupart C (1981) Ocean and continents: Similarities and differences in the mechanism of heat loss. J Geophys Res 86(B12):11535-11552

Thiede J (1979) History of the North Atlantic Ocean: Evolution of an asymmetric zonal paleoenvironment in a latitudinal ocean basin. Maurice-Ewing-Ser (AGU) 3:275-296

Thiede J, Eldholm O (in press) Speculations about the paleodepth of the Greenland-Scotland Ridge during late Mesozoic and Cenozoic times. In: Bott MHP et al (eds) Structure and development of the Greenland-Scotland Ridge — New techniques and concepts. Plenum Press, New York London

Vail PR, Mitchum RM, Shipley TH, Buffler RT (1980) Unconformities of the North Atlantic. Philos Trans R Soc London Ser A 294:137-155

Mesozoic and Cenozoic Paleogeographic Maps

A. M. Ziegler, C. R. Scotese, and S. F. Barrett[1]

1. Introduction

A knowledge of paleogeography is critical in computing tidal friction and its effect on the Earth's rotation through time (Krohn and Sündermann, this Vol.). Important factors are the dimensions of ocean basins, the widths of continental shelves, and the details of coastal configurations. Paleogeographic maps have been greatly improved in recent years, although the certainty with which geologists can address these factors is highly variable. Seven appropriately detailed paleogeographic maps have been published for the Paleozoic periods (Scotese et al. 1979) and seven Mesozoic and Cenezoic maps have been prepared in the same style for the present paper. Before discussing the details of these new maps, we describe some of the general problems facing the paleogeographer, with particular reference to how these might affect the tidal computations of the hydrodynamicist.

General ocean basin geometry has been reliably established for the Mesozoic and Cenozoic periods, but is very tentative for the earlier maps, because undeformed ocean crust older than 200 m.y. is unknown. The linear magnetic anomalies of the younger ocean floor have now been dated to the extent that the widths of ocean basins can be determined to the nearest 200 km (see Sclater et al. 1981). In the Paleozoic, however, errors of thousands of kilometers are possible, as these distances are based on such vague considerations as biogeographic patterns, and modern rates of plate motion, which are known to be variable in space and time (Ziegler et al. 1979). Since resonance is such an important factor in tidal dissipation calculations, it is doubtful at this stage that the Paleozoic maps would be accurate enough for this purpose.

Shelf widths are, for most areas and times, reasonably portrayed on the paleogeographic maps. During many periods these shelves were considerably wider than at present, although a systematic bias exists due to the fact that paleogeographers prefer to work on periods with wide shelf expanses. These intervals provide the most extensive fossil

1 University of Chicago, Department of Geophysical Sciences, 5734 S. Ellis Avenue, Chicago, IL 60637, USA

Tidal Friction and the Earth's Rotation II
ed. by Brosche/Sündermann

record, as well as the most cosmopolitan faunas, and therefore the most easily correlated rocks. Sea level has fluctuated on a variable time scale in the past (Vail et al. 1977) so the shelf widths as portrayed should be regarded as representing one extreme of a wide spectrum for each time period. For instance, our mid-Silurian map (Scotese et al. 1979, Fig. 15) shows a maximum 45% of the continental crust covered with shallow seas, while it is known that just prior to this period, a glacio-eustatic lowering of sea level (Sheehan 1973) resulted in perhaps as little as 5% of the continental area being covered with marginal seas. Such lowered sea levels result in unconformities, or a "negative" geologic record, and are not attractive time periods for paleogeographers.

Another problem with respect to tidal calculations involves the depth of water covering the continental shelves. The present marginal seas may be anomalously deep on the geologic scale because of the rapid sea level rise due to the melting of the ice sheets just 11,000 years ago. This effectively drowned the shelves up to 150 m (Schopf 1980). The usual processes of sedimentation have not had time to adjust to this rapid rise. Major glaciations, and their extreme effect on sea level, are relatively rare events in geologic history, having occurred just three times since the beginning of the Paleozoic (Late Ordovician, Carboniferous to Permian, and Pleistocene). This constitutes less than 25% of this 580 million year interval. Normally, indigenous carbonate production in the lower latitudes would have maintained extremely shallow depths across the shelf. Moreover, reef growth near the continental margin in many cases would have effectively isolated the broad continental seas from the effects of ocean generated tides. Even in the cooler latitudes, terrigenuous sediments would prograde across the shelf and generally result in shallower conditions than seen today. It may very well be, then, that tidal dissipation in the shallow seas was not as important during most periods in the geologic past. In any case, the present depth of the shelf margin is, in part, an historic accident. Also, it cannot be assumed that the gradient from shelf margin to shore would have been relatively constant in the past. Our paleogeographic maps will, in the future, show both reef tracts and a bathymetric subdivision of the shelf regions, and this should help with this particular problem.

Finally, special coastal configurations, such as the Bay of Fundy and the North Sea, can create very high tides, and account for a disproprotionate amount of the earth's tidal friction. Such local situations are probably beyond paleogeographic resolve in most cases. Again, however, the present may be atypical of the geologic past. During the recent rapid sea level rise, valleys eroded by rivers or carved by glaciers have become estuaries and bays, resulting in coastlines that are unusually irregular. Of course, paleogeographic maps are always going to appear to be simpler than geographic maps because of the incompleteness of the geologic record, but we believe this simplicity approaches the real situation.

In the following pages we present the seven Mesozoic and Cenozoic paleogeographic maps (Figs. 1-7) and briefly discuss the geometric, paleomagnetic, and tectonic considerations we used in their preparation. These maps are preliminary to more elaborate maps being compiled for the "Atlas of Paleogeographic Maps" at the University of Chicago.

2. Continental Fits

The Mesozoic and Cenozoic maps presented in this paper are based on the recent plate tectonic models of the Atlantic (Sclater et al. 1977) and the Indian oceans (Norton and Sclater 1979). There are uncertainties regarding the early breakup history of the southern continents prior to the early Cretaceous, and the tectonic histories of the Gulf of Mexico, Mediterranean, Arctic and Antarctic regions.

The early spreading history of the Indian Ocean is not known, and there is still considerable debate as to how the eastern and western halves of Gondwana must be reassembled. In Du Toit's original reconstruction of the southern continents, Madagascar was placed in the northern notch along the eastern coast of Africa, that is, adjacent to Kenya. Paleozoic paleomagnetic poles cluster most tightly when this fit is used (Klootwijk 1979). It has been proposed that the alternate fit of Madagascar adjacent to Mozambique results in a better alignment of the Tasman, Trans-Antarctic, and Cape fold belts (Barron et al. 1978). A third reconstruction of the Indian Ocean (Johnson et al. 1980), though similar to Du Toit's fit, displaces India approximately 500 km further south. This reconstruction matches Jurassic shorelines in Africa, Madagascar and India, and may better explain the pattern of magnetic anomalies from the Mozambique Basin. The reconstruction used in this set of maps to link eastern and western Gondwana is modified after Smith and Hallam (1970).

The early tectonic history of the Gulf of Mexico is similarly constrained by the "fit" used to reconstruct the surrounding continental blocks. The reconstruction we have chosen requires that Mexico, Honduras and the Yucatan peninsula be rotated to the northwest and out of the way of South America (Van der Voo et al. 1976). There is some evidence for major sinistral faults in Mexico and southwestern United States along which this motion may have taken place (Silver and Anderson 1974, McKee and Jones 1979). Other reconstructions of the Gulf of Mexico displace Mexico and Central America in a similar manner (Pilger 1978).

The tectonic histories of the Arctic and Antarctic regions must remain speculative at this time. The northern portions of eastern Siberia must be included with the Alaskan north slope, and in all

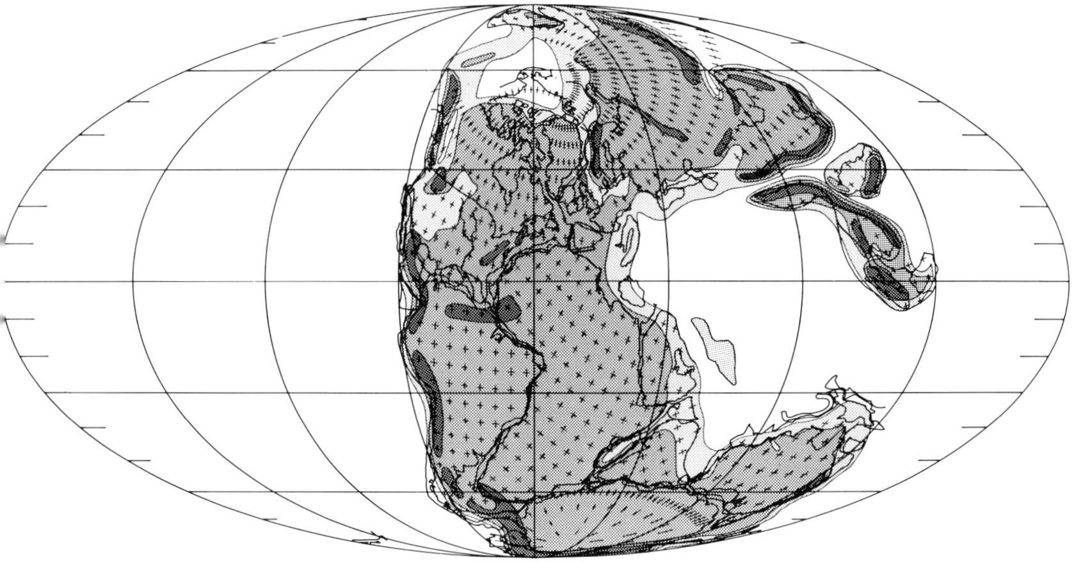

Fig. 1. Earliest Triassic (Induan Stage) paleogeography. Deep oceans, unshaded; shallow seas, light shading; low land, intermediate shading; mountains, dense shading

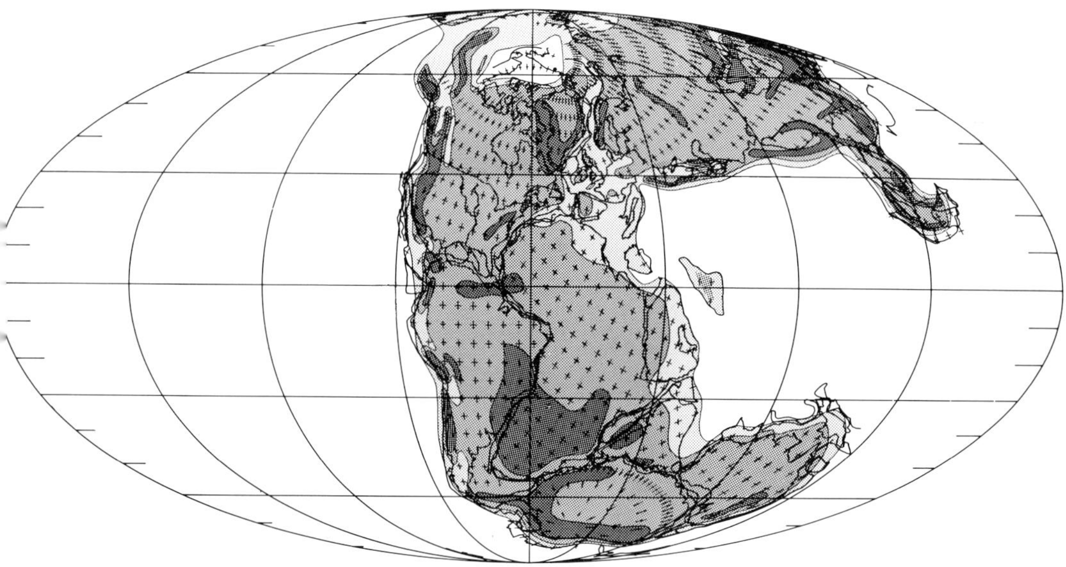

Fig. 2. Early Jurassic (Pliensbachian Stage) paleogeography. Deep oceans, unshaded; shallow seas, light shading; low land, intermediate shading; mountains, dense shading

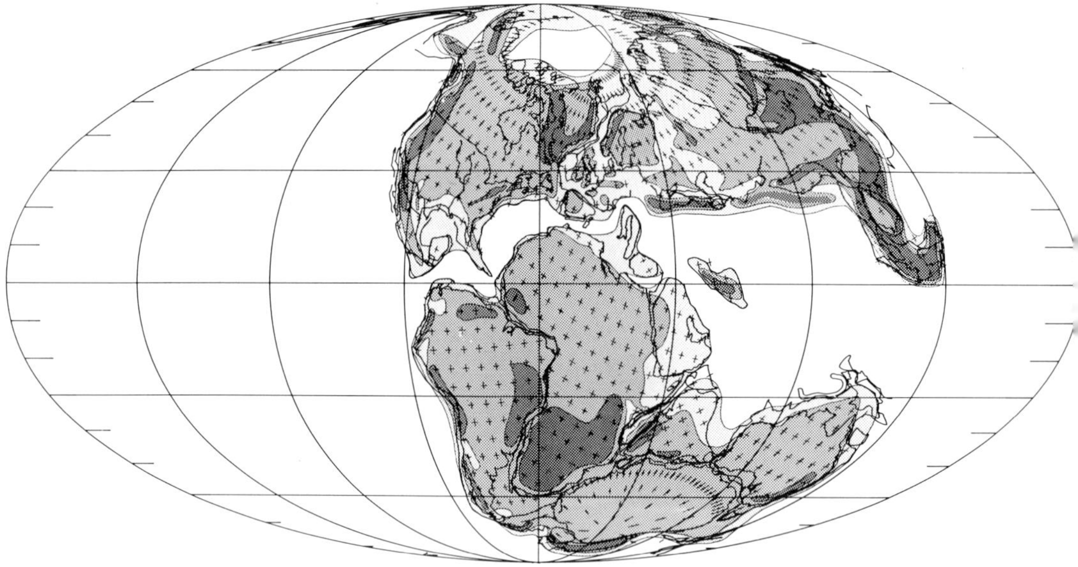

Fig. 3. Latest Jurassic (Volgian Stage) paleogeography. Deep oceans, unshaded; shallow seas, light shading; low land, intermediate shading; mountains, dense shading

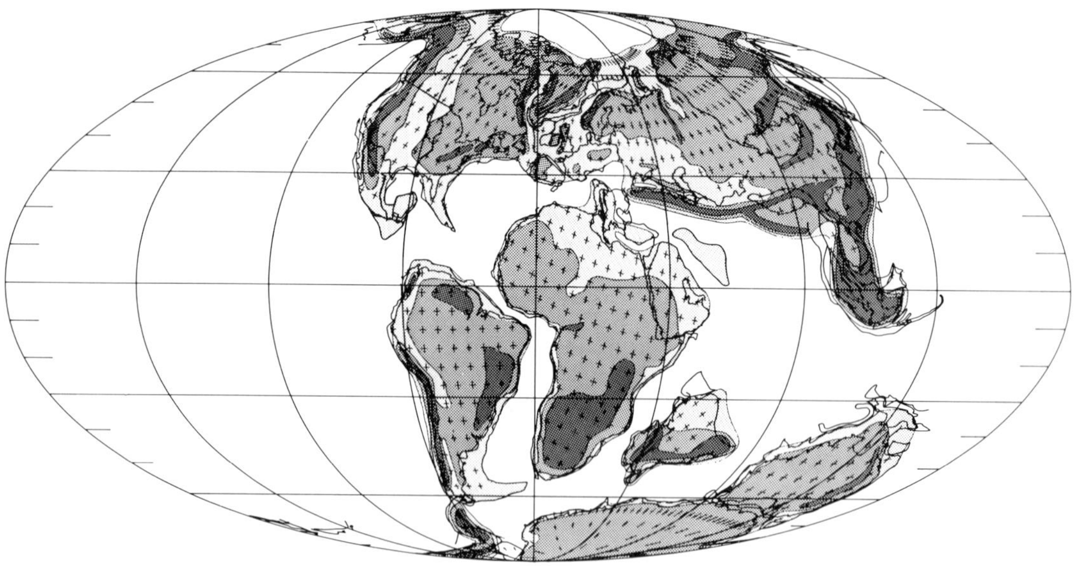

Fig. 4. Middle Cretaceous (Cenomanian Stage) paleogeography. Deep oceans, unshaded; shallow seas, light shading; low land, intermediate shading; mountains, dense shading

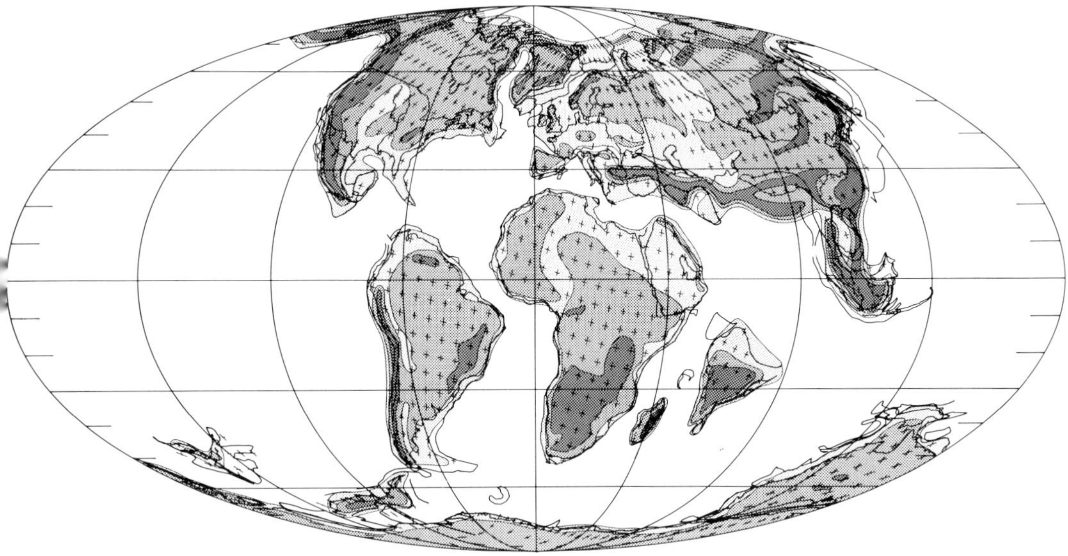

Fig. 5. Latest Cretaceous (Maestrichtian Stage) paleogeography. Deep oceans, unshaded; shallow seas, light shading; low land, intermediate shading; mountains, dense shading

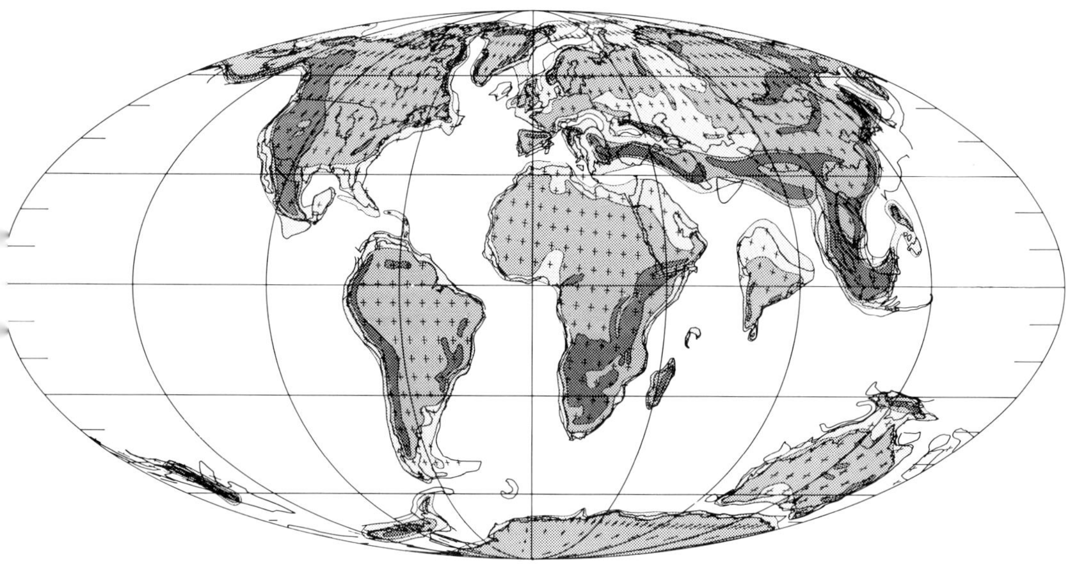

Fig. 6. Middle Eocene (Lutetian Stage) paleogeography. Deep oceans, unshaded; shallow seas, light shading; low land, intermediate shading; mountains, dense shading

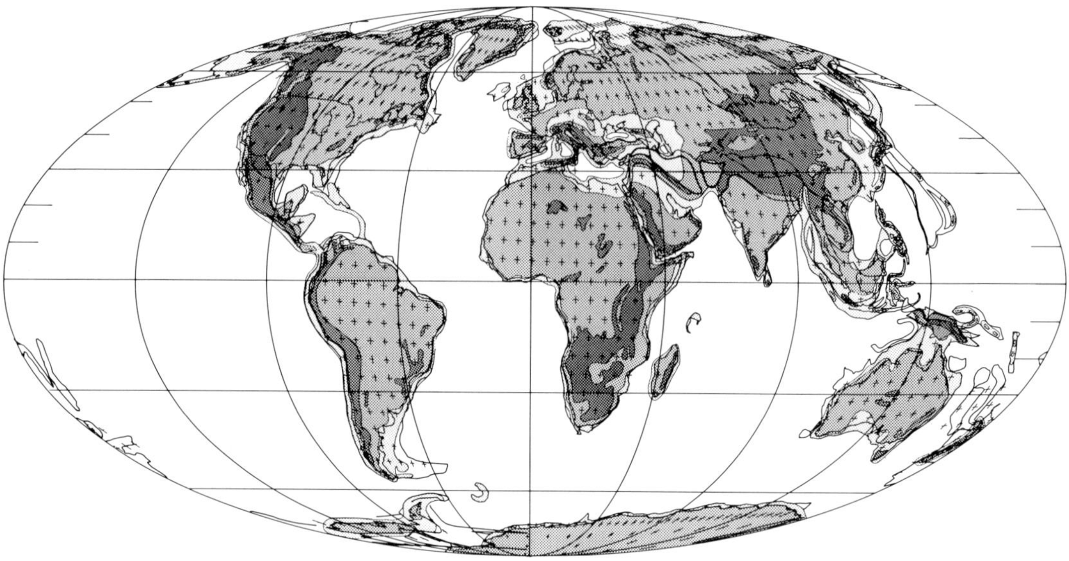

Fig. 7. Middle Miocene (Vindobonian Stage) paleogeography. Deep oceans, unshaded: shallow seas, light shading; low land, intermediate shading; mountains, dense shading

our maps, these terrains are kept fixed with respect to North America and Siberia. Paleomagnetic evidence no longer supports the counter-clockwise rotation of the northern portion of Alaska away from Arctic Canada.

Western Antarctica is probably composed of several micro-continents. In addition to the Western Antarctic peninsula, the Ellsworth mountains and Marie Byrdland must be accounted for in any reconstruction of the southern oceans. The reconstruction used in these maps is similar to the reconstruction of De Wit (1977). Marie Byrdland is fitted against East Antarctica, and the Ellsworth mountains are rotated clockwise so that the folded Paleozoic terrains line up with the main trend of the Trans-Antarctic mountains. The Western Antarctic peninsula is fitted against the Falkland plateau, providing a continuation of the Andean Cordillera. An alternate reconstruction based on paleomagnetic data (Alley and Watts 1979) rotates the peninsula clockwise against eastern Antarctica, closing the Weddell Sea.

3. Paleomagnetic Orientations

Mesozoic and Cenozoic paleomagnetic data were compiled from the summaries of Irving (1960-1965) and McElhinny (1968-1978) and selected for reliability. Pole positions that were obtained from samples that were insufficiently demagnetized, that were from unstable tectonic areas, or that did not exhibit a primary remanence, were rejected. Mean poles for each continent were calculated for the Early and Late Triassic, Early and Middle Jurassic, Early, Middle and Late Cretaceous, Paleocene, Eocene, and Miocene, using the paleomagnetic data which meet the reliability criteria. These mean paleomagnetic poles were then rotated into superposition by applying the rotation required to close the ocean basin, according to the plate tectonic model discussed in the previous section. Global mean poles, in African coordinates, were calculated for each time interval and plotted to form a "global apparent polar wander path" (Fig. 8). In making each reconstruction, the entire continental assembly was oriented with respect to the original geographic north poles for each time interval by superimposing the global means pole on the rotation axes.

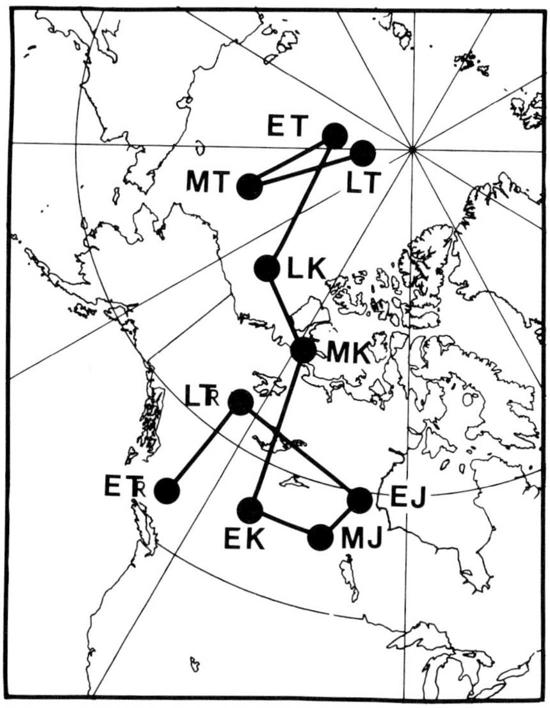

Fig. 8. Global apparent polar wander path in African coordinates. LT = Late Tertiary (Miocene); MT = Middle Tertiary (Eocene); ET = Early Tertiary (Paleocene); LK = Late Cretaceous; MK = Middle Cretaceous; EK = Early Cretaceous; MJ = Middle Jurassic; EJ = Early Jurassic; LT = Late Triassic; ET = Early Triassic

The individual continental mean paleomagnetic poles clustered well after the continents were restored to their correct relative position. Confidence circles of the means (α 95) were less than 11°, with the exception of the Early Triassic (α 95 = 20°). A measure of the scatter of the data, the dispersion parameter K, was similarly at acceptable levels ($K > 50$). The apparent dispersion of paleomagnetic data for the Early Triassic may be the result of a "misfit" between north and south Pangaea. Morel and Irving (1981) and Smith et al. (1981) have proposed alternate Pangaea reconstruction. In their reconstructions, Europe, rather than North America, is adjacent to northwest South America. Though these reconstruction satisfy the paleomagnetic information, they have yet to be reconciled with geologic, tectonic, and biogeographic data.

The Tertiary maps are based on 86 poles and the Cretaceous on 69 poles. The Jurassic and Triassic are based on 46 and 48 paleomagnetic poles, respectively. Because of the sparse amount of data available for the Late Jurassic, a global means pole was not calculated, but the position of the pole for the Late Jurassic (Volgian) was interpolated between the Middle Jurassic and Early Cretaceous poles along the global apparent polar wander path.

4. Convergent Zones

Paleomagnetic data is the primary means of orienting continental pieces. However, construction of detailed maps showing shorelines, paleobathymetry, and distribution of sediment types requires information from stratigraphy and structural geology. This is especially true for compressive, collisional, or other deformed areas.

The main problem areas on our maps lie in the belt of compressive and collisional tectonics that spans the globe from the Tethyan areas of the Caribbean through the Alpine and Himalayan belt to Southeast Asia, and from there north around the Pacific to include eastern Asia and western North America. We feel comfortable with our treatment of the Alpine region, which differs somewhat from published accounts. The action began in the late Jurassic (Fig. 2) with the opening of the Central Atlantic. Paleomagnetic data indicate that Italy (and presumably other Adriatic areas) remained with Africa (Channell et al. 1979). The relative motion of Europe and Africa is described by a rotation pole in northern Europe (Dewey et al. 1973, p.3157). This means that spreading in the Atlantic was transformed into compression in the Himalayan belt, with a zone of shear in between. We believe that the Pontides of northern Turkey and adjacent areas of the Lesser Caucasus to the east, and the Rhodope Massif and Moesia to the west were involved in this sinsitral shear zone, which resulted in the opening of the Black Sea. The situation is similar to the late Tertiary opening of the Andaman Sea by the northward (sinistral) motion of the Indian Ocean plate Curray et al. 1979). Some have suggested that northern Turkey and adjacent areas were derived from Gondwana (Sengor 1979), but we are dubious about this because these areas have a European flavor both biogeographically (Sengor and Yilmaz 1981, p. 203) and tectonically (having been involved in Hercynian tectogenesis). Paleomagnetic evidence quoted by the above authors in support of a Gondwanian origin for these terrains comes from rocks that are poorly dated, and is now thought to be inconclusive (Van der Voo pers. comm. 1981).

Innovations we have made in our maps with respect to the Himalayan zone are significant. It has been necessary to palinspastically widen ("undeform") the Himalayas and Tibetan plateau a total of 1500 km because it is known that the Indian continent collided with Asia in the late Eocene (Molnar and Tapponnier 1975) and this amount of convergence of India and Siberia has taken place since then. Justification for this comes from the observation that the crust in this part of the world is 70 km thick (Chang and Cheng 1973, p. 261), about double normal thickness, and this doubled thickness extends over a north-south distance of 1500 km. All the maps have this palinspastic restoration, except the Middle Miocene map, since collision then was well underway. We believe that a substantial convergence took place throughout the whole Himalayan-Alpine belt with decreasing amounts to the west. The lowering of sea level during the late Tertiary must in part be due to this decrease in area of continental crust and concommitant increase in the width of the ocean basins. We note that the "Old Red" and "New Red" continents of earlier times also followed periods of continental collision.

The peripheral belts of the north Pacific are not adequately dealt with on our maps. Much palinspastic work remains for areas as diverse as south China, Japan, Kamchatka and adjacent parts of eastern Siberia, south Alaska, and parts of the Cordilleran belt of Canada, the U.S.A., and Mexico. All of these areas were surely south of, and probably separate from the main continental nuclei of Siberia and North America in the early Mesozoic. The paleomagnetic directions, sediments (carbonates), and faunas (tropical) for these peripheral areas all indicate a southerly provenance (Ross 1978, Monger and Price 1979). Only the first two maps of our sequence are affected, for by the Latest Jurassic these terrains were accreted to the continental nuclei. We note that this convergence corresponds in time with the initial opening of the central Atlantic. In other words, there was a major change in world plate geometry about the end of the Jurassic.

The concept of Pangaea, the megacontinent of late Paleozoic and early Mesozoic times, must therefore be modified. To be sure, the major continental nuclei, with the notable exception of south China (Feng and Zhu 1980), were together during this interval, but significant marginal terrains around the north Pacific were well separated from

Asia and North America until the late Jurassic. These terrains may have existed as fringing arcs, similar to the situation in the western Pacific today. Pangaea, as an agglomeration of major continental nuclei existed from the late Triassic, the time of the addition of south China, to the mid Jurassic, the time of the opening of the central Atlantic. As a unified mass of all continental crust, it never really existed, as several of the marginal terrains of the north Pacific are based on crust as old as late Precambrian.

5. Conclusions

A. Published Paleozoic maps are not yet accurate enough for tidal dissipation calculations, because the widths of the ocean basins are so uncertain.

B. Paleogeographic maps generally represent the extreme times of sea coverage of continental crust, so tidal computations based on them should not be treated as representing the long-term average.

C. Shelf depths cannot be assumed to be constant through time, and in general may have been much shallower, even though shelf widths were usually greater in the past.

D. The shallower shelf depths and less irregular coastlines of most geologic periods would suggest that tidal dissipation associated with marginal seas may have been less during most geologic intervals than at the presesent time.

Acknowledgments. The maps were drafted by Carol S Kazmer and Rebecca L. Curtis. Technical assistance was provided by Cal C. Plotter and Z. "Rocks" MacHine. General financial support for the Paleogeographic Atlas Project at the University of Chicago comes from the National Science Foundation (Grant EAR-7915133), the Shell Development Company, the Amoco International Oil Company, Mobil Exploration and Producing Services, Inc., the Exxon Production Reserach Company, the Chevron Oil Field Research Company, the Cities Service Company, the Marathon Oil Company, the United States Geological Survey, the British Petroleum Company, Ltd., and the Pennzoil Exploration and Production Company.

References

Alley R, Watts DR (1979) Paleomagnetic investigation of the northern peninsula. Eos Trans Am Geophys Union 60:240

Barron EJ, Harrison CGA, Hay WW (1978) A revised reconstruction of the southern continents. Trans Am Geophys Union 59:436-449

Chang CF, Cheng HL (1973) Some tectonic features of the Mt. Jolmo Lungma Area, southern Tibet, China. Sci Sin 14:247-265

Channell JET, D'Argenio B, Horvath F (1979) Adria, the African Promontory, in Mesozoic Mediterranean palaeogeography. Earth Sci Rev 15:213-292

Curray JR, Moore DG, Lawver LA, Emmel FJ, Raitt RW, Henry M, Kieckhefer R (1979) Tectonics of the Andaman Sea and Burma. In: Watkins JS, Montadert L, Dickerson PW (eds) Geological and geophyiscal investigations of continental margins. AAPG Mem 29. AAPG, Oklahoma

Dewey JF, Pitman WC III, Ryan WBF, Bonnin J (1973) Plate tectonics and the evolution of the Alpine System. Geol Soc Am Bull 84: 3137-3180

De Wit MJ (1977) The evolution of the Scotia Arc as a key to the reconstruction of Gondwanaland. Tectonophysics 37:53-81

Feng YM, Zhu BQ (1980) Melanges and tectonic development of the west Qinling Mountains. Acta Geol Sin 1980:34-43

Irving E (1960-1965) Paleomagnetic directions and pole positions. Parts I-VII. Geophys J R Astron Soc 3:96-111 (1960a); 3:444-449 (1960b); 5:72-79 (1961); 6:263-267 (1962a); 7:263-274 (1962b) (with Stout PM); 8:249-257 (1963); 9:185-194 (1965)

Johnson BD, Powell CM, Veevers JJ (1980) Early spreading history of the Indian Ocean between India and Australia. Earth Planet Sci Lett 47(1):131-143

Klootwijk CT (1979) A review of palaeomagnetic data from the Indo-Pakistani fragment of Gondwanaland. In: Farah A, DeJong KA (eds) Geodynamics of Pakistan. Geol Surv Pakistan, Quetta, pp 41-80

McElhinny MW (1968-1978) Paleomagnetic directions and pole positions. Parts VIII-XV. Geophys J R Astron Soc 15:409-430 (1968a); 16:207-224 (1968b); 19:305-327 (1969); 20:417-429 (1970); 27:237-257 (1972a); 30:281-293 (1972b); 49:313-356 (1977) (with Cowley JA); 52:259-276 (1978) (with Cowley JA)

McKee JW, Jones NW (1979) A large Mesozoic fault in Coahuila, Mexico. Geol Soc Am, Abstr Programs 11:476

Molnar P, Tapponnier P (1975) Cenozoic tectonics of Asia: Effects of a continental collision. Science 189:419-426

Monger JWH, Price RA (1979) Geodynamic evolution of the Canadian Cordillera-progress and problems. Can J Earth Sci 16:770-791

Morel P, Irving E (1981) Paleomagnetism and the evolution of Pangaea. J Geophys Res 86:1858-1872

Norton IO, Sclater JG (1979) A model for the evolution of the Indian Ocean and the breakup of Gondwanaland. J Geophys Res 84:6803-6830

Pilger RH Jr (1978) A closed Gulf of Mexico, pre-Atlantic Ocean plate reconstruction and the early rift history of the Guld and North Atlantic. Trans Gulf Coast Assoc Geol Soc 28:385-393

Ross JRP (1978) Biogeography of Permian Ectoproct Bryzoa. Palaeontology 21:341-356

Schopf TJM (1980) Paleooceanography. Harvard Univ Press, Cambridge p 341

Sclater JG, Hellinger S, Tapscott C (1977) The paleobathymetry of the Atlantic Ocean from the Jurassic to the present. J Geol 85:509-552

Sclater JG, Parsons B, Jaupart C (1981) Oceans and continents: similarities and differences in the mechanism of heat loss. J Geophys Res 86:11535-11552

Scotese CR, Bambach RK, Barton C, Van der Voo R, Zielger AM (1979) Paleozoic base maps. J Geol 87:217-277

Sengor AMC (1979) Mid-Mesozoic closure of Permo-Triassic Tethys and its implications. Nature (London) 279:590-593

Sengor AMC, Yilmaz Y (1981) Tethyan evolution of Turkey: A plate tectonic approach. Tectonophysics 75:181-241

Sheehan P (1973) The relation of late Ordovician glaciation to the Ordovician-Silurian changeover in North American brachiopod faunas. Lethaia 6:147-154

Silver LT, Anderson TH (1974) Possible left-lateral early to middle Mesozoic disruption of the southwestern North American craton. Geol Soc Am, Abstr Programs 6:955

Smith AG, Hallam A (1970) The fit of the southern continents. Nature (London) 225:139-144

Smith AG, Hurley AM, Briden JC (1981) Phanerozoic Paleocontinental World Maps. Cambridge Univ Press, Cambridge, p 102

Vail PR, Mitchum RM, Thompson S (1977) Seismic stratigraphic and global changes of sea level, Part 4: global cycles of relative changes of sea level. IN: Payton CE (ed) Seismic stratigraphy-applications to hydrocarbon exploration. AAPG Mem 26, pp 83-97

Voo Van der R, Mauk FJ, French RB (1976) Permian-Triassic continental configurations and the origin of the Gulf of Mexico. Geology 4:177-180

Ziegler AM, Scotese CR, McKerrow WS, Johnson ME, Bambach RK (1979) Paleozoic paleogeography. Annu Rev Earth Planet Sci 7:473-502

Movements of the Continental Crust and Lithosphere-Aesthenosphere Systems in Precambrian Times

J. D. A. Piper[1]

1. Introduction

Since publication of the first volume of Tidal Friction and the Earth's Rotation, considerable advances have been made in our understanding of the behaviour and relative movements of the Earth's surface lithosphere in Precambrian times. In particular, an improved understanding of the temperature-pressure phase relationships of crustal and upper mantle materials has refined the models of lithosphere-aesthenosphere interactions during earlier phases of the Earth's history (see for example Baer 1977 and Tarling 1980), and a widening of the palaeomagnetic data base has confirmed that these interactions have been radically different during the two major divisions of post-Archaean (<2700 m.y.) times. The present-day scheme of "plate tectonics" incorporating six large rigid plates and a number of smaller plates growing at constructive margins along the mid-ocean ridges, moving relative to one another along transform faults (conservative margins), and being consumed at destructive margins is seen to be only one stage in an evolving system of surface tectonics. It should more specifically be described as "multi-continent" plate tectonics with the Proterozoic (2700-570 m.y.) régime distinguished as "single-continent" plate tectonics. As will be discussed in this paper, these evolutionary changes appear to have been brought about by two important factors: the first factor is the temporal change in the aesthenosphere convection system, and the second factor is the change in the relative strengths of the continental and oceanic parts of the lithosphere.

The transition from the Proterozoic to the Phanerozoic eras at about 570 million years ago, although identified on a faunal basis by the diversification and shelly secretions of the metazoa (multicellular animals), coincided closely with the change in tectonic style. Indeed, these two events seem to have been intimately related because the change in tectonic régime consequent upon continental breakup and dispersal produced environmental and chemical changes favourable to the faunal diversification. This evolutionary development of the

1 Sub-Department of Geophysics, University of Liverpool, Liverpool L69 3BX, Great Britain

lithosphere tectonics also had some important bearings on the distribution and characteristics of shallow seas and hence on the magnitude of tidal friction. The observation that the present tidal retarding couple extrapolated backwards in geological time predicts a critical (and clearly hypothetical) approach of the Earth and Moon at about 1500 m.y. provides one of the major challenges to historical geology. This is namely to provide a qualitative, and as far as possible quantitative, explanation for the change in the tidal retarding couple.

2. The Proterozoic Supercontinent

During the 1970's the extent to which the present-day system of plate tectonics could be extrapolated back into geological time was being very actively debated. One school, impressed by the ability of the theory to incorporate so many diverse aspects of Phanerozoic (570-0 m.y.) geology, has considered it to be generally applicable to Archaean (3800-2700 m.y.), Proterozoic and Phanerozoic times (e.g., Burke and Dewey 1973a, Burke et al. 1976). The opposing school was impressed by the evidence for evolutionary changes in the geological record and argued that the prevailing tectonic regime has changed with time (e.g., Engel and Kelm 1972, Shackleton 1973, Sutton and Watson 1974). These discussions have been resolved, in part, by the results of palaeomagnetism which have shown that the time sequences of palaeopoles from the diverse fragments of the Precambrian continental crust (shields) satisfy a unique apparent polar wander (a.p.w.) path prior to the beginning of Cambrian times and back to the end of Archaean times (which is effectively the beginning of the palaeomagnetic record). However, owing to the statistical nature of this subject and difficulties which are inherent in its practical application, it is not possible to define differences in palaeopoles to better than 5-10° or about 1000 km and thus the scale of internal deformation of the Precambrian crust cannot be well constrained by this method.

Although the results of palaeomagnetism have demonstrated that the continental (sialic) crust formed a single integral unit in Proterozoic times and not a range of dispersed units as at the present time, it does not appear to be in conflict with the major tenets of plate tectonic theory: the driving force of aesthenosphere convection (with probably a contribution from gravity sliding of the oceanic lithosphere) was still the motivator of the tectonic events taking place in the sialic crust and resulted in the consumption of oceanic lithosphere into the mantle. To a limited extent the signature of this consumption is identified around the margin of the Proterozoic crust although the dynamics of this process appear to have precluded the obduction of the oceanic (simatic) crust onto the sialic crust prior to about 1000 m.y. It is argued here that it is the scale of this convection which has changed, while the declining radiogenic heat production within the Earth has resulted in an increase in the thickness and rigidity of the continental and oceanic components of the lithosphere; although the strengths of these two components have both increased progressively, they have increased a different rates so that the interactions between them have changed.

Because the simatic component of the lithosphere was entirely consumed into the lithosphere prior to late Precambrian times, its nature cannot be directly inferred. However, the shape of the deep oceans can be indirectly deduced from the palaeomagnetic evidence relating to the continental crust, and some conclusions about the shelf seas are possible from the record of the sedimentary rocks deposited on the continent; this latter evidence has been discussed in a contribution to the earlier volume (Piper 1978). Here I first outline the palaeomagnetic evidence for Precambrian times because it is this evidence which defines quantitatively the distribution and movements of the continental crust with respect to the ancient magnetic poles. The implications of the combined geologic and palaeomagnetic data to inferring the types of lithosphere and aesthenosphere systems operating during these times are then evaluated. The collective data provide the data base for estimating the importance of the tidal retarding couple in Precambrian times.

The distribution of the sial during the Proterozoic era is illustrated in Fig. 1. This has been derived from comparisons of the palaeomagnetic data between the African and Laurentian (North America-Greenland) shields (Piper 1975, 1976a,b), the Laurentian and Fennoscandian (Baltic) shields (Piper 1976a, Patchett et al. 1978, Piper 1980a) and the Australian and India-Antarctica shields (Piper 1982a). The configuration of the Siberian shield between the African and Laurentian shields was proposed on the basis of similarities in the trends of geological structures and the similarity of the geological histories of the adjacent margins by Sears and Price (1978); it appears to satisfy the palaeomagnetic constraints although the data base is rather weak (Piper 1980b). There are very few palaeomagnetic results from the remaining Precambrian shield of North and South China and Kazakhstania with which to determine their positions with respect to the other shields. They are inserted here in an attempt to satisfy three constants namely (1) to preserve the continuity of the sial, (2) to retain a proximity to the Australian and Siberian shields respectively, with which they are associated in the later transient supercontinents of Pangaea and Gondwanaland, and (3) the continuity of structural trends. The position of a few minor fragments of Precambrian crust can only be speculated on. The most important of these is Madagascar for which the Phanerozoic position in Gondwanaland has been disputed (cf. Embleton and McElhinny 1975, Kent 1973); this includes several relicts of pre-3000 m.y. terrain and a position adjacent to the ca. 600 m.y. Mozambique belt is

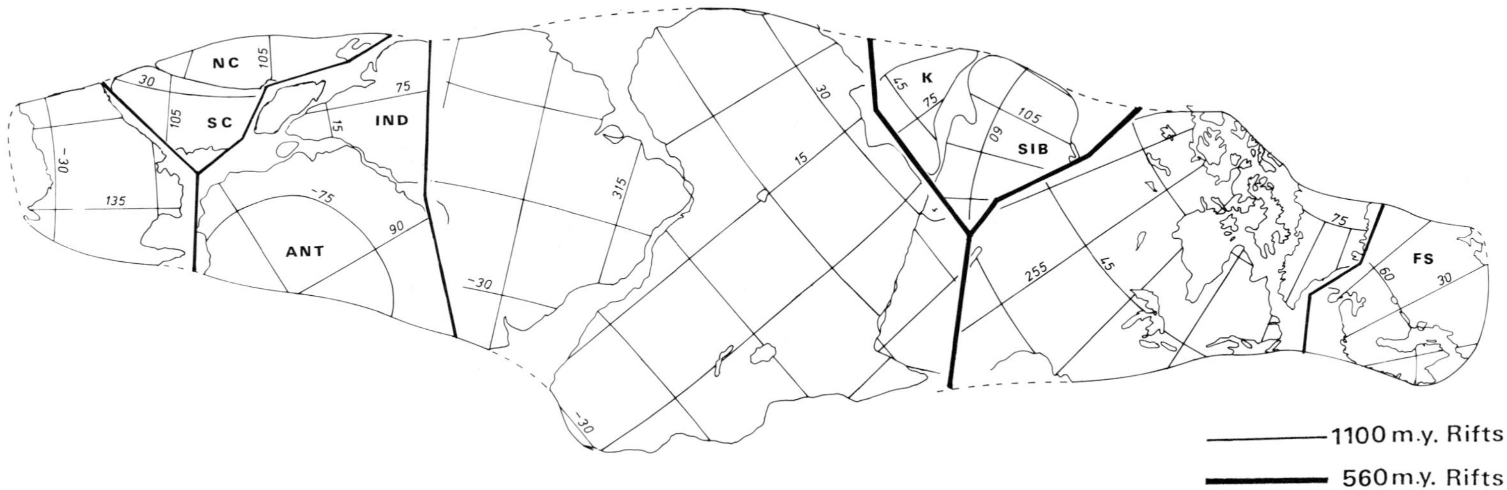

Fig. 1. The Proterozoic Supercontinent defined predominantly from palaeomagnetic evidence (see text). The positions of the North China (NC), South China (SC) and the Kazakhstania (K) Shields are inserted in the positions shown to preserve the continuity of the sialic crust; these positions are supported by some geological data but no palaeomagnetic evidence. Other shields lettered in this figure are: IND India; SIB Siberia, and FS Fennoscandia. The lines of latitude and longitude are shown at intervals of 15° and the figures are in degrees north and east respectively. The positions of these Precambrian Shields on present day coordinates are illustrated in Scotese et al. (1979), Fig. 1

Table 1. The proterozoic supercontinent-Euler poles and angles of rotation with respect to the Laurentian Shield

Operation[a]	Euler pole °E	°N	Rotation angle (°)
Fennoscandia to Laurentia (pre. 1100 m.y.)	8	21	- 41
Fennoscandia to Laurentia (post 1100 m.y.)	274	80½	- 66½
Siberia to Laurentia	309	66	+ 86
Africa to Laurentia	138	73	-146
South America to Laurentia	329½	44	+ 57
and	138	73	-146
India to Laurentia	238	50½	+137
Antarctica to Laurentia	235	-75	-110
Australia to Laurentia	281	45	+116
Kazakhstania to Laurentia	182	78	+161
North China to Laurentia	192	21	+ 96
South China to Laurentia	204	17	+ 98

Rotations are indicated as clockwise (negative) or anticlockwise (positive). The late Precambrian reconstruction of India-Antarctica and Australia with respect to Africa are given by Smith and Hallam (1970).
[a]This operation is the rotation of the Precambrian Shield into its Proterozoic position relative to the Laurentian Shield and yields the reconstruction of the Proterozoic crust shown in Fig. 1.

probably inappropriate so it is included here adjacent to India as in the reconstruction of Smith and Hallam (1970).

The Euler poles and angular of rotation have been calculated in Table 1 relative to the Laurentian shield which is retained in present day coordinates (following closure of the Davis Strait by rotating Greenland according to Bullard et al. 1965). In Figs. 2-9 the remaining shields and their palaeopoles are rotated according to these operations.

3. The Precambrian a.p.w. Path

The a.p.w. path is the resultant path incorporating the palaeomagnetic poles from the continental shields, and it is shown here to define a single integral swathe on the reconstruction of Fig. 1. The movement of the supercontinent is thus illustrated here in terms of the a.p.w. path; this is equivalent to a sequence of palaeogeographic maps, but in place of keeping the rotation axis fixed and illustrating the migration of the continent, the movement of the palaeomagnetic pole (and hence presumably the rotation axis) is shown with respect to the continental crust reconstructed according to the operations of Table 1. The palaeomagnetic poles are derived from a variety of igneous, metamorphic and sedimentary rocks with a very variable control over the age estimates of the magnetisations. Generally the most reliable reference points are provided by those igneous rocks intruded into cold crust at ages defined by Rb-Sr isochron, U-Pb and Pb-Pb studies, and sometimes, by palaeomagnetic contact or fold tests to suggest the primary nature of the remanence. Metamorphic rocks have less well constrained ages of magnetisation in general, but they frequently define sequences of high to low blocking temperature remanence directions acquired in a time sequence which correlates crudely with K-Ar and Rb-Sr mineral ages; they are most useful for defining the magnitude and direction of a.p.w. movements between points defined by first rank dated poles from igneous units. Sedimentary rocks have less direct use in a Precambrian context because the origin of their magnetic remanences is often obscure with the primary depositional and later components proving difficult to isolate as such. In addition, comparable problems relate to radiometric studies of these rocks because they may have become isotopically closed at a variety of post-depositional stages associated with lithification, thermal or tectonic events, and it is frequently only possible to fix the maximum age of the remanence by reference to the age of the basement on which the sediment is deposited; for these reasons data from sedimentary rocks are normally of ancillary value only. The properties and limitations of the palaeomagnetic data have been extensively discussed elsewhere (e.g., Irving and McGlynn 1976, Piper 1978, Dunlop 1981) and need not be further noted here. The limitations have probably been overstressed because there proves to be a remarkable conformity of the majority of the palaeopoles with a single a.p.w. path (Figs. 2-9). In part this is because the multicomponent magnetisation record identified in many Precambrian rocks will lie somewhere along the overall a.p.w. path, although not necessarily at the ages inferred from the regional thermal history or in the sequence inferred from the blocking temperature spectra. In particular, the age sequence of the magnetisations may not be in the high to low blocking temperature sequence if both hematite and magnetite are carriers of remanence because members of the hematite-ilmenite series may form as primary igneous phases or as secondary alteration phases.

3.1 2850-2200 m.y.

The oldest palaeomagnetic data come from relict cratonic terrains that have not been reworked by later tectonic activity (Fig. 2) but which have, in general, been variably subjected to later reheating events. For this reason the nature of some of the magnetic remanences assigned to this interval has been questioned (see Dunlop 1981 for discussion), but key palaeopoles with confirmatory tests come from the Pilbara and Yilgarn cratons of Western Australia, the Kaapvaal and Zimbabwe cratons of southern Africa and the Superior and Slave (Dogrib dykes) cratons of North America. In the Superior craton Irving and Naldrett (1977) and Schults and Dunlop (1981) identify remanence directions predating the intrusion of the ca. 2630 m.y. Matachewan dykes. The path connecting these older poles with the palaeopoles from the dykes is the same as that defined by a sequence of $\geq$2600 m.y. palaeopoles from the Pilbara craton in western Australia (McElhinny and Senanayake 1980). The oldest African pole position from the Modipe gabbro in Botswana conforms with the Laurentian Shield data at ca. 2600 m.y. and the path connecting pre- and post-Matachewan poles from the Laurentian Shield is identified between African poles >2450 m.y. (1-4,7) and <2450 m.y. (5,6) in age. African palaeopoles 2 and 3 from the Great Dyke and related dyke swarms in Zimbabwe define a small loop in the a.p.w. path at 2500 m.y.; Indian poles 1 to 5 from ca. 2500 m.y. rock units also fall on this earlier part of the path but may be younger magnetisations because this area was subjected to regional metamorphism at about 2000 m.y. The rapid shift at ca. 2450 m.y. carries the a.p.w. path close to the present poles. Morris (1977) identified a loop here beginning with palaeopoles from ca. 2350 m.y. volcanics, extending to the pole from the Gowganda "A" remanence (the Gowganda is part of the Huronian sedimentary succession deposited on the uplifted and eroded Superior terrain prior to 2100 m.y. and includes glaciogenic deposits) and back to the Gowganda "B" remanence. This loop includes the palaeopoles from Australian units 4-11, of which four are from dyke swarms specifically assigned to this interval, and African pole 5.

The recognition that these oldest data conform to a single narrow a.p.w. path has shown the continental crust nucleated and stabilised in Archaean times had already been aggregated into the Proterozoic Supercontinent by 2600 m.y. It also constrains the scale of relative motions between the relict cratonic nucleii: although the larger part of the continental crust has been subjected to tectonic reworking in later times, this has not been on a scale sufficient to destroy the conformity of these older palaeomagnetic data with the single a.p.w. path. It would appear to limit the resultant compression and/or extension of the continental body during the Proterozoic era to $\leq$10% (Piper 1982a).

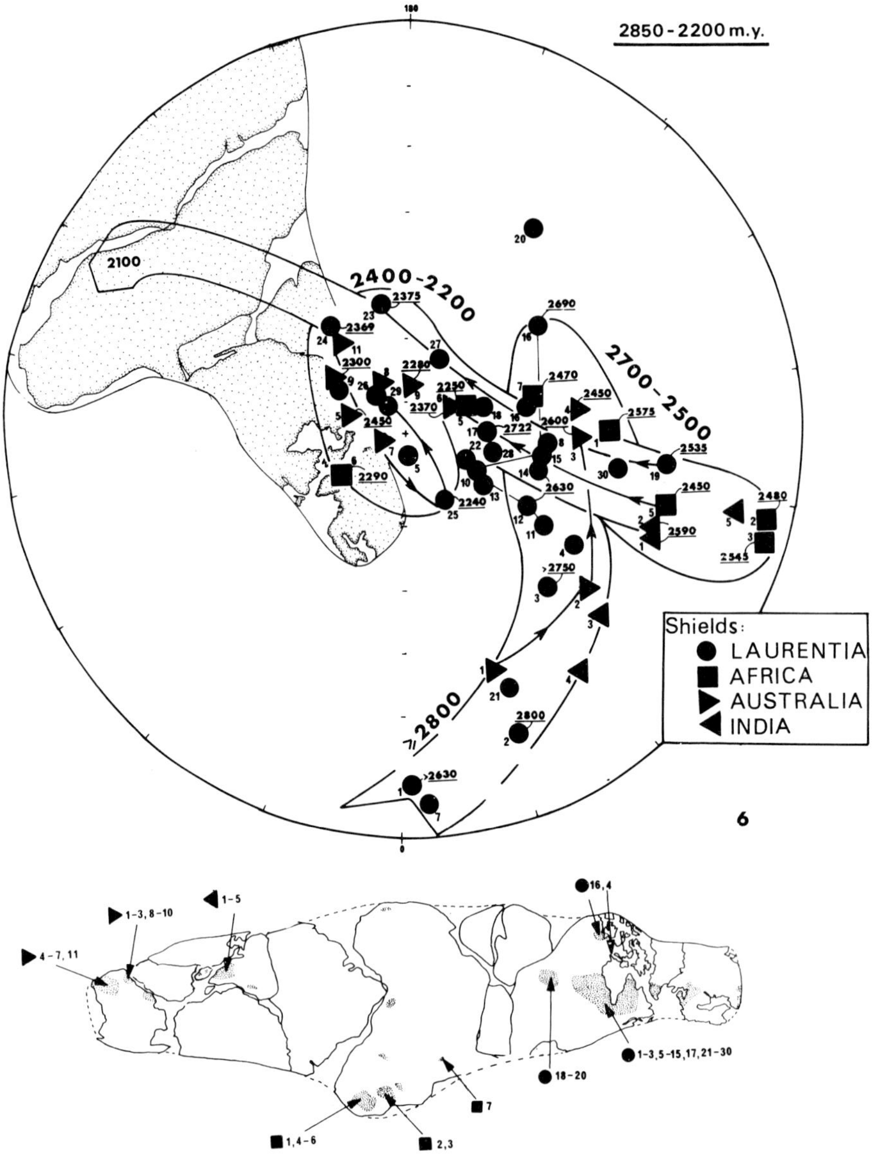

Fig. 2. Palaeomagnetic poles assigned to pre-2200 m.y. times in approximate order of age (where known). Laurentian Shield (•): 1, Kamiskotia Complex (>2630*); 2, Dundonald Sill (~2800); 3, Skead Group (>2750**); 4, Unmetamorphosed Kaminak dykes (2370^{+}); 5, Frechville greenstone; 6, Archaean gabbro; 7, Archaean basalt-gabbro; 8, Pike's Hill lava; 9, Archaean gneiss; 10, Poohbah alkaline complex (2706-2556^{+}); 11-15, Matachewan dykes (2630**); 16, Dogrib dykes (2690**); 17, Chibougamau greenstone sills (2722*); 18, Stillwater Complex (2690*,**); 19,20, Beartooth dykes (2765*, 2450*);

(Legend to Fig. 2, cont.)
21,22, Shelley Lake granite, group 1 and group 2 (2595++); 23, Thessalon volcanics (2325*); 24, Mugford basalts (2320*); 25, Gowganda prefolding "A" remanence (2240*); 26, Gowganda Formation; 27, Firstbrook Formation; 28, Coleman Member (>2100*); 29, Gowganda postfolding "B" remanence (~2200); 30, Cobalt Formation (2650-2100).
African Shield (■): 1, Modipe gabbro, Botswana (2575*); 2, Great Dyke of Zimbabwe (2480*); 3, Satellite dykes of the Great Dyke (2545*); 4, Lower Ventersdorp lavas (2450*); 5, Upper Ventersdorp lavas (2252*); 6, Gaberones Granite (2290*); 7, West Kenya granites (2470*-2580+).
Australian Shield (▸): 1, Duffer Formation (3452**); 2, Duffer Formation "B"; 3, Duffer Formation "C" (>2600); 4, YE dykes (2450*); 5, Ravensthorpe dykes (2450*); 6, Widgiemooltha dykes (2370*); 7, YA dykes (2500-1700*); 8, Mt. Goldsworth lode ore MG3 (3000-2000); 9, BR dykes (2280*); 10, CD dykes (2300*); 11, Koolyan obbing-Dowd's Hill ore (2750-2200).
Indian Shield (◂): 1,2, Quartz-magnetic metamorphic rocks (2590*); 3,4,5, metamorphosed dykes at Chitaldurg, Dharwar and Hyderabab (~2500).
The palaeopoles utilised in this and subsequent diagrams are summarised in the tables to number XVI published at intervals in the Geophysical Journal of the Royal Astronomical Society or in references in Piper (1982a). All poles are plotted after the rotations listed in Table 1 keeping the Laurentian Shield in present coordinates. The method of age determination where known is given as: +K-Ar, ++^{39}Ar/^{40}Ar or K-Ar isochron age, *Rb-Sr mineral or isochron age, **U-Pb; for clarity only some of these ages are shown on the figures. The lines link palaeopoles from the same rock unit and the arrows illustrate the older to younger age relationships indicated by the field relationships or suggested by the blocking temperature spectra. (Latitude markers are in divisions of 20 degrees and the perimeter is the 40° south line of latitude; azimuthal equidistant polar projection). In Figs. 2 to 9 the one and two digit numbers are the pole reference, the underlined four digit numbers are assigned ages in millions of years and the large boxed numbers are general age assignments to those parts of the a.p.w. path

3.2 2200-2000 m.y. (Fig. 3)

At about 2200 m.y. an important change in the character of a.p.w. movements is recognised: after this time movements are considerably more rapid and up to about 800 m.y. they typically define flattened and closed loops (Figs. 3-8) with similar outward and return paths separated by sharp changes in direction or "hairpins" at the extremities. Because these movements are rapid they have proved comparitively more difficult to define, but it is clear that a single a.p.w. path is applicable because the scale of movements is now recognised (with the exception of a few short intervals) in the Laurentian Shield alone. At 2100 m.y. for example, the scale of movements

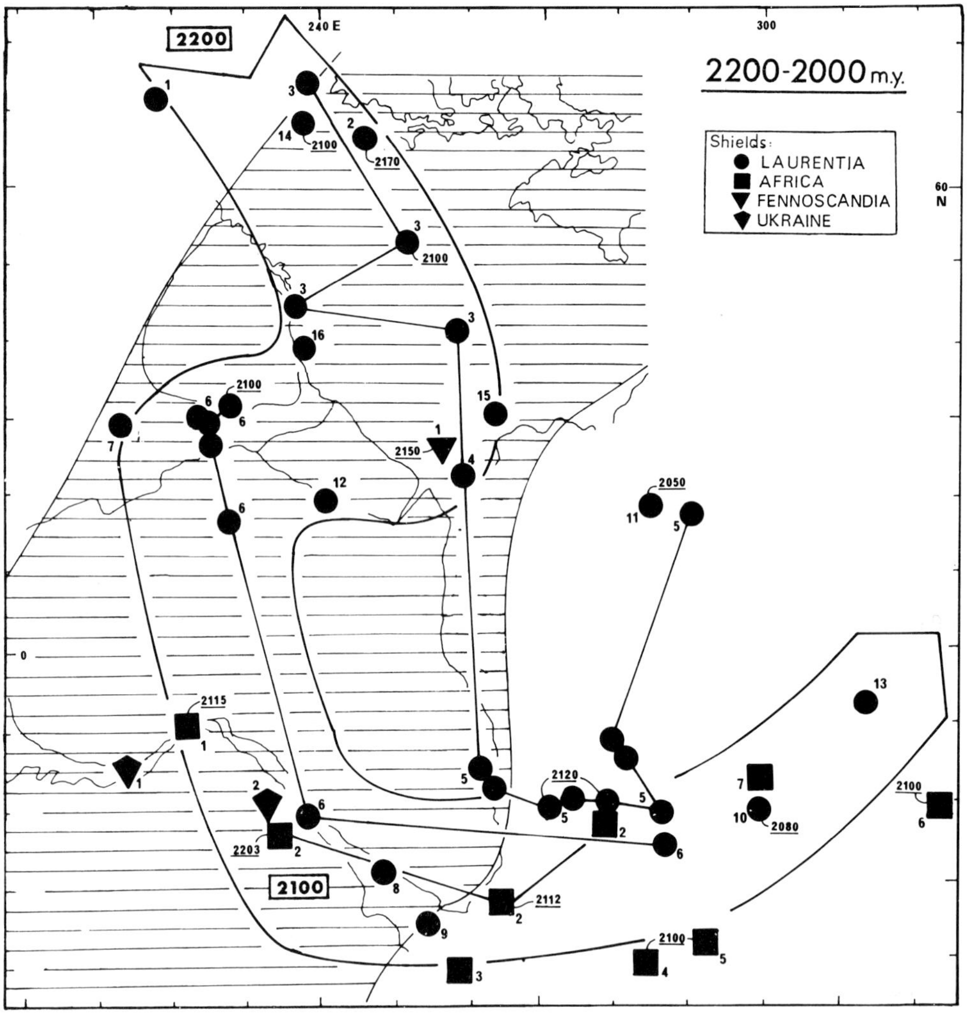

Fig. 3. Palaeomagnetic poles assigned to the interval 2200-2000 Ma. Laurentian Shield (o): 1, Gowganda "B"; 2, Big Spruce Complex "D" (<2171*, 2067*); 3, Nipissing "C" (2120*, 2110*); 4, Cobalt Formation, Nipissing Contact; 5, Nipissing "D" (2120*, 2110*); 6, Abitibi dykes (2210*-2102*); 7, Marathon dykes (2172*); 8, Owl Creek dykes (2090+-1910+); 9, Sokoman iron formation, pre-folding remanence (2200-2100); 10, Indian Harbour dykes (2080+); 11, Indian dykes (2050*); 12, "X" dykes (2160+); 13, Duck Lake sill (2090+-1490+); 14, Otto Stock (2100**); 15, Gunflint Formation; 16, Soudan iron ores and volcanics.
African Shield (■): 1, Tonalites, NW Sahara (2115*); 2, Cunene anorthosite complex (2112*); 3, Transvaal System, lavas (~2203*); 4, Syntectonic igneous rocks, NW Sahara (~2115*); 5, Tarkwaian intrusions, Ghana (~2100+,**); 6, Obuasi greenstone body (~2100+,**);

(Legend to Fig. 3, cont.)
Orange river lavas (2100).
Ukrainian Shield (◆): 1, Amphibolitised diabases (2240^{+}-2180^{+}); 2, Taratash Complex (2080).
Fennoscandian Shield (▼); 1, Jatulian igneous rocks (2150^{*}).
(In Figs. 3 to 8 a simple square plot is used with divisions of latitude and longitude in 10 degree intervals numbered in degrees north and east respectively. The Laurentian Shield is in present coordinates and other shields and poles are rotated according to the operations of Table 1)

is defined by the ca. 2100 m.y. Abitibi dyke swarm and the Nipissing diabase in the Superior craton. Two predominant directions of magnetisation are recognised in the Nipissing diabase; the first (but not necessarily the oldest) is westerly and defines poles numbered 3 in Fig. 3 conforming with the predominant direction identified in the Abitibi dykes, and the second is a more common southerly direction which yields poles numbered 5 in this figure. Contact tests strongly suggest that both of these sets of directions are primary (Roy and Lapointe 1976, Irving and Naldrett 1977, Morris 1979) and indicative of protracted intrusion at ca. 2100 m.y. Although the complete data from these rock units suggest that a continuous a.p.w. path is represented here (Fig. 3), the direction of movement is uncertain. Stability evidence and the wider data base from elsewhere in the Laurentian Shield have suggested the sequence westerly to southerly as plotted in Fig. 3 (Roy and Lapointe 1976), but Symons (1971) recognised the southerly direction in sills and the westerly direction in dykes cutting them. Morris (1981) has since shown that the southerly direction predates intrusion of the Sudbury complex (1960-1780 my.), and it is possible that the westerly Nipissing directions plotted in Fig. 3 belong to the later part of the curve illustrated in Fig. 4, although the Abitibi results confirm that this position is also relevant to 2100 m.y. The path plotted in Fig. 3 is represented in Africa by data including the sequence of palaeopoles in the ca. 2100 m.y. Cunene complex in the Kasai (Congo) craton, poles from the Birrimian terrain at the southern margin of the West Africa craton, and from rock units in the 2150-1950 m.y. Dorsal Eglabs terrain at the northern margin of this craton.

3.3 2000-1750 m.y. (Fig. 4)

At about 2000 m.y. the Archaean Slave terrain at the northern margin of the Laurentian Shield had been elevated close to present surface levels and was being rifted along its margins with the formation of deep rifts (aulacogens) intermittently filled with sediments. Rocks from the Coronation Geosyncline and the Authopuscow aulacogen here define much of the resultant a.p.w. motion between ca. 2000 m.y. and 1750 m.y. and data are plotted as poles 3 to 14 in Fig. 4 (see Evans

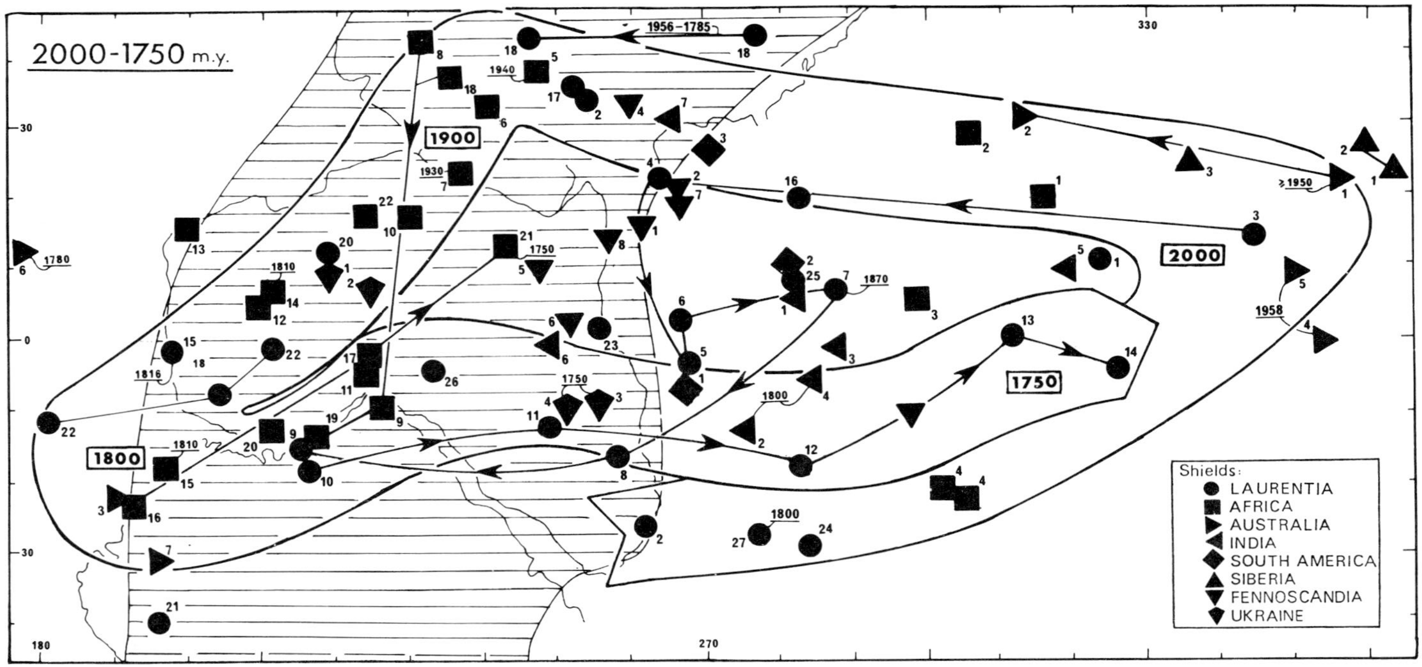

Fig. 4. Palaeomagnetic poles assigned to the interval 2000-1750 Ma. Laurentian Shield (●): 1-2, Nagaunee iron ores (≥1890⁺); 3, Western River Formation; 4, Hornby Bay Group; 5, Akaitcho River Formation; 6, Seton Formation (1794-1770*); 7, Kahocella Group, primary (1870**); 8, Douglas Peninsula Formation; 9, Stark Formation; 10, Tochatawi Formation; 11, Takiyuak Formation; 12, Pearson Formation "A"; 13, Et-Then Group primary; 14, Pearson Formation "B"; 15, Murdoch, Thompson Lake and Menihek Association (1816*); 16, Indin dikes (2050⁺); 17, Otis-gabbro (1930⁺); 18, Sudbury Complex (1956-1785*); 19, Big Spruce "X" (1900-1600); 20, Late N2 diorties, West Greenland, high blocking temperature component (≥1690*); 21, Laxfordian high blocking temperature component, NW Scotland; 22, Nipissing "E" Hudsonian overprint; 23, Cameron Bay

(Continuation of Legend to Fig. 4 on p. 265)

(Legend to Fig. 4, cont.)
Group porphyry (1733^*); 24, Retty periodotite, Labrador "A"; 25, Slave craton overprint (SE directions); 26, Nordfjord gneisses, E. Greenland; 27, Peninsula Sill (1800^{**}).
African Shield (■): 1, Dolerite intrusion, Ivory Coast (1730^+); 2, Cunene anorthosite complex, younger facies (2030^*); 3, Orange River lavas (2100-1850); 4, Aftout plutons ($1940-1910^*$); 5, Aftout gabbro ($1940-1910^*$); 6, Losberg intrusion (1910^*); 7, Bushveld Complex (1910^*); 8, Palabora Complex, pyroxenite + Vera Hill syenite; 9, Palabora Complex, eastern syenite, cross cutting dyke; 10, Vredefort ring complex (1930^*); 11, Diorite intrusions, NW Sahara ($<1910^*$); 12, Main dyke swarm related to southerly extension of Great Dyke, Limpopo overprint (ca. 1900-1800); 13, Great Dyke extension, Ruri + Crystal Springs dyke swarm, (overprint, ≤ 1810); 14, Sebanga dyke ($\sim 1810^*$); 15, Mashonaland dolerites (1810^*); 16-17 Waterberg Formation, lower sites ($1910-1740^*$); 18-20 Palabora Complex groups 1-3; 21, Rust der Winter, acid volcanics (1752^*); 22, Limpopo gneisses "A". Results 18-22 are listed by Morgan and Briden (1981).
South American Shield (◆): 1, Supamo-Pastara granite-greenstone terrain ($2000-1900^+$); 2-3, Imatuca granulite complex ($2055-2000^*$).
Australian Shield (▶): 1-2, Mt. Goldsworthy lode ores MG1 and MG2 (>1950); 3, Koolyanobbing "A" deposit; 4, Mt. Tom Price ore (1958^*-1800); 5, Mt. Newman ore (1958^*-1800); 6, Hart dolerite (1780^*); 7, YF dykes (~1700).
Indian Shield (◀): 1-4, Gwalior Traps (1800^*); 5-6 Viskhapatnan charnockites (~1800); 7, Kondapalli Charnockites.
Fennoscandian Shield (▼): 1, Central Karelia dolerites; 2-3, Central Karelia sandstones (1870-1610); 4, Northern Karelia sandstones (1870-1610); 5, Kola Peninsula, gabbro-norite intrusions; 6, Omega gabbro-dolerite intrusions; 7, Shoksha Group, South Karelia; 8, Pedalsel'sk and Pukhta Groups.
Siberian Shield (▲): 1-2 Khapchan Series; 3, Quartz-gabbro stock ($2085-1850^+$).
Ukrainian Shield) (◆): 1, Bug-Podolia Series gneisses (2000^+); 2, Oktyabr'sk alkaline massif (2060^+); 3, Korosten Complex (1750); 4, Korosten granite (1750)

---

et al. 1980, McGlynn and Irving 1978). The earlier part of this path incorporates ca. 2000 m.y. data and the sequence of magnetisations in the Sudbury complex. The path connecting Laurentian poles 17 and 18 with the highest blocking temperature components in ca. 1800 m.y. Husonian mobile belts (20,21), a probable overprint of this age in the elevated Superior craton (23), and the extremity of the Coronation-Authapuscow loop (Stark and Tochatwi poles 9 and 10) is identical to that defined from the African Shield between 1900 m.y. intrusive complexes of southern Africa (6-10) and the 1810 m.y. Mashonaland dolerites (15). This part of the path is also defined by uplift magnetisations related to ca. 2000 m.y. metamorphism in the Limpopo mobile belt separating the Zimbabwe and Kaapvaal cratons, and the sequence of magnetisations in the Palabora complex of South Africa (Morgan and Briden 1981). A common change in ambient field polarity from poles to antipoles (as plotted in Fig. 4) is represented here in the data from the Laurentian, African and Australian Shields (Piper 1982a).

The return part of the Coronation-Authapuscow loop links the highest blocking temperature magnetisations in the Hudsonian mobile terrains (Piper 1981a) with the lower blocking temperature and uplift-related magnetisations in these terrains, and is also defined by the sequence of magnetisations in the lower part of the ca. 1760 m.y. Waterberg sediments in southern Africa; the collective age assignments of the data plotted here suggest that this part of the path was executed rather rapidly at about 1750 m.y. (Fig. 4). The interval represented here also includes data from ca. 1800 m.y. metamorphic terrains in India and South America, and from pre-Svecokarelian (>1800 m.y.) terrains in the Fennoscandian Shield.

3.4 1750-1520 m.y. (Fig. 5)

The palaeopoles defining the a.p.w. motion during this interval come predominantly from the large tracts of crust affected by tectonic and thermal reworking at about 1800 m.y. and correlating with the "Hudsonian" event in the Laurentian Shield. Many of the magnetisations were acquired during uplift and cooling following this event, and in West Greenland for example, the Laurentian sequence 1-6 in Fig. 5 represents a time-sequence acquired during cooling of a ca. 1 km thick section of metamorphic terrain (Piper 1981a); this path would appear to represent an a.p.w. rate of 1-2 ^{o}/m.y. and a local uplift rate of about 10-20 m/m.y. at ca. 1700 m.y. The same sequence is represented in outline by magnetisations in the Labrador Trough (Park 1977) and the Churchill terrain (Park 1975) elsewhere in the shield, and the path incorporates a large number of individual determinations from other parts of the Hudsonian terrain (Morgan 1976). The youngest part of the West Greenland sequence just overlaps (Fig. 5) with the sequence of ca. 1650 m.y. magnetisations from the analogous terrain in N.W. Scotland (corrected for Phanerozoic drift) which appear to define the extremity of the 1750-1600 m.y. loop illustrated in this figure. It is possible that the collective path illustrated here should be extended 30° to the SE to incorporate the antipoles of the Laurentian data points 47, 53 and 54 as plotted because tests suggest the pole 47 at least, is primary and related to a 1750 m.y. age (Schmidt 1980), and because the polarities are mostly opposite to the Lewisian poles 43-45. This consideration would not influence the inter-shield comparisons because it applies to data from one shield alone, but the a.p.w. movement illustrated here should be regarded as a minimum.

The bulk of the data from elsewhere relevant to the interval come from the Svecofennian and Svecokarelian terrains of the Fennoscandian

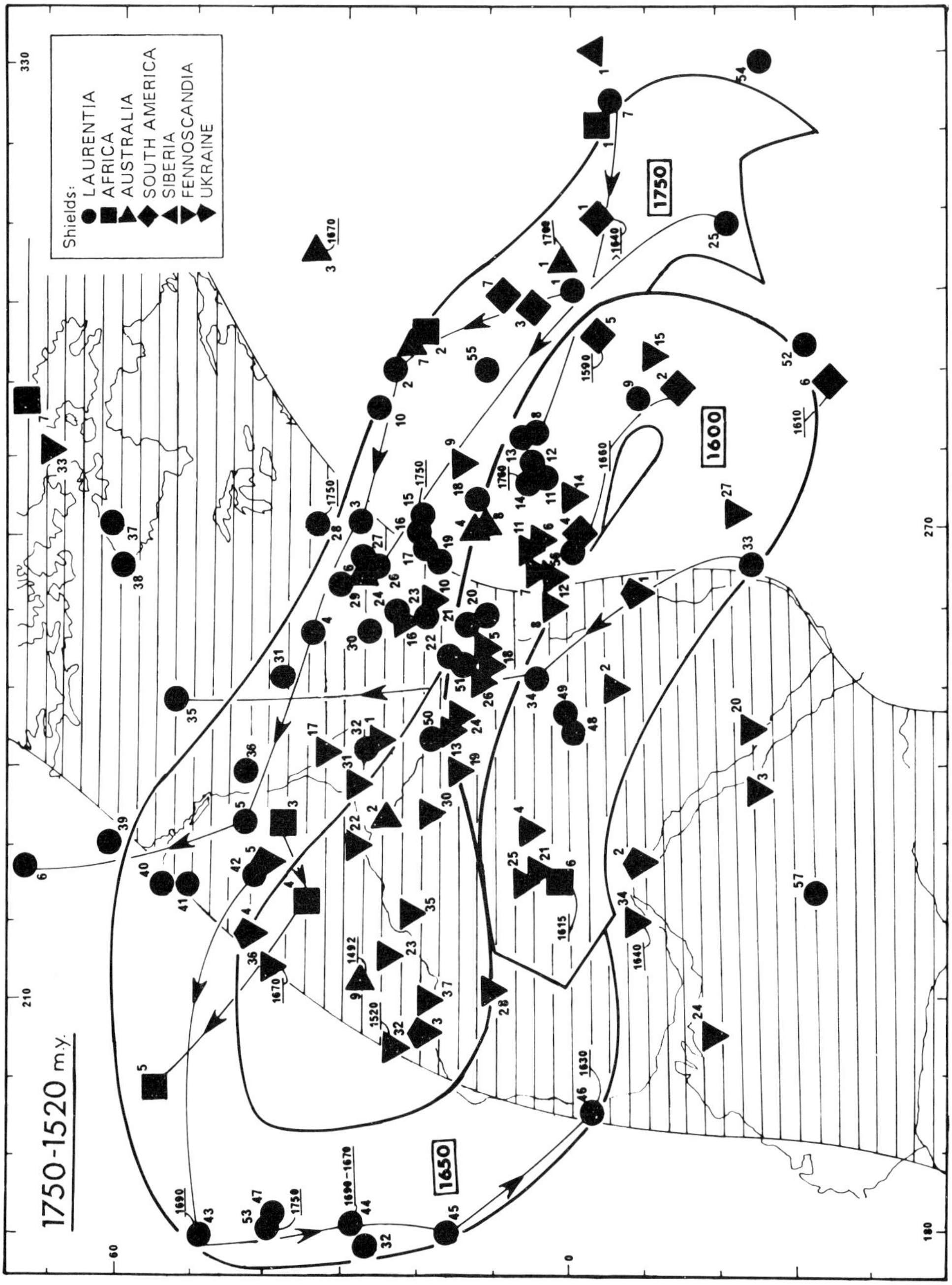

Fig. 5 (Legend see page 268)

Fig. 5. Palaeomagnetic Poles assigned to the interval ca. 1750-1500 Ma.

Laurentian Shield (•): 1-6, Sequence of uplift magnetisations, Nagssugtoqidian terrain, W. Greenland (~1700+); 7, Pearson Formation "B"; 8, Pearson Formation "C"; 9, Martin Formation (1830-1650+); 10, Kaellinghaetten gneiss, W. Greenland; 11, Retty peridotite, Labrador "B"; 12, Menihek Formation, Labrador "B"; 13, Big Spruce Complex, "Y"; 14, Dubawnt Group (1760*); 15, Itivdleq dykes (~1750+); 16, Coronation geosyncline overprint; 17, Metamorphosed Kaminak dykes (1900-1620+); 18, Nonacho Group (1850-1750+); 19, Kahocella secondary; 20, Daly Bay metamorphics "A" (1690-1590+); 21, Sokoman iron formation, post-folding remanence; 22, Itivdleq dykes (1790-1650+); 23, Kangamiut dykes south of Nagssugtoqidian Front; 24, Amitsôq gneisses; 25, Flin-Flon secondary (~1760*); 26, Flin-Flon Tertiary (1700-1600+); 27, Gunflint Formation (≥1600*); 28, Spanish River Carbonatite (1750*); 29, Nordre Strømfjord gneisses; 30, Angmagssalik gneisses (1800-1600+); 31-32, Molson dykes (1650-1280+); 33, Seward Formation "A"; 34, Seward Formation "B"; 35, Seward Formation "C"; 36, Wind River dykes (1880-1680+); 37, Sagdlerssuaq dykes (1620++); 38-39, Ketilidian metavolcanics (1700-1600+); 40, Amphibolitised dykes, north of Nagssugtoqidian Front; 41, Sarfanguaqland amphibolites; 42, Lewisian "A2" (1650-1625+); 43, Lewisian "A3" (1690-1680+); 44, Lewisian "A4" (1690-1670+); 45, Lewisian "A5" (1500-1470+); 46, Lewisian "B" (1630*); 47, Eskimo Volcanics, Northwest Territories (1750*-1625+); 48, Flaherty Volcanics (<47, 1690+-1640*); 49, Haig intrusions (1790*-1620+); 50, Eskimo Volcanics, secondary; 51, Cape Smith Metavolcanics (1650-1450+); 52, Lewisian Outer Hebrides (1720-1500*+); 53, Richmond Gulf Volcanics; 54, Richmond Gulf-Manitounuk Island Volcanics; 55, Sparrow dykes (1550-1390+, 1700++); 56, Beartooth dykes, 7 and 8 (1500+); 57, Iron Mountain ore. It is possible that the lobe in the a.p.w. path at ca. 1750 m.y. should be extended 30° to the SE to accommodate the antipoles of 47 and 53 (see text).

African Shield (■): 1, Dykes intruding Gueld el Hadid Series (1760-1350+); 2, Dykes intruding d'Akileb Deilel Series (1760-1350+); 3-5, Waterberg Formation, higher sites (<1760**Ma); 6, Van dike Mine, dolerite dyke, South Africa (1615*); 7, Limpopo overprint "B".

South American Shield (◆): 1, Cudivero-Caicara, acid metavolcanics (1960-1640*); 2, Roraima Sandstone (1660*); 3-4, Roraima intrusive suite (1660*); 5, Parguaza pluton (1590 -1524**,+); 6, Telles Pires (1610*); 7, Beneficente Group (1480*).

Fennoscandian Shield (▼): 1-2, Vuollerim gabbro; 3, Harads amphibolites; 4, Radmanso gabbro-diorite; 5, Tärendö Gabbro; 6, Tärendö acid rocks; 7, Ylivieska Gabbro; 8, Pohjanmaa gabbro-diorite; 9, Tammela gabbro-diorite; 10, Mikkeli gabbro-diorite; 11, Hyvinkää gabbro-diorite; 12, SW Finland gabbro-diorites; 13-16, Kallax Gabbro; 17, Korstrask Gabbro; 18, Niemisel Gabbro; 19-20, Sangis Gabbro; 21-24, Gallivare basic rocks; 25-26, Stora Lulevatten gabbro; 27-29, Jokkmokk basic rocks; 30, Loftahammar gabbro (1660-1470*+); 31, Uppsala metabasites "A" (≥1570*); 32, Hälleforsnäs giant dyke

(Legend to Fig. 5, cont.)
"A" (1520*); 33, Hame dolerties (1800-1640*); 34, Kumlinge dolerites (1640**); 35, Föglö dolerite; 36, Åva intrusive (1700-1670*); 37, Upper Dala Volcanics (≤1634*).
Australian Shield (►): 1, YD dykes (~1700); 2, YC dykes (≤1500,<YD); 3, GB dykes (1670*); 4, GA dykes (1470*); 5, Iron Monarch, positive group ore (1750-1510); 6, Iron Monarch, negative group ore (1750-1510); 7, Iron Prince ore (1750-1510); 8, Gawler Range Volcanics (1525*); 9, Lunch Creek lopolith (1492*).
Siberian Shield (▲): 1, Diabase and diorite, Aldan region (1590+).
Ukrainian Shield (◆): 1, Uman granite; 2, Turchinka gabbro; 3, Korosten gabbros (1550+-1650+); 4, Ukrainian diabases 1550+-1650+); 5, Olivine diabases and lamprophyres (1770+); 6, Taratash Complex, secondary (1550+)

Shield which were mobilised at about the same time as the Hudsonian terrains of the Laurentian Shield, and the magnetisations are linked to the interval between late tectonic events at deeper crustal levels dated ca. 1780 m.y. and igneous intrusion into stabilised crust at higher levels (in an "anorogenic" environment) between 1640 and 1530 m.y. There is a close correlation between these data and the Laurentian record at 1650-1600 m.y. notably between the path defined by Fennoscandian poles 36(1700-1670 m.y.), 34(1640 m.y.) and 30(~1660 m.y.) to poles from northern Sweden (13-29) linked to the interval 1690-1530 m.y., and coinciding with the younger Laurentian polar path (43→46, 48-50, 53 and 55) plotted in this figure. Data from the other shields are rather few, but the single path includes the African sequence from the upper part of the Waterberg succession (3→5) and dyke swarms (1,2 and 6), Australian dyke swarms (1→2,3) and hematite ores (6 and 7), and in South America by a range of ca. 1600 m.y. poles from Suriname and Venezuela (Onstott and Hargraves 1981).

3.5 1450-1125 m.y. (Fig. 6)

It is not yet possible to define the a.p.w. path between 1520 and 1450 m.y. in detail because few data points can confidently be assigned to this interval. It presumably commences with Fennoscandian pole 32 (1520 m.y.) and Australian pole 9 (1492 m.y.) in Fig. 5, and may incorporate Australian pole 4 (the error estimate on the age of the latter is large) and Laurentian pole 23 in Fig. 6. At 1450-1400 m.y. however, it is closely constrained by poles from anorogenic complexes intruded in the Laurentian Shield (poles 1-9 and 13), and the path moves via the slightly-younger poles from comparable Fennoscandian complexes (2-5, 1390-1290 m.y.), Australian poles 1 and 2, and African pole 5 to African pole 1 at ca. 1280 m.y. The direction of movement along this part of the path is also defined by the magnetisations in Jotnian sediments and lavas of Sweden, and probably by Ukrainian sediments although the ages of the latter are

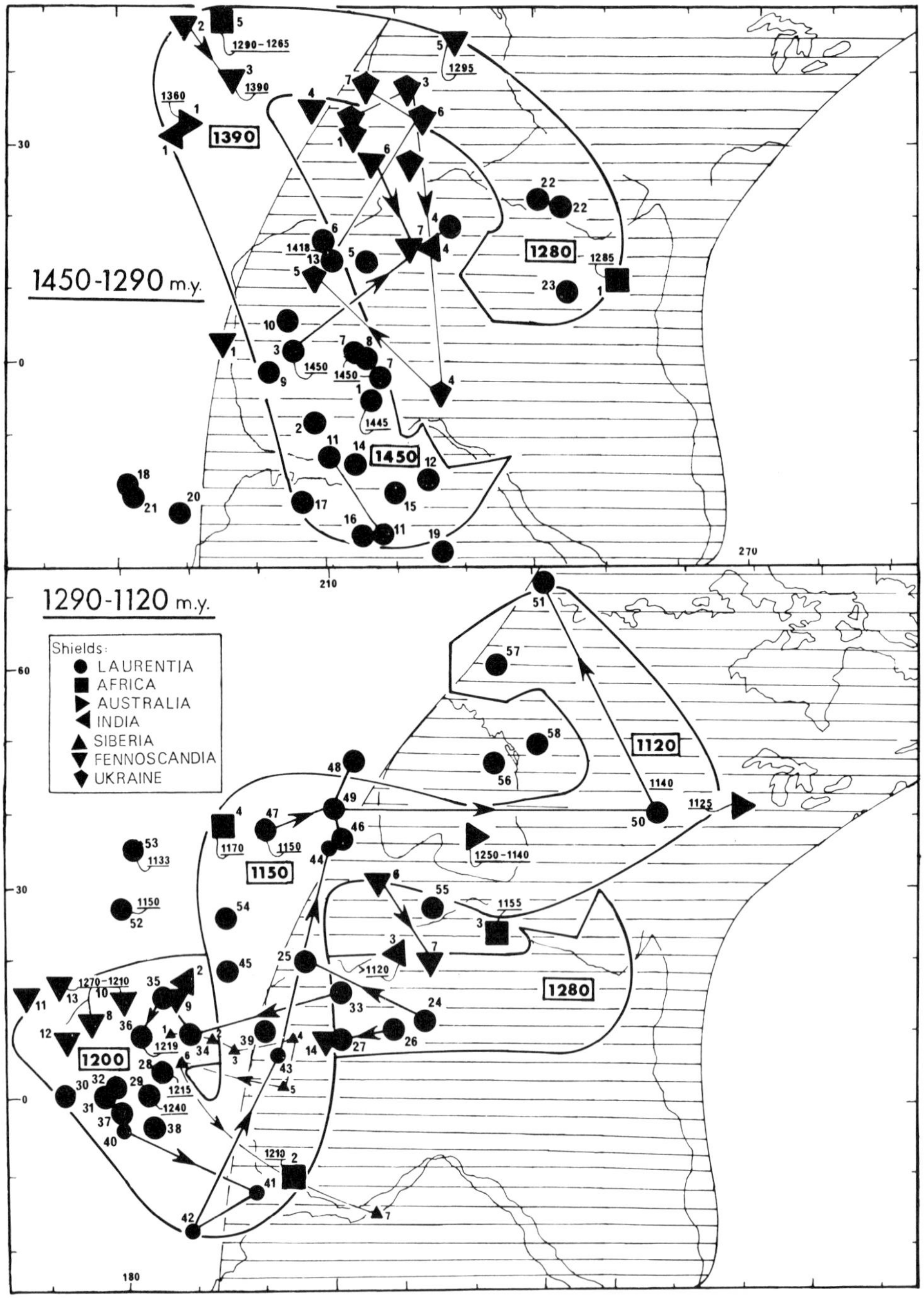

Fig. 6. (Legend see opposite page)

Fig. 6. Palaeomagnetic poles assigned to the interval 1450-1120 m.y. Laurentian Shield (•): 1, Croker Island Complex (1445*); 2, Sherman granite (1380*); 3, Harp Complex (1450*); 4, Harp dykes (<3); 5, Harp country rocks; 6, Arbuckle granites (1400-1320*); 7, Michikamau anorthosite (1450-1400^{+}); 8, St. Francois rocks (1380-1250*); 9, Mistatin pluton (1310*); 10, Seal Lake Group redbeds (1320); 11, Sibley Group (1340*); 12, Spokane Formation argillites (1300?); 13, Nain anorthosite (1418*); 14-16, Upper Belt Supergroup, Klintla, Shepard and Purcell Formations; 17, Missoula Group, Montana; 18, Grinnell argillite, Montana; 19, Pioneer shale, Apache Group, Arizona; 20, Uinta Mountain Group, Utah; 21, Big Cottonwood Formation, Utah; 22, Baron-quartzite, Wisconsin and Sioux quartzite, Minnesota; 23, Western Channel Diabase (1392*); 24, 25, Lower, Upper Gardar lavas (~1283*); 26, 27, Lower, Upper Powdermill lavas (1220-1200); 28, Muskox intrusion (1214*); 29, Coppermine Group (1240*); 30, Mackenzie diabases (1250 -1230*); 31, Laramie Range anorthosite; 32, Rama diabase (1200-850^{+}); 33, BDO dykes 1251*-1212*); 34, 35, Mid-Gardar lamprophyre dykes (1249 -1228*) and Gardar NNW dolerite dykes (<34); 36, Kûngnât ring dyke (1219*, <35); 37, Sudbury dykes (1205*-1189*); 38, Savage point sills; 39, South Trap Range lavas (1190*); 40-44, Grand Canyon Supergroup, stratigraphic succession of poles from lower formations; 45, Seal Group "B"; 46, Narssaq gabbro; 47, Hviddal Giant Dyke (1150*); 48, Gardar gabbro giant dykes; 49, NE dykes Tugtotôq (1150*-1140*); 50, 51, Ilîmaussaq marginal syenites and fractionates (1140*,1120*, <49); 52, Gila Country diabase sills (1140^{+}, 1150*); 53, Logan diabase (Canada, 1133); 54, Logan diabase (U.S.A.); 55, Aillik dykes; 56, Logan dykes reversed; 57,Alona Bay lavas; 58, Baraga County dykes.
Indian Shield (<): 1, Newer dolerites, Singbhum; 2, Bhima sediments (1300-1200); 3, Kaimur Sandstone (>1120*); 4, Cuddapah Traps (1400-1200).
African Shield (■): 1, Pilanesberg dykes, (1282*); 2, Kisii Series lavas (1210^{++}); 3, National kimberlite (1155*, 1140*); 4, Premier Mine kimberlite (1220-1170*); 5, Barby Formation lavas (≤1290*, 1265*).
Fennoscandian shield (▼): 1, Jotnian sediments, Finland; 2, Nordingrå gabbro-anorthosite (1385*); 3, Nordingrå granite (1385*); 4, Gavle granite; 5, Ragunda complex (1295*); 6, Jotnian (Dalarna) sediments; 7, Jotnian (Dalarna) lavas; 8, Nordingrå Complex, Jotnian remagnetisation; 9, Vaasa dolerites, Finland (1270* , 1225*); 10, Satakunta dolerties, Finland (1236*, 1225*); 11, Market dolerties, Finland (1270**); 12, Post-Jotnian dolerites, East Sweden (1254^{++}); 13, Post-Jotnian dolerites, West Sweden (1290 -1210*); 14, Post-Ragunda complex dykes (<1295*). (More recent Rb-Sr dating of unit 2 suggests that it may be as old as 1500 m.y. and relevant to the position of Australian pole 9, and Fennoscandian pole 32 in Fig. 4).
Ukrainian Shield (◆): 1, Avzyan Group, Malvinzer and Kataske sequences; 2, Avzyan Group, Green and Ushakovo sequences; 3, Avzyan Group, combined result; 4, Avzyan Group Revet sequence; 5, Zil' Merdak Group, combined result; 6, Katav and Zil' Merdata Groups; 7, Katavsh Group. These sedimentary sequences are constrained by K-Ar

(Continuation of Legend to Fig. 6 on p. 272)

(Legend to Fig. 6, cont.)
ages in the range 1350-940 Ma.
Australian Shield (▶): 1, Morowa lavas (1360*); 2, Giles Complex (1250+-1140+); 3, Lake View dolerite (1125*).
Siberian shield (▲): 1, Oslgansk Series; 2-4, Tingusik Series; 5-7, Sukhopit Series. (These sediments are constrained within a maximum age range of ca. 1300-930 m.y. and may not belong to this sector of the a.p.w. path)

not closely defined. At about this time the a.p.w. path is again well defined by results from the extensive igneous events related to rifting of the Laurentian and Fennoscandian Shields which ultimately culminated in breakup at about 1000 m.y. The early Gardar Province (Laurentian poles 24 → 25) and Keweenawan (26 → 27) palaeopoles define the return path of the loop illustrated in Fig. 6 with the sequence of poles from intersecting dyke swarms 1250-1200 m.y. in age from southern Greenland (33 → 36). At 1220-1190 m.y. the palaeopole position is defined by the extensive dolerite magmatism collectively grouped as the Mackenzie event in the Canadian Shield where it includes an acurate dyke swarm spanning the shield from its northern margin (poles 28 and 29 are derived from lavas and intrusions at this extremity) to its southern margin (pole 37), the mid-Gardar event in southern Greenland, and the post-Jotnian event in the Fennoscandian Shield (poles 8-13). With the exception of a single African pole (2) and (probably) palaeopoles from sediments of the Siberian Shield (1-7), this part of the path does not appear to be represented in the record from the other shields.

Over this general time interval the western margin of the Laurentian Shield was also being subjected to rifting and aulacogen formation with deposition of thick sedimentary sequences. Palaeopoles from these successions (12, 14-23) fall along the 1400-1200 m.y. path in conformity with a general assignment to this interval, and the lower part of the Grand Canyon succession (Elston and Grommé 1974) defines a sequence of palaeopoles beginning at ca. 1220 m.y. (Fig. 6) and defining a migration towards the Laurentian Shield which is also defined by 1150-1120 m.y. poles (46 → 50) from the last phases of Gardar magmatism in Greenland.

3.6 Continental Breakup and Relative Movements at ca. 1100 m.y.

The palaeopoles 9 to 14 from the post-Jotnian dolerites in the Fennoscandian Shield dated by reference to age determinations in the range 1290-1180 m.y., are the youngest palaeopoles to conform with the primitive reconstruction of Fig. 1 (Patchett et al. 1978) and post-1050 m.y. poles from this shield are found to conform to a different reconstruction (Figs. 22, 23 and Table 1). The resultant motion within this interval was a clockwise rotation of the Fennoscandian

Shield by about 110° relative to the Laurentian Shield so that they were again contiguous in a different configuration (Patchett and Bylund 1977 and Fig. 22): this configuration was retained until Lower Cambrian times (Fig. 9). The breakup of this part of the supercontinent thus followed closely on the extensive 1250-1150 m.y. magmatism characterised by widespread intrusion of dolerites as dyke swarms and other types of intrusion, the development of central alkaline complexes, and rifting, with the latter being associated with deposition of coarse fluviatile deposits and eruption of lavas (Bridgwater and Windley 1973). It is certain that this rifting, which commenced prior to 1280 m.y., is associated with this first brittle breakup of the continental crust because igneous activity occurred intermittently and with increasing intensity until disruption was complete. In the Keweenawan rift zone extending from the Greenville mobile belt via Lake Superior into the Laurentian Shield (Fig. 8) volcanism continued until about 1090 m.y. and sedimentation continued for some time after this. This rift zone is now seen as a failed arm of a larger triple junction (Burke and Dewey 1973b) with the successful arm being the rift between Greenland and Fennoscandia along which breakup ultimately took place. The Torridonian sediments of N.W. Scotland were deposited as a coarse clastic wedge derived from the north west on this subsiding coastal margin between 1100 and 1000 m.y. (see Fig. 7), and encroachment of the sea onto this shelf is identified by a change from fluviatile to marine conditions at the top of the succession (Williams 1966).

It is regretable that there are few data points from Australia and India relevant to the interval 1450-1120 m.y. plotted in Fig. 6 because the primitive reconstruction of Fig. 1 is also inapplicable to the Late Precambrian data from these shields (McElhinny and Embleton 1976). Australian poles 1-3 in Fig. 6 appear to conform to the primitive reconstruction until 1125 m.y., and Indian poles may be relevant until about 1150 m.y., but since no precise age estimates are available for these data it is not possible to be sure of this. There are also very few data points from these shields until 750 m.y. to constrain the post-1100 m.y. movements but minor evidence (Antarctica pole 1, Indian pole 1, and Australian pole 1 in Fig. 7) suggest that the Gondwanaland reconstruction may have been approximately achieved by about 1000 m.y. The geological record for the breakup of these shields is also less well defined than it is between the Laurentian and Fennoscandian Shields, and the only conclusions that can be drawn from the present data (cf. Figs. 6 and 9) are that the Australian and India-Antarctica shields broke off from the remainder of the supercontinent after about 1150 m.y. and prior to 750 m.y., and moved into the conventional Gondwanaland reconstruction (Smith and Hallam 1970). We have as yet, nosignificant information bearing on the size of the oceans which opened and closed during these ca. 1100 m.y. movements, but comparison of the continental geometries before and after these events suggests that they essentially involved a "shuffling" of the dismembered blocks with relative movements taking the form of rotations and strike-slip motions rather than a complete decoupling with the formation of intershield oceans. Since the signatures of subduction beneath the continents were already becoming evident by this time (Sect. 5), this inference would be supported by the absence of obducted ophiolites at 1100 m.y. The resultant effect of this first continental breakup was to shorten the long axis of the continental body from about 220° to about 180° of arc and produce a clustering of the Precambrian Shields.

3.7 1125-1000 m.y. (Fig. 7)

Rocks associated with the Keweenawan rifting and dated 1150-1050 m.y. define the a.p.w. path shown in Fig. 7 which illustrates a reversal of magnetisation specifically defined from igneous units (DuBois 1962, Pesonen and Halls 1979). The normal and reversed axes of magnetisation in these units are significantly non-antiparallel, a phenomenon also observed in the older Gardar lavas (Laurentian poles 24 and 25 in Fig. 6). Because the stratigraphic relationships of the Keweenawan lava flows illustrating this phenomenon are not completely resolved, the contribution of a non-axially geocentric dipole source to this effect cannot be separated at present from the contributions of rapid a.p.w. movements and secondary magnetic components (Palmer et al. 1981, Pesonen and Nevanlinna 1981). However, any complex dipole model for this period must accomodate the observation that the sequence of 1150-1050 m.y. palaeopoles from southern Africa in Fig. 7 falls precisely along the path defined by the Laurentian data and illustrates the same change in ambient magnetic field polarity; this latter correlation between the African and Laurentian data is the best observed since the pre-1750 m.y. data in Fig. 4 and clearly illustrates that the same reconstruction is applicable to these younger data.

The direction of a.p.w. motion along this path is identified by the sequence of poles from a number of igneous units in the Keweenawan rift (4 → 36, 5 → 24, 6 → 35, and 10 → 11) and from the Jacobsville sedimentary succession (43 → 46, Roy and Robertson 1978) deposited near the top of the rift succession. In north west Scotland two segments of this path at about 1100 and 1040 m.y. are defined by the sequence of poles in the Lower (16-20) and Upper (47-54) Torridonian succession (Smith et al. 1982). The highest parts of this succession define the hairpin in the a.p.w. curve at about 1040-1020 m.y. and the migration towards the post-1020 m.y. loop illustrated in Fig. 8. The upper part of the Grand Canyon sedimentary succession defines a larger part of this same path and the same migration towards this loop (Elston and Grommé 1974).

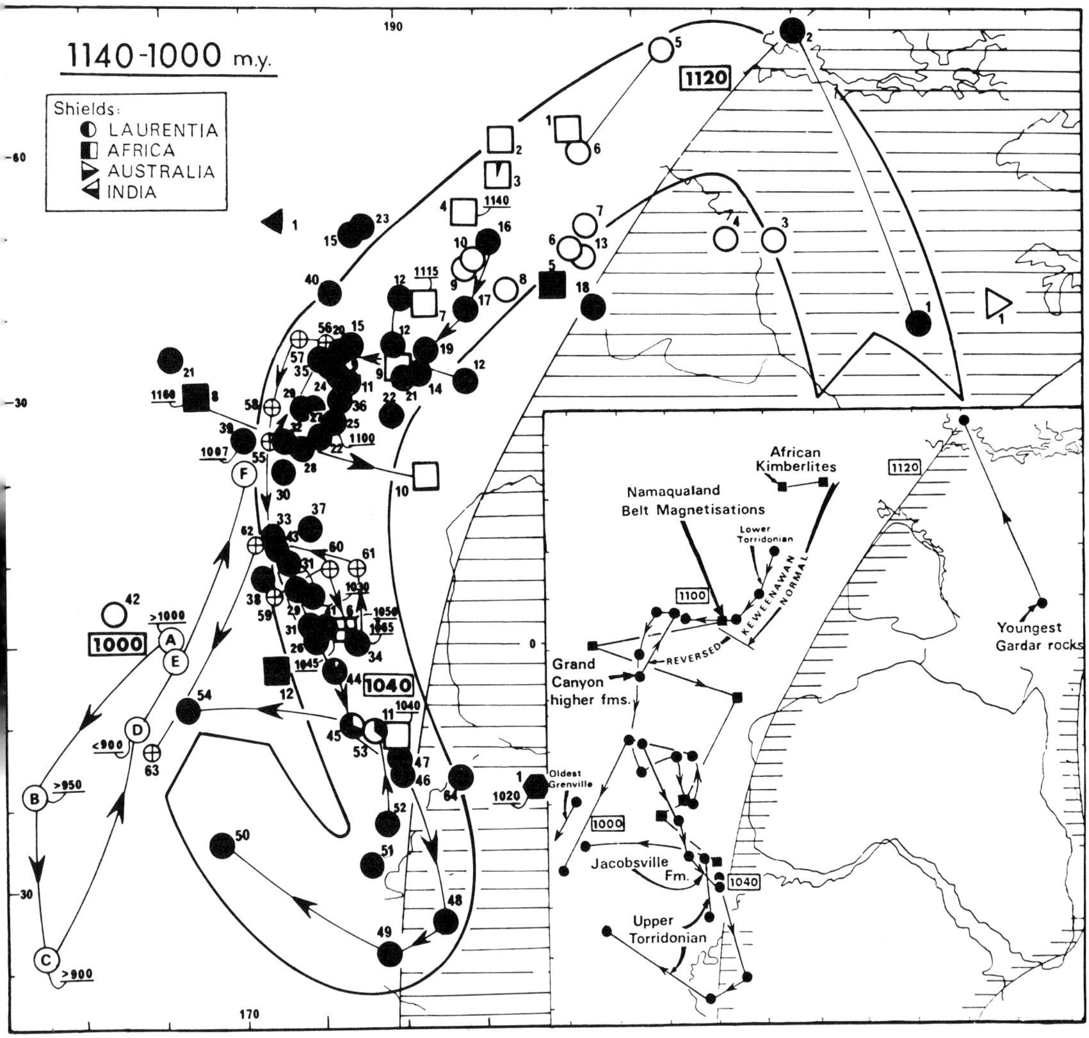

Fig. 7. Palaeomagnetic poles assigned to the interval 1120-1000 Ma. Laurentian Shield (●): 1-2, Ilîmaussaq intrusion (1140*, 1120*); 3, Baraga County dykes; 4, Mamainse lavas (1040*); 5, Lower Cape Gargantua volcanics; 6, Logan dykes (reversed group); 7, Marquette County dykes; 8, Duluth Complex (reversed sites); 9, Osler lavas (1115+); 10, North Shore Volcanics (reversed group, 1135); 11, North Shore volcanics (normal group, 1105); 12, Duluth complex (three poles, normal group, 1115+); 13, Thunder Bay dykes (reversed 1135+, normal 1170-1150*); 14, Keweenawan intrusive (1120+); 15, Keweenawan Gabbro Wisconsin, Minnesota; 16-20 Stoer Group (in approximate stratigraphic order); 21, Endion Sill (1150+) and Lester River Sill; 22, Beaver Bay Complex, basalts (1090+); 23, Hazel Formation, New Mexico; 24, Upper Cape Gargantua Volcanics (1090+); 25, Portage Lake lavas (1100+); 26, Freda Sandstone; 27, Michipicoten Island lavas (952+);

(Continuation of Legend to Fig. 7 see p. 276)

(Legend to Fig. 7, cont.)
28, Copper Harbour lavas (1060*); 29, Pikes Peak granite (1030+*); 30, Isle Royale lavas; 31, Nonesuch Shale (1065 -1025*); 32, Basalt dykes Minnesota; 33, Nankoweap Formation (1000); 34, Cardenas lavas (1065*); 35, Logan dykes (normal group 1050+); 36, Mamainse Lavas (1050*); 37, Copper Harbour conglomerate, secondary; 38, Fond du Lac Formation; 39, Kearsage rhyolite (1007); 40, Nemagonsenda carbonatite (1035+); 41, Freda and Nonesuch Sandstone (1045*); 42, Seal and Croteau igneous suite (~950); 43-46 Jacobsville JIC, JIB, JIA, JISA; 47, Diabaig Formation; 48, Torridonian, Cape Wrath; 49, Applecross Formation, Lower Sites; 50, Applecross Formation, upper sites; 51, Applecross, basal sites, Loch Broom; 52, Applecross Formation, Loch Maree; 53, Aultbea Formation, lower part; 54, Aultbea Formation, higher part; 55-60, 34, 61-63, Grand Canyon Supergroup, poles from higher formations in stratigraphic order after Elston and Grommé (1974); 64, Bass Limestone is an earlier result from these rocks. A to F are mean thermochron poles from the Greenville Province (data for this interval are covered in detail in Fig. 8).
Note that the Keweenawan rocks defined as normal are reversed according to the interpretation of Fig. 8 (see inset).
African Shield (◧): 1, Kanangono kimberlite, Ivory Coast; 2, Montrose kimberlite, South Africa; 3, Waterberg dolerites; 4, Umkondo dolerites (1140*); 5, Group de Char Formation, Mauretania (1019); 6, Chela Group Angola (1100-1050*); 7, Umkondo lavas (1090*); 8, Guperas lavas (1250**); 9, Koras Group, Kalkpunt Formation (1200-1000); 10, Auborus Formation (~1020, <8); 11, O'okiep intrusions (1020*, 1090-1050**); 12, Nosib Group (<1010**,>844*).
Australian Shield (▷): 1, Lake View dolerite (1125*), (primitive reconstruction).
Indian Shield (◁): 1, Mundwara complex (1000-950). (Gondwanaland reconstruction).
Antarctica Shield (◁▷): 1, Vestfold Hills dykes (1020**). (Gondwanaland reconstruction).
(In this figure the symbols have been subdivided according to the percentage of magnetisations in the palaeomagnetic study defining the pole (*solid part* of the symbol) or antipole (*open part* of the symbol) as plotted. This is to show the reversal along this track observed in both the Laurentian and African data). After Piper (1982a)

3.8 1020-840 m.y. (Fig. 8)

The definition of this path has been largely achieved by palaeomagnetic studies of the Grenville and Sveconorwegian mobile belts (Fig. 8) which developed along the margins of the Laurentian and Fennoscandian Shields respectively as a thermal-tectonic lineament immediately prior to the continental breakup at 1100 m.y. Although this belt has been considered in the context of continental collision at this time (Burke et al. 1976, Irving and McGlynn 1976), it has long been clear that the frontal zones demarcating this belt from the

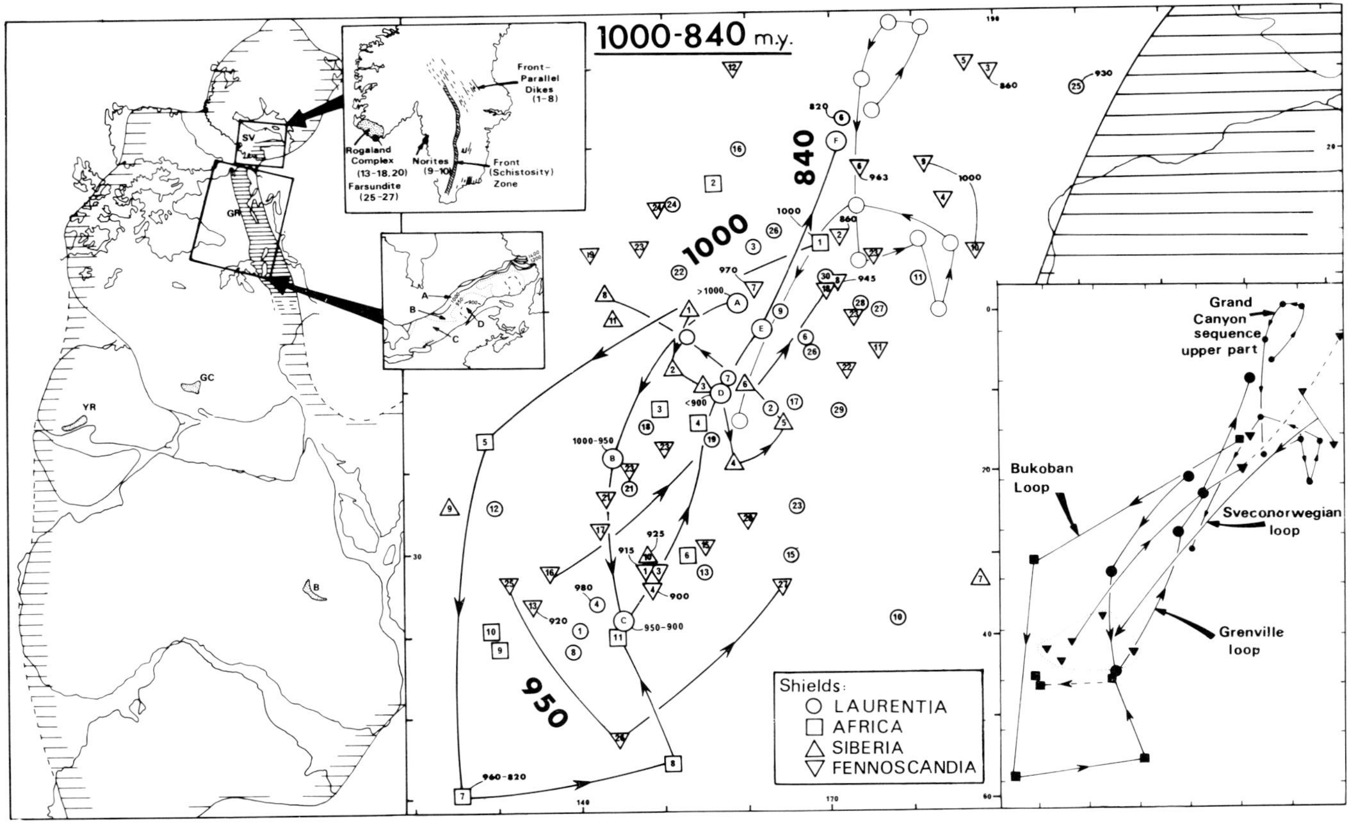

Fig. 8 (Legend see page 278)

Fig. 8. Palaeomagnetic poles assigned to the interval 1000-800 Ma. Laurentian Shield (o): (Palaeomagnetic poles derived from the Grenville mobile belt and its margins are listed in alphabetic order because their ages are not directly known.) 1, Allard Lake anorthosite; 2, Frontenac dykes (this pole may belong to the a.p.w. path at ca. 650 Ma); 3, Grenville Front anorthosite; 4,5,6, Haliburton intrusion HA (960^{++}), HB, HC (820^{++}); 7, Indian Head anorthosite; 8,9, Morin anorthosite M1, M2; 10,11, Mealy Mountain E, NW; 12, Magnetawan metasediment; 13, Ottawa intrusions; 14, St. Urbain; 15,16 Thanet gabbro complex, A, B; 17, Umfraville gabbro; 18, Wilberforce pyroxenite; 19,20,21, Whitestone anorthosite, W, Y, Z; 22, Seal-Croteau igneous suite (~950); A-F are the mean thermochron poles from the Grenville province as calculated by McWilliams and Dunlop (1978); 23, Keweenawan shock induced remance B; 24, Eileen Sandstone; 25, El Paso rocks (930^{*}); 26, Michael Gabbro; 27, Orienta Sandstone, primary; 28, Rama diabase (935^{+}, 850^{+}); 29, Shabogamo gabbro; 30, Freda and Nonesuch Sandstone Formations, secondary; 31, Rapitan Formation "X" component ($900-825^{+}$); 32, Steel Mountain anorthosite. Small open circles are the poles stratigraphic order from the Grand Canyon succession.
Fennoscandian Shield (▽): 1, Karlshamn dyke (915^{*}, 855^{*}); 2, Tärnö dyke (860^{*}); 3, Bräkne-Hoby dyke (860^{*}); 4, Väby dyke; 5, Fäjö dyke; 6, Nilstorp dyke (965^{*}); 7, Arby dyke (974^{*}); 8, Falun dyke (945^{*}); 9, Bratton norite (1000^{+}); 10, Algon norite (1000^{+}); 11, Dolerite, Bratton; 12, Listed and Bolshavn "hyperites"; 13, Aana-Sira Massif (920^{+*}); 14, Rogaland Complex (900^{+*}); 15, Garsaknett body; 16, Rogaland Complex north part; 17, Egersund-Ogna Massif; 18, Rogaland, "Y" component; 19, Amphibolites and hyperites, Bamble and Kongsberg ($1100-950^{+}$); 20, Amphibolites, norites and anorthosites, south Rogaland; 21, Hunnedalen dolerites; 22, Tuve dolerite, Gothenburg; 23, "Hyperite" dykes of south Sweden; 24, "Hyperites" of Kragero district; 25, Farsundite complex; 26, Farsundite, south part; 27, Farsundite, north part.
African Shield (□): 1, Bukoba Sandstone (>990-1000); 2, Abercorn Sandstone; 3, Klein Karas dykes (890^{+}); 4, Marico River intrusions; 5, Kigonero Flags (>890); 6, Florida Formation secondary; 7, Gagwe Amygdaloidal lavas (960^{+}, 820^{+}); 8, Manyovu Red Beds; 9, Bukoban dolerites (860^{+}, 810^{+}); 10, Mbala dolerites; 11, Malagarasi Sandstone; 12, Ikorongo Group.
Siberian Shield (Δ): 1-8, Yenisei Ridge sedimentary sequences in stratigraphic order; 1, Gorbilok Group; 2, Udere Group; 3, Pogoryui Group; 4, Potoskui Group; 5, Shuntar Group; 6, Kirgitei Group; 7-8 Nizhneangara Group; 9, Yenisei Group, Kuznets Alatau. (The maximum age range of these sediments is 1320-750 Ma). 10, Burovaya Suite (925).
The figure on the *left* illustrates the continental reconstruction of the major shields (with the reservation that the Gondwanaland reconstruction may not have been reached, Sect. 3.6) applicable to this time interval together with the locations of the main data sets defining the a.p.w. path. These are the Sveconorwegian mobile belt in the Fennoscandian Shield with data derived from within the belt and

(Legend to Fig. 8, cont.)
from the front parallel dykes to the east, the Grenville mobile belt with representative mean palaeopoles derived from uplift chrontour zones A to D and low blocking temperature remanences, the sediments of the Grand Canyon Supergroup (GC), the Yenesei Ridge in the Siberian Shield (YR), and the Bukoban System in Tanzania (B). These various units define part or whole of the same elongate a.p.w. loop (*right hand inset*)

---

earlier-stabilised parts of the shield could not be the sites of a suture because critical structures continue across these zones and are reworked into the mobile belts (e.g., Baer 1976). It is now generally accepted that the a.p.w. record from this zone is an integral part of the whole-shield path (e.g., Buchan and Dunlop 1976, Patchett and Bylund 1977, McWilliams and Dunlop 1978).

At least the majority of the magnetisations in these belts were acquired during the protracted phase of post-tectonic cooling and are thus related to the uplift history. A simplified assessment of the collective data can be made by calculating mean poles from zones defined by K-Ar ages on micas which presumably became closed systems to the migration of argon in the same time sequence (although not, in general, at the same actual times) as the magnetisations. These K-Ar zones (Fig. 8, inset) yield mean palaeopoles A→F in Fig. 8 defining a closed loop. This has been more precisely calibrated by $^{39}Ar:^{40}Ar$ studies (e.g., Berger et al. 1979) and the sequence of differing blocking temperature remanences in individual rock units (e.g., Buchan and Dunlop 1976). The uplift history of the Sveconorwegian belt (Fig. 8) in the Fennoscandian Shield is less well understood than that of the Grenville belt, but the same loop is defined by the general sequence of palaeopoles from ca. 1000 m.y. magnetisations near the margins of the belt (9,10,19) to ca. 950 m.y. Magnetisations from major intrusive complexes in the interior of this belt (13-17, 20,25-27), to late magnetisations in this latter zone (18, Stearn and Piper 1982). The return path is also defined by ca. 950-850 m.y. dykes (1-8) intruded into the already-stabilised shield parallel to the mobile belt and linked to uplift-related post-tectonic fracturing of this margin (Fig. 8 and Patchett and Bylund 1977). The collective data suggest that the entire closed loop in Fig. 8 represents a time interval of about 1020-840 m.y. with the hairpin at the apex being about 950 m.y. in age. The same path is defined by the sedimentary-volcanic Bukoban succession deposited around the western margin of the Tanzania craton in central Africa and, in part, by the sedimentary successions of the Yenesei Ridge in the Siberian Shield (Fig. 8).

3.9 840-500 m.y. (Fig. 9)

The youngest part of the loop illustrated in Fig. 8 is defined by poles linked to age estimates between 860 and 820 m.y. (Patchett and Bylund 1977, Berger et al. 1979, Stearn and Piper 1982) and these data indicate an upward migration of the a.p.w. path as illustrated in this figure (McWilliams and Dunlop 1978). Unfortunately, there are few data points applicable to the interval 820-750 m.y. from which to constrain the subsequent path, and the assessment of a.p.w. during Late Precambrian times is also less satisfactory because the palaeopoles are derived predominantly from sedimentary rocks where the age and origin of the magnetic remanences is often obscure. However, in spite of this deficiency the collective data from this interval yield a coherent model for a.p.w. using the composite reconstruction derived from the older data (South America-Africa, Siberia, Laurentia and Fennoscandia) and from the Gondwanaland reconstruction (Australia and India-Antarctica, see Smith and Hallam 1970 and McElhinny and Embleton 1976). This is illustrated in Fig. 9 where the collective data are seen to conform to a single path until about the beginning of the Cambrian.

Although this path defines a loop in common with the older data back to 2100 m.y., it is different in form by being an open feature rather than flattened (cf. Fig. 9 and Figs. 4-8). Poles ca. 750-700 m.y. in age from India, Africa and Australia lie close to the present poles, while the migration away from this position in late Riphean and early Vendian ($\leq$710 m.y.) times is consistently suggested by magnetisation sequences in a number of sedimentary successions from Namibia and South Africa (African poles 2,4,7,14,15 $\rightarrow$ 16) which carry the path through sequences of magnetisations from a number of sedimentary successions in the Ukrainian sector of the Fennoscandian Shield which are specifically assigned a Vendian age (710-570 m.y.). The hairpin at this point also incorporates Siberian data while the path between this point and the latest Precambrian hairpin at 600 m.y. is defined by palaeopoles in the Adelaide Geosyncline of Australia, the Ukrainian, African and Laurentian Shields. Although absolute age assignments relevant to this path are few, they do plot in correct sequence along this path and include the La Tinta Formation of Argentina (690 m.y.), the Franklin lavas (625 m.y.), diabases and stills (675-625 m.y.) which represent a major episode of dolerite intrusion along the northern margin of the Laurentian Shield, and intrusions from northern and central Africa (10,11) dated 610 and 617 m.y. respectively. Also plotting along this path is a low blocking temperature remanence from amphibolite-facies terrain in the Grenville belt (1) in common with a 670 m.y. intrusion from this belt; in common with other Laurentian poles of probable Late Precambrian age (2-6, 8-10), they conform to the path defined by data from elsewhere although the age data cannot confirm this assignment.

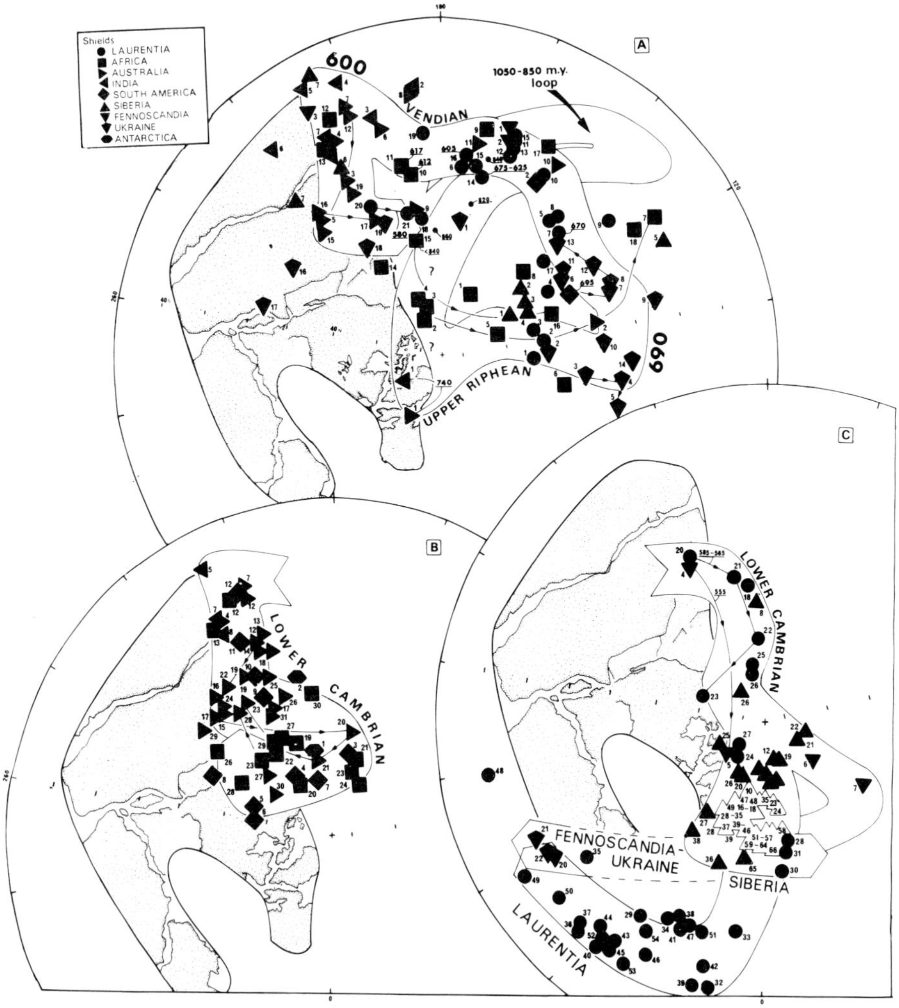

Fig. 9. Palaeomagnetic poles from the major shields assigned to the interval 800 Ma Lower Palaeozoic times.
Laurentian Shield (•): 1, Cordova "A"; 2-6, Colorado intrusive (LPc-C, 704-485^{+}); 7, Tudor Gabbro (670^{++}); 8, Iron Mountain-Maclure Mountain Complex (LPc-C); 9, Frontenac dykes (817-751^{+}); 10, Grenville dykes (LPc); 11, Franklin lavas etc. (625^{++}); 12, Franklin diabases (675-625^{+}); 13, Aston Bay dykes; 14, Coronation sills (647^{+}); 15, Johnnie Formation (LPc); 16, Cloud Mountain basalt (615^{+}); 17, Lamotte Formation (1, LPc-C); 18, Lamotte Formation (E, LPc-C); 19, Quebec Lavas, A; 20, Ikertôq shear belt pseudotachylites (>23); 21-23, West Greenland alkaline province (585-570^{+*}); 24, Chequamegon Sandstone, secondary; 25, Rapitan Formation, "Y" component; 26, Orienta Sandstone, secondary; 27, Jacobsville Sandstone, J2 component; 28, Lamotte Sandstones, 2a component (C); 29, Lamotte Sandstone, 2b component (C);
(Continuation of Legend to Fig. 9 on page 282)

(Legend to Fig. 9, cont.)
30, Lodore Formation (C); 31, Bonnetere Dolostone (C); 32, Bradore Sandstones (C_1); 33,34, Quebec lavas, B and C components; 35, Ratcliffe Brook Formation (C_1); 36, Carrara-Bonanza King Formation (C_{1-m}); 37, Carrara Formation (C_{1-m}); 39, Waynesboro Formation (C_{1-m}); 40, Rome Formation (C_{1-m}); 41, Tapeats Sandstone (C_{1-m}); 42, Ophiolite Complex, Quebec (550^+); 43, Hickory Sandstone (C_{m-u}); 44, Cap Mountain Limestone (C_{m-u}); 45, Lion Mountain Limestone (C_{m-u}); 46, Welge Sandstone/Morgan Creek Limestone (C_{m-u}); 47, Point Peak Shale (C_{m-u}); 48,49, Colorado intrusives (525^+); 50-54, Wichita granites (525); 55, Quebec dykes (497^+).
Fennoscandian Shield (▼): 1,2, Egersund dykes (LPc); 3, Batsfjord dykes (650^+); 4,5, Alnö Complex, A (553^*, 554^+), B; 6, Fen Rødberg ($600\text{-}530^{+*}$); 7, Nexo Sandstone (LPc-C).
Ukrainian Shield (▼): 1, Lower Katav Group (LPc); 2, Katavsk Group (LPc); 3-5, Kuk-Karauk Group (LPc(V)); 6-8, Basinsk Group (LPc(V)); 9, Inzer Group (LPc); 10, Katav Group, Podinzer sequence; 11, M-U Katav Group (LPc); 12,13, Basinsk Group, higher formations (LPc(V)); 14-19, Man' Insk Nyorovei and Kokpel Groups (LPc(V)); 20, Middle Urals intrusions (O_{m-u}); 21, Karachan Group sediments (O_1); 22, Bardym Group.
Siberian Shield (▲): 1, Omninsk Group (LPc); 2, Malginsk Group (LPc); 3, Tsipanda Group (LPc); 4, Lakhanda Group (LPc); 5, Ust' Kunda Group (LPc-C_1); 6, Podkrasno tsvetnaya Group (C_1); 7, Karagasski Suite (LPc(V)); 8, East Aldan sediments (C_1); 9, Charsk Group (C_1); 10, Ust' Agul'sk Group (C_1); 11, Emyaksa Group (C_1); 12, Chernoles Group (C_m); 13-15, Ust'Maisk and Amga Groups (C_m); 16-17, Olenek River Groups (C_m); 18, Silisir Dzhakhtar Groups; 19,20, Ust' Maya Group (C_m); 21, Amga River sediments (C_m); 22, Ust'Botoma Group (C_m); 23, Dzakhtar Group (C_m); 24, Siligir Group (C_m); 25, Akcha Group volcanics (C_m); 26, Nadak Group (C_m-C_u); 27-29, Upper Lena Group (C_u); 30-34, 37-38, 42-45, Verkholensk Group (C_u); 35, Evenkiisk Group (C_u-O_8); 36, Gornaya Altai Group (C_u-O_1); 39, Chukuk and Markha Groups (C_u); 40, Lena River Sills (C_u); 41, Iglinsk Group (C_u); 46, Lena River sills (O_1); 47, Ust' Kut Group (O_1); 48, Ahakit River Limestones (O_1); 49, Ust' Kutsk Group (O_1); 50, Ust'Kutsk and Kazimirovsh Groups (O_1); 51, Chertovsk Group (O_m); 52, Lena River sandstones and clays (O_m); 53-56, Lena River sediments; 57, Lower Makarovsk Group (O_{m-u}); 58-59, Makarovsk Group (O_u); 60, River Nepa, combined result (O_u); 61, Makarovsk Group and Chertovsk horizon (O_u); 62, Bratsk and Makarovsk Groups (O_u); 63, Bratsk Group, combined result (O_u); 64, Taratash complex (Ou); 65, Lower Bratsk Group (Ou); 66, Lena River sediments (Ou).
African Shield (■): 1, Pre-Nama dykes ($640 \pm 70^*$); 2, Blaubeker Formation ($\leq 670^*$); 3, Mbozi Complex (750^+); 4, Lower Nama Group, N1 (<2); 5, Plateau Series A (LPc); 6, Dokhan Volcanics ($665\text{-}605^+$); 7, Upper Nama Group N2 (LPc); 8, Group de Char Formation, secondary; 9, Klipheuval Formation (LPc-C); 10, Adma diorite (610-590); 11, Ntonya ring structure ($617 \pm 24^*$); 12, Nama Group, N3 (Lpc); 13, Sijarira Group (LPc-C); 14, Nosib Group, NQ2 ($679\text{-}660^*$); 15, Otavi Group, DC1 (<840, >651, $684\text{-}618^*$); 16, Otavi Group DC2-3 (<DC1); 17, Nosib

(Legend to Fig. 9, cont.)
Group, NQ3 (<NQ2); 18, Mulden Group (560-550*); 19, Plateau Series, B (LPc-C); 20, Nama Group, secondary; 21, Sabaloka ring structure (530*); 22, Moroccan lavas (C_{1-m}); 23, Ben Azzer volcanic sediments (520*); 24, QS dykes (530-480+); 25, Dokhan Volcanics (500-465+); 26, Doornpoort Formation (550-500); 27,28, Adras de Mauretainie CO_{10}, CO_8 (C-O); 29, Tassili sediments (C-O); 30, Table Mountain series (O).
Australian Shield (►): 1, YB dykes (735*); 2, Wooltana Volcanics ($\leq$900-850); 3, Copley Quartzite (LPc); 4, Merinjina Tillite (PLc); 5, Tapley Hill Formation (LPc); 6, Angepena Formation (LPc); 7,8, Brachina Formation (LPc); 9, Bunyeroo Formation (LPc); 10, Arumbera Sandstone (LPc-C); 11, Antrim Plateau Volcanics (LPc-C_1); 12, Lower Arumbera sediments (LPc-C_1); 13, Upper Arumbera sediments (LPc-C_1); 14, Todd River sediments (C_1); 15, Aroona Dam sediments (C_1); 16, Basal Lake Froome Group (C_m); 17, Lake Froome Group, lower (C_m); 18, Hawker Group (C); 19, Red beds, Kangaroo Island (C_1); 20, Hudson Formation (C_1); 21, Hugh River Shale (C_{1-m}); 22, Billy Greek Formation (C_{1-m}); 23, Lake Froome Group, upper (C_{m-u}); 24, Panta pinna Formation (C_m); 25, Dundas Group (C_u); 25, Jinduckin Formation (O_1); 27, Stairway sandstone (O_m); 28, Tumblegooda Sandstone (O); 29, Mereenie Sandstone (S-D); 30, Merinjina Tillite, secondary; 31, Brachina Formation, secondary.
Indian Shield (◄): Malani rhyolites (730*); 2,6, Rewa Sandstone (LPc-C); 3-5, Bhander Sandstone (LPc-C); 7, Khewra Sandstone (C_1); 8, Baghanwala Formation (C_m).
South American Shield (◆): 1, La Tinta Formation (695*); 2,5, Purmamarca sediments (C); 3, North Tilcara sediments (C); 4, South Tilcara sediments (C); 6, Abra de Cajas sediments (C); 7, Salta sediments (O); 8, Suri Formation (O); 9, Urucum Formation (O); 10,11, Salta and Jujuy sediments (O); 12, Bolivian sediments (O).
Antarctica Shield (⬬): 1, Mirnyy Station charnockites (495*); 2, Sør Rondale intrusions (O_{1-m}, 485)

The hairpin at 600 m.y. appears to be defined by a number of sedimentary magnetisations of probable latest Precambrian age from Gondwandland, but the Proterozoic-Phanerozoic transition has been specifically studied in Central Australia (Kirschvinck 1978a,b) where sedimentation appears to have been practically continuous at this time. The palaeopoles from these units are in a similar position to poles of late Middle and Upper Cambrian age from the same continent (Fig. 9) but the scale of a.p.w. during the intervening period is less well established: the collective palaeopoles of Lower-Middle Cambrian age from Gondwanaland appear to define a large closed loop (Fig. 9, McElhinny and Embleton 1976, Piper 1976a) but since the nature and age of some of these magnetisations have been disputed (Klootwyk (1980), the form of the loop remains unclear. Taken at their face value the Lower to early Middle Cambrian poles from Australia (20,21) and Africa (14,18,19) and South America (3) carry the Gondwana path to the west, after which it retraces a path to poles of Middle (Australian poles 16,23 and 24, African pole 21, Indian pole 8) and Upper (Australian pole 25) Cambrian age. In a more general way this path is also defined by Cambrian poles from South America (2-6). Antarctica (1,2) and Africa (22-24) for which there are no specific age assignments.

In contrast, the a.p.w. paths of the Laurentian, Siberian and Fennoscandian Shields specifically diverge from the Gondwanaland path at this time (Fig. 9). The latest Precambrian to early Cambrian poles from the Laurentian Shield define a rapid movement towards the present pole (20-23, with a number of probable Late Precambrian magnetisations 24-27 utilised by Morris and Roy 1977 to define the Hadrynian Track and belong to the path either here or at ca. 820-750 m.y., see also Fig. 9); this path (Piper 1981b) incorporates Fennoscandian poles of Lower Cambrian age up to pole 7 prior to which the paths for the two shields diverge. The late Lower Cambrian to early Ordovician palaeopoles (i.e., ca. 540-470 m.y.) from the Laurentian and Siberian Shields illustrate relatively small a.p.w. movements in comparison with the older data but lie in contrasting areas on the Proterozoic reconstruction and show that it is no longer applicable to these times.

Thus these differing a.p.w. paths show that at least four major continental blocks were moving relative to one another by the end of Lower Cambrian times. These were the Laurentian, Fennoscandian, Siberian and Gondwanaland continents of Palaeozoic times and it is possible that the China Shields had broken off from Australia and the Kazakhstania Shield had broken off from Siberia as part of these movements. The combined a.p.w. path for late Precambrian to Ordovician times is a classic example of continental breakup and dispersal as defined by palaeomagnetic data (Fig. 10). It shows that part of the supercontinent in nothern Africa lay astride the South pole in latest Precambrian-early Cambrian times, while the continental breakup in Lower Cambrian times caused a rapid dispersal which carried northern Africa via an excursion into lower latitudes,the scale of which is in dispute (see Klootwyk 1980),back to the South pole by late Middle Cambrian times. In contrast the Fennoscandian, Siberian and Laurentian Shields moved rapidly across the South pole in Lower Cambrian times and into low latitudes by Middle Cambrian times (i.e., after 530 m.y.). These motions are summarised by the palaeogeographic maps of Fig. 11 which complement the maps applicable to post-Cambrian times given by other workers (Scotese et al. 1979, Ziegler this Vol.). It should be noted that the palaeomagnetic method provides no control over palaeolongitude, and therefore because the Siberian, Laurentian and Fennoscandian Shields lay in low latitudes after Lower Cambrian times it is not possible to constrain the distances between them; the separations shown in Fig. 11 are arbitrary.

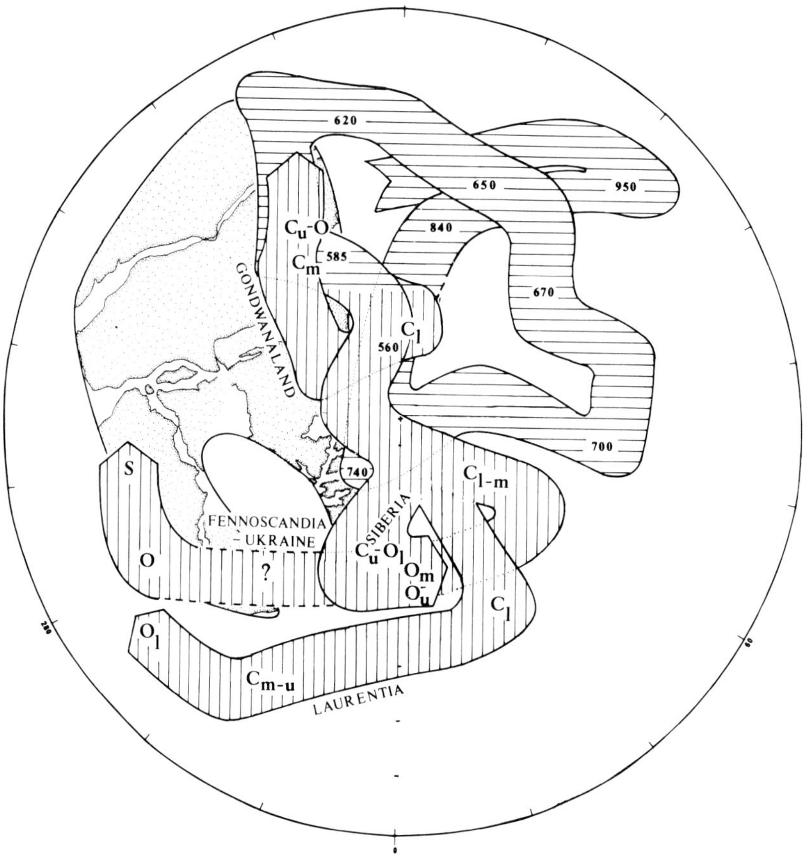

Fig. 10. Summary of the 840 to 450 m.y. a.p.w. paths on the Late Precambrian supercontinent reconstruction (the data are plotted in full in Fig. 9). Note that a single path is applicable to this reconstruction until Lower Cambrian times when the paths from the Gondwanaland, Laurentian, Fennoscandian and Siberian diverge in sequence on this reconstruction. It is evidently no longer applicable by Middle Cambrian times. C Cambrian; O Ordovician; S Silurian; l,m.u, lower, middle, and upper divisions. Note that there are no data relevant to the Fennoscandian-Ukrainian block between Lower Cambrian and Lower Ordovician times (After Piper 1982b)

3.10 The Continental Breakup and Dispersal in Lower Cambrian Times

The relative movements in Lower Cambrian times led to the growth of large new ocean basins which were ultimately consumed by subduction and obduction of this ocean crust as the continents moved together again to form the Caledonian and Hercynian orogenic belts. This contact between the four major blocks illustrated in Fig. 11 was achieved between mid-Silurian and mid-Carboniferous times to form the transient

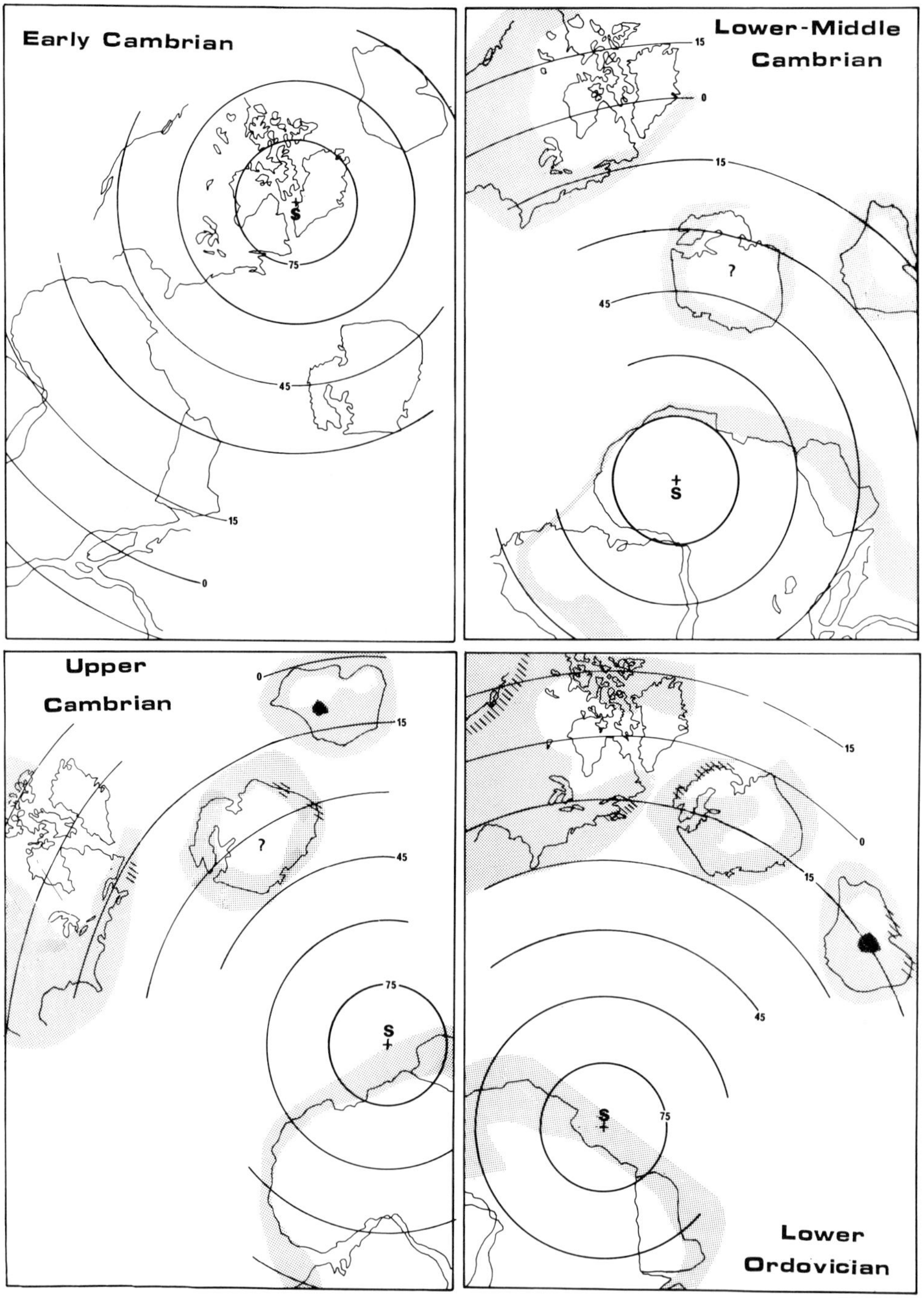

Fig. 11. The palaeogeographic implications of the Cambrian palaeomagnetic data plotted in Figs. 9 and 10 illustrating the break up of the Proterozoic Supercontinent in Lower Cambrian times and the stabilisation of large ocean basins between several major shields by Middle Cambrian times. The *stippled areas* are shallow shelf seas and the cross hachured areas are evaporites. The *inclined lines* show areas characterised by the emplacement of obducted ophiolites as the ocean crust between these shields was consumed when they began to move together during formation of the Caledonian orogenic belts (ca. 450-350 m.y.)

supercontinent of Pangaea. The formation of these new ocean basins in Lower Cambrian times was a major event and brought about the faunal diversity and provincialisation which distinguish the Cambrian faunas from the much less provincial and more uniform Late Precambrian faunas (Cowie 1971, Holland 1974, Vidal 1977, 1979). It is recorded by a number of chemical and geological events (Piper 1982b) which distinguish it from the first continental breakup at about 1100 m.y. (Sect.3.6). These are:

1. The $Sr^{87}:Sr^{86}$ isotope ratio of seawater shows a gradual increase during the Proterozoic era indicating a progressive reworking of older continental crustal material (Viezer 1976), but at the Late Precambrian-Cambrian boundary this increase is interrupted and succeeded by lower values in the Phanerozoic which are more typical of Archaean times. This observation implies that there was a rapid influx of Upper Mantle-derived strontium into the marine system at this time which is an expected consequence of the formation of new ocean basins.
2. Other chemical indicators which define a scarcity of oceanic crustal signatures during the Proterozoic but show a marked change at the beginning of Phanerozoic times to values more typical of the Archaean permobile regime are K:Na and Rb:Sr (Engel et al. 1974). The large-ion-lithopile elements also tend to be higher in Proterozoic silicic igneous rocks suggesting the predominance of ensialic melting processes (Glikson 1982). A sympathetic change in the $K_2O:Al_2O_3$, $Na_2O:Al_2O_3$ and $K_2O:Na_2O$ ratios of marine shales in Phanerozoic times (Viezer 1973) is another signature of this change in the marine environment.
3. Alkaline magmatism is a ubiquitous indicator of continental rifting and breakup (Bailey 1974) and took place extensively in the continental crust immediately prior to the Lower Cambrian breakup. A major alkaline igneous province incorporates the Fennoscandian, Laurentian, Siberian and north eastern part of the African Shields (Doig 1968, Harris and Gass 1981) with activity commencing at about 610 m.y. and continuing until about 555 m.y.; it was associated with widespread rifting some of which developed into the lines of continental separation at this time while others were activated by the Mesozoic-early Tertiary breakup of Pangaea. Elsewhere a major change in Upper Mantle environments which may have motivated this breakup is indicated by a change in the characteristic magmatism; along the Arabian margin of the African Shield, accretion related to subduction was taking place between ca. 1000 and 600 m.y. and is indicated by high P_{H_2O} calk-alkaline magmatism (Greenwood et al. 1976); after 600 m.y. however, these conditions ceased and subsequent magmatism took the form of low P_{H_2O} peralkaline magmatism (Harris and Gass 1981).
4. Change in tectonic style. The integrity of the a.p.w. record from the individual shields has led several workers to conclude that the Proterozoic moble belts developed in an ensialic environment (Piper et al. 1973, McElhinny and Embleton 1976, McWilliams and

Kröner 1981) and this conclusion is inherent in the conformity of the palaeopoles to a single path in Figs. 2-9. Critical contrasts between these older mobile belts and Phanerozoic orogenic belts (Kröner 1976, Piper 1978) are the paucity of volcanics (especially those which can be related to the subduction of ocean crust), the lack of ophiolitic characteristics in mafic igneous units, and a continental origin for most, or all, of the granites (Wynne-Edwards 1976, Glikson 1981). By late Precambrian times however, there is increasing evidence for limited continental separation in some mobile belts and subduction around the margins of the continental crust; miogeosynclines were developing (Fig. 22) and calc-alkaline volcanic suites appear here in the latter part of Precambrian times; elsewhere ophiolitic fragments are present in some parts of the Pan-African domain and appear to require limited crustal separation and ocean development not accompanied by subduction (the delamination process of McWilliams and Kröner 1981). By Palaeozoic times the convergence of the major continents along the Caledonian, Hercynian and Uralian belts was characterised by orogenic tectonics (e.g., Windley 1977) with widespread emplacement of ophiolite slices into fold belts, subduction of ocean crust along the borders of these belts leading to widespread development of calc-alkaline igneous suites and development of wedges of flysch sediments in marginal basins before and during collision, followed by molasse sedimentation at syn- and post-tectonic phases. This change in tectonic style is an important consequence of the change from single- to multicontinent tectonics at the Proterozoic-Phanerozoic transition.

5. Changes in the chemistry of the oceans and the characteristics of shelf seas coincident with this transition are outlined in a later section.

3.11 Summary of Continental Movements in Precambrian Times

The changes in the palaeomagnetic vectors yielding the a.p.w. paths of Figs. 2-9 are composed of changes in magnetic inclination (sensitive to changes in palaeolatitude) and changes in magnetic declination which could be achieved by the in situ rotation of the crustal block. Since the palaeomagnetic method is only sensitive to changes in palaeolatitude it can only define rates of change in this parameter and not the gross movements of the continental crust; however, when both changes in declination and inclination are large it is probable that large translations of the continental crust took place. In Fig. 12 the rate of a.p.w. movement is plotted as a function of time and shows that these rates were significantly lower before 2150 m.y. than between 2150 m.y. and Lower Cambrian times; as discussed in Sect. 5, this correlates with an important change from small to large scale tectonics in the continental crust. For short periods during the interval 2150-540 m.y. rates of movement were

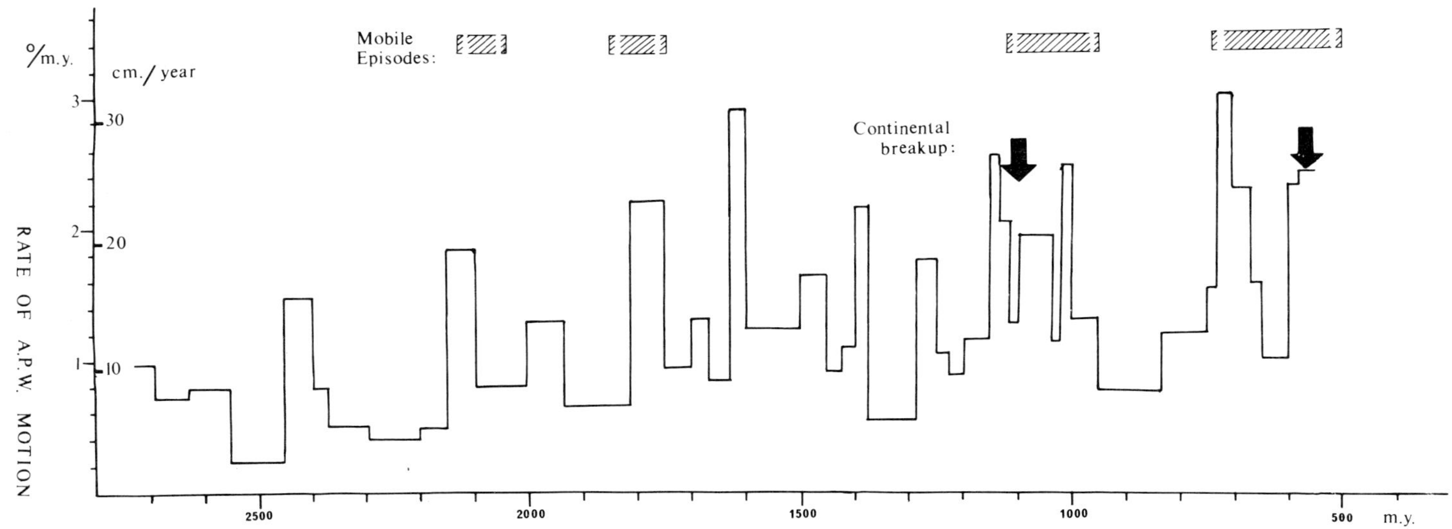

Fig. 12. The rate of apparent polar wander movement represented by the paths of Figs. 2 to 9 plotted as a function of time. Note that this does not represent the total movement of the Proterozoic Supercontinent but only that part to which palaeomagnetism is sensitive

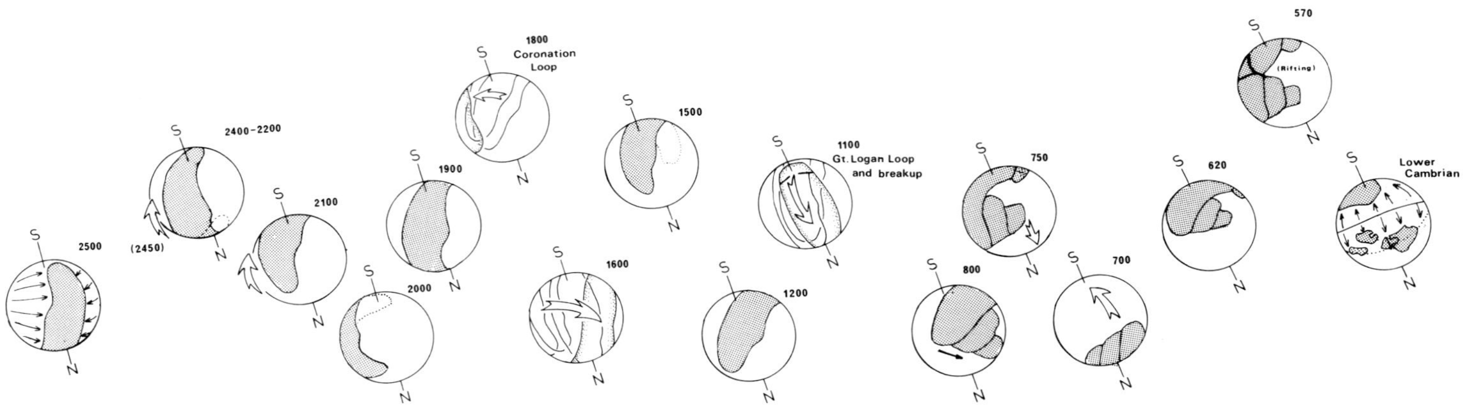

<u>Fig. 13.</u> Cartoon representation of movements of the continental crust during Proterozoic times. Note the repeated tendency for the continental crust to return to a mean stable position astride the south pole between 2100 and 600 m.y.

Table 2. Mean palaeomagnetic poles for the interval 2800-600 m.y.

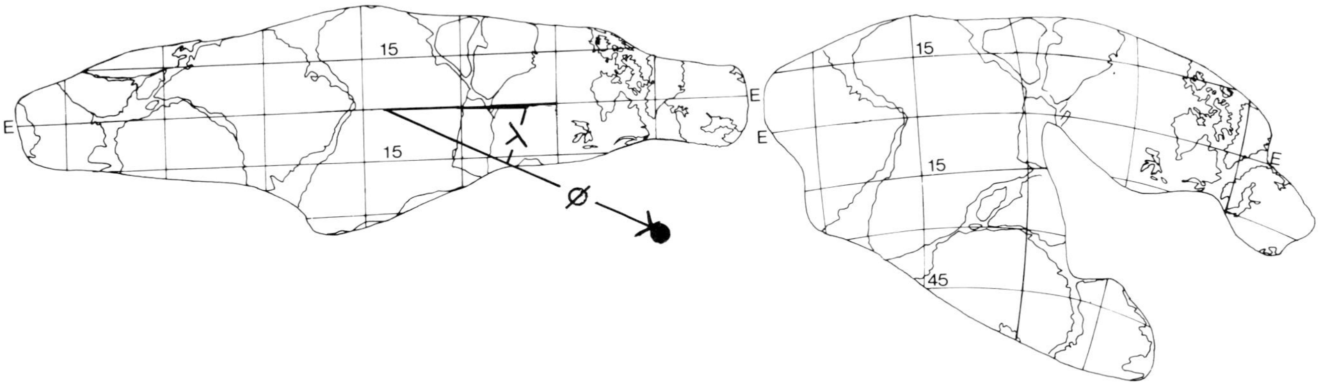

Age	Pole		φ	λ	Age	Pole		φ	λ	Age	Pole		φ	λ
2800	200	0[a]	25	242	1725?	260	20	38	134	1120	240	60	60	342
2750	230	-25[a]	28	145	1690	210	50	52	324	1100	185	30	47	286
2700	270	-65[a]	75	140	1630	205	5	21	258	1060	180	0	45	247
2600	270	-40[a]	58	116	1600	290	-10	67	79	1040	190	-20	41	212
2500	255	20	36	31	1520	210	20	23	298	1000	160	0	66	245
2350	205	60	60	326	1450	220	-15	18	175	950	145	-40	84	215
2250	210	-70[a]	73	164	1400?	210	15	19	289	840	175	10	52	259
2200	240	75	74	342	1375?	195	40	47	300	740	310	75	88	343
2150	230	30	30	353	1280	245	15	23	30	710	280	-40[a]	65	115
2100	270	-30	51	110	1250	210	15	19	289	690	300	-20[a]	77	87
2000	165	-15[a]	60	225	1220	175	10	50	259	650	160	0	67	246
1930	245	35	39	0	1200	200	-10	28	223	615	190	5	36	251
1810	195	-15	33	216	1160	200	35	43	304	600	205	-20	30	198
1750	150	5[a]	70	254	1140	260	40	52	13	570	205	20	26	292

(Legend to Table 2 see p. 292

(Legend to Table 2)
The palaeomagnetic pole is given in degrees east and degrees north as defined by the a.p.w. paths of Figs. 2 to 9. [a]Indicates a north pole; the remainder are south poles. ϕ = angular distance of the geometrical centre of the supercontinent (15°E, 10°N on present coordinates) from the ancient pole, λ = angle (clockwise) between the Fennoscandian end of the major axis of the continent and the direction to the ancient pole measured in degrees clockwise from the long axis of the continental crust as indicated below. Lines of latitude and longitude are not indicated on the palaeogeographic reconstructions presented in this paper because they were changing more or less continuously during Precambrian times. In the reference figures above the long axis of the Proterozoic Supercontinent is aligned along the equator and lines of latitude and longitude are indicated at 15° and 30° intervals respectively. The primitive reconstruction is applicable until approximately 1000 m.y. and the modified reconstruction is applicable to Late Precambrian times

very high and there is a tendency for these rapid movements to occur during, or just after, major mobile episodes in the continental crust at ca. 2100, 1800, 1100 and 600 m.y. In Phanerozoic times this régime has been succeeded by low rates of continental drift interrupted by rapid movements in mid-Devonian, late Permian and late Cretaceous times (Briden 1967, Embleton 1969, McElhinny 1973).

A remarkable feature of the a.p.w. path between 2150 and 600 m.y. was the tendency for the continental crust to return repeatedly to a similar position following the large excursions which executed the a.p.w. loops (Figs. 4-9). This preferred position of the south pole was near to the geometrical centre of the Proterozoic supercontinent in the vicinity of northern Africa and gives the tendency for many of the Proterozoic palaeomagnetic poles to plot in low latitudes between 180 and 270°E in these figures. This characteristic of the continental crust to return repeatedly to a position astride the south pole is illustrated diagrammatically in Fig. 13, where it will be noted that this system contrasts with early Proterozoic times (the small scale crustal tectonic regime, Sect. 5) when the continental crust was distributed approximately uniformly between the poles. The explanation for the large a.p.w. loops of the 2150-540 m.y. interval is not completely clear; they presumably represent large scale instabilities in the mantle convection system driving the continental crust, but since it was only in Lower Cambrian times that these movements achieved continental dispersal, they cannot be the signatures of collisional events (Irving and Park 1972).

In Table 2 the mean palaeomagnetic poles for the interval 2800-600 m.y. are listed for the continental crust reconstructed according to the operations listed in Table 1 (with the Laurentian Shield in present coordinates). The third and fourth columns in this table list the angular distance of the centre of the continent from the pole and the angle between the major axis and the direction to the pole at each of these times.

4. Geophysical Implications of the Precambrian Palaeomagnetic Record

The discovery that the Proterozoic palaeomagnetic record defines a single coherent a.p.w. path yielding a contiguous reconstruction of the Precambrian shields with a strong confirmation in observed geology (Sect. 5) has several important geophysical implications. In particular, it precludes major changes in the dimensions of the Earth since 2700 m.y. and defines the gross changes in the behaviour of the magnetic field since this time; the latter considerations are relevant to models for the geomagnetic dipole field.

4.1 The a.p.w. Correlation Paradox and Limits to Change in the Earth's Palaeoradius

Embleton and Schmidt (1979) observed an impressive correspondence between the palaeopoles lying within the range 2000-1600 m.y. from the African, Australian and Laurentian Shields on present day coordinates. This observation is very difficult to explain in the context of conventional Plate Tectonic theory because it requires that these shields have now reached the same relative positions that they occupied in mid-Proterozoic times. It prompted these workers to propose that these observations would result if global expansion had taken place from a palaeoradius of 0.55 times the present value at 1600 m.y. to the present value by 400 m.y. (Schmidt and Embleton 1982). This proposal has revived interest in geological models for global expansion, advocated in particular by Carey (1976), and has been used to explain the absence of accreted Precambrian crust (Glikson 1979).

The Proterozoic Supercontinent analysed here demonstrates a close correspondence of the a.p.w. data from the shields assuming a different continental configuration from the present but no significant change in the Earth's palaeoradius. Clearly both this model and that of Schmidt and Embleton cannot be correct, and since it is important to clarify any changes in the radius of the Earth, this subject merits some further discussion here. The positions of the Australian and Laurentian Shields utilised in the supercontinent reconstruction (Fig. 1) are not widely different from the present-day coordinates used by Embleton and Schmidt, and the close correspondence in field polarities between the two shield is strong confirmation that the same a.p.w. paths are being correlated in both models. Inevitably there is subjective judgement about some of the poles which should be included in this comparison, but accepting those Australian poles which can with confidence be assigned to the interval considered by Embleton and Schmidt, namely the Mt. Newman and Tom Price ores (ca. 1800 m.y.), the Hart dolerite (1780 m.y.), YF dykes (1700 m.y.), Edith River Volcanics (1700 m.y., the original quoted pole is used here for this unit), the Iron Monarch positive and Iron Prince ores, the GB and YD dykes (1700 m.y.) and the GA dykes (1500 m.y.), it is seen in Fig. 14 that (whilst there is some uncertainty about the part of the E-W equatorial track at ca. 1750 m.y. which some of these poles correlate with) there is little to choose between the two configurations. However, for the 2600-2000 m.y. interval for which there are an equal number of Rb-Sr dated poles from Australia, the data defining the pre-2000 m.y. path (the Track 6 of Irving and Naldrett 1977) are clearly distinct from the contemporaneous data for the Laurentian Shield on present coordinates (Fig. 14) and the overlap of the a.p.w. paths between 1800 and 1500 is thus only transient.

There are some subjective differences between the ca. 2100-1900 m.y. African data used in the two analyses which apply to poles such as the Aftout dolerites for which no precise age estimates are available, but Fig. 14 illustrates those data with the more definitive estimates. It can be seen here that, whilst there is some correlation between the African and Laurentian data at the top of Track 5 (Fig. 3 and Roy and Lapointe 1976) on present coordinates, the African palaeopoles tend to be of opposite polarity to the Laurentian data falling in the same area. Also the African poles falling on the apex of the ca. 1800 m.y. loop are more that 100 m.y. older than the possible age of the Laurentian poles falling here. In addition, if the African path is extended to the 1810 m.y. position, it is seen to be remote from any Laurentian data with this possible age. More critical is the absence of correlation between pre-2200 m.y. data from the two shields (Fig. 14).

The analysis of Fig. 14 suggests that the correlation paradox identified by Embleton and Schmidt (1979) results from 1. the Australian and Laurentian Shields occupying similar relative positions at the present time to that pertaining in Proterozoic times, and 2. a crossing of the antipolar a.p.w. paths of the African and Laurentian Shields at ca. 2000 m.y.; pre-2100 m.y. data cannot be used to support a model for global expansion. The palaeopole correlations of Figs. 2-9 are achieved by the standard assumption of a constant global radius, and the supercontinent model is an alternative assessment of inter-shield correlations supporting the case for an approximately constant global radius since at least the end of Achaean times. This is consistent with the data from the other terrestrial planets (McElhinny et al. 1978).

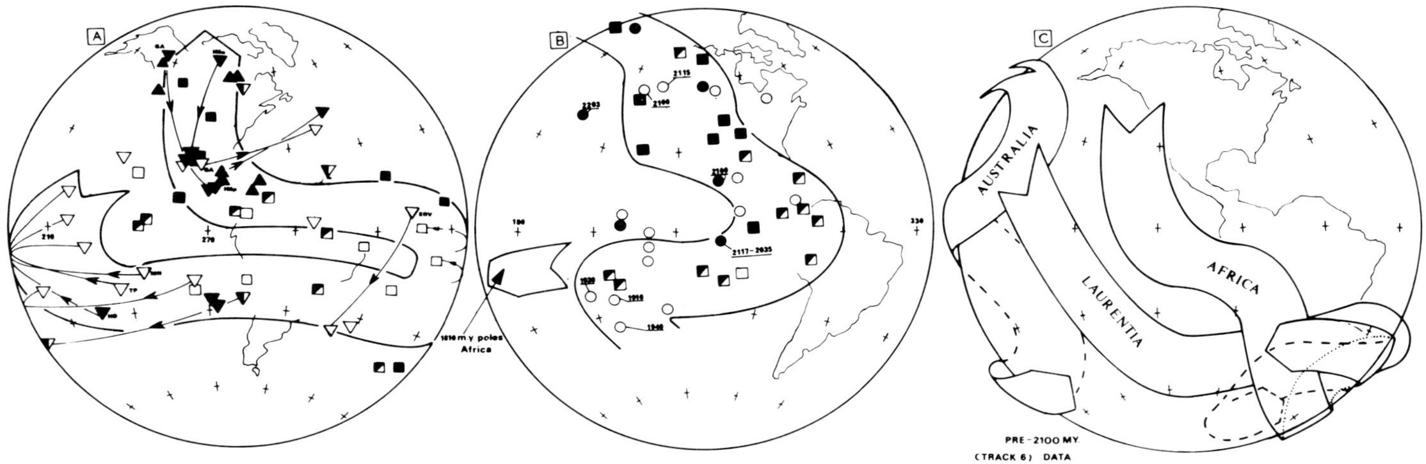

Fig. 14. The a.p.w. correlation paradox after Embleton and Schmidt (1979) showing the apparent conformity of the African and Laurentian palaeopoles at ca. 1800-1600 m.y. (A), and African and Laurentian palaeopoles at ca. 2000-1900 m.y. (B); closed symbols are poles, open symbols are antipoles and divided symbols are mixed polarities. Because of the large equatorial shift at ca. 1800-1750 m.y. (Fig. 4), it is not clear which part of the a.p.w. path at this time many of the Australian data points belong to, but their positions before and after correction for the supercontinent reconstruction are not widely different (A) because Australia and Laurentia now have a similar spacing and orientation to that pertaining in Proterozoic times. Note in (B) that the African and Laurentian poles correlating on present day coordinates are mostly of opposite polarity and differ in age by 100-200 m.y. (cf. Figs. 3 and 4). There is no evidence in the pre-2100 m.y. palaeomagnetic record for this paradox because the a.p.w. paths do not overlap significantly on present day coordinates (C)

4.2 The Axial Geocentric Dipole Assumption

The working model used in palaeomagnetism assumes that the geomagnetic field source is a geocentric dipole with an axis aligned along the Earth's spin axis. It was originally adopted because historial data averaged over periods of several hundreds of years approximately averaged out the components of secular variation. This general assumption needs to be refined in detail during some intervals of geological time: during Upper Tertiary times palaeopoles are typically displaced by a few degrees to the far side of the geographic pole with respect to the areas from which they are derived. This observation could result from an axial dipole source shifted to the north of the centre of the Earth (Wilson 1970, 1971); the source appears to have shifted closer to a geocentric source during the last few million years (McElhinny 1973). Examples have been identified in ca. 1280 and ca. 1100 m.y. lava successions from Greenland (Piper 1978) and North America (Palmer 1970) respectively where successive normal and reversed axes of magnetisation are non-antiparallel (Sect. 3.7). This effect could result if the dipole source were non-geocentric at these times, but in both instances the effects of other complicating factors have yet to be resolved (Palmer et al. 1981). The description of some older precise reversals (e.g., Bingham and Evans 1975, Piper 1981b) shows that the duration of any non-dipole effects was limited, but the findings of Wilson (1970, 1971) imply that small non-dipole effects may persist for several tens of millions of years and show that even very accurate determinations of the ancient pole position may be displaced from the actual pole of rotation by up to about 5°.

The test that the ancient field source was axial must depend primarily on palaeoclimatic evidence, and several palaeoclimatic indicators confirm that the axial geocentric model is applicable to Phanerozoic times (Briden and Irving 1964). For Precambrian times there are very few palaeoclimatic indicators that can be used to confirm the axial nature of the field, and the sedimentary record cannot be dated with the precision required for this test. The distribution of the Late Precambrian (ca. 800-570 m.y.) glaciogenic deposits appears to follow the sequential movement of the continental crust over the poles (Piper 1982b), but poor age control during this interval of rapid a.p.w. motions (Fig. 9) preclude a more definitive conclusion.

However, as Figs. 2-9 illustrate, the assumption of an axial geocentric dipole source for the geomagnetic field source throughout Proterozoic times has led to a particularly simple result, and it would appear that the extrapolation of this assumption based on Tertiary and Recent data back into remote geological times has been justified.

4.3 The Intensity of the Precambrian Magnetic Field

Recent studies have shown that the strength of the dipole source fluctuates by 2-3 times during periods of the order of 10^3-10^4 years (Games 1980), so that individual determinations of the ancient field intensity are of limited value unless they represent a time-average. Most of the data relating to the intensity of the pre-Tertiary field were acquired in the 1960's using techniques and assumptions that are now regarded as unsatisfactory; however, they give an overall picture which is consistent and presumably reflects the gross variation in the field strength, although little emphasis should be placed on individual determinations. Determinations of the Virtual Dipole Moment (VDM) relevant to Phanerozoic times exhibit a progressive fall off to about 10 percent of the present value by 500 m.y. ago (Smith 1967 and Fig. 15). This effect is not caused by a decay in the natural remanence because it does not extend back into the Precambrian where values up to three times the present value have been recorded. Some of the palaeointensity studies from Precambrian rocks have not reported the method of derivation fully and no quality selection is made here. These studies have usually reported a value for the equivalent field intensity at the palaeoequator and this has been converted into the equivalent VDM (M) using the relationship:

$$B = \frac{M \times 10^{-7}(1 + 3 \sin^2\lambda)^{\frac{1}{2}}}{R^3}$$

where λ is the latitude (= 0 at the equator) and R is the radius of the Earth; age estimates have been adjusted where more reliable data are now available.

Note that values in excess of 15×10^{22} A.m^2 depend on one authority only, but values at least as high as the present field (8×10^{22} A.m^2) were common during Precambrian times. There is no systematic decline during these times, and although there is a dramatic fall in VDM in Cambrian times when the geological environment would suggest a marked increase in the tidal retarding couple, there has been a progressive increase as the Earth and Moon have receded in Phanerozoic times (this volume). Hence this dramatic change in VDM is more likely to be due to a subdivision of the mantle convection system (viz. the continental dispersion, Sect. 3.10) with consequent effects on temperature changes at the core mantle interface. The present intensity record shows that any link between the Earth-Moon distance and the VDM, if it exists, must accomodate an *increase* (prominent in the Phanerozoic, less obvious in the Precambrian) in this parameter as the Moon has receded.

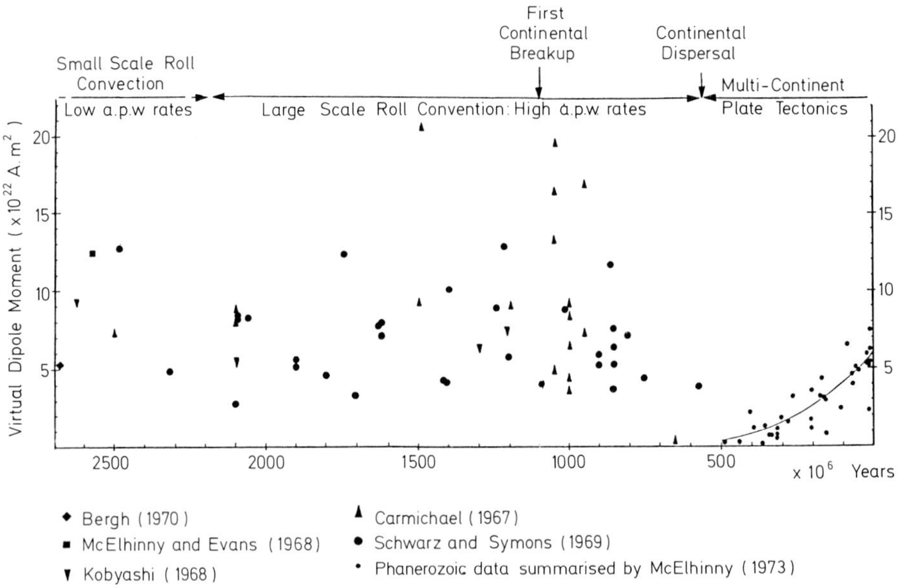

Fig. 15. The Virtual Dipole Moment as a function of time before the present and derived from palaeointensity determinations. The Precambrian data have been converted to their equivalent dipole moment from information given in the references listed and the age estimates have been adjusted where more reliable information is now available. The Phanerozoic data are summarised by McElhinny (1973) and are taken directly from that reference

4.4 The Polarity History of the Magnetic Field

McElhinny (1971) has investigated the polarity bias in the Phanerozoic palaeomagnetic data and demonstrated an apparent periodicity of 350×10^6 years (Fig. 16): the Lower Palaeozoic field appears to have been marginally more reversed (opposite to the present geomagnetic configuration) than normal (same configuration as the present geomagnetic field), while the Upper Palaeozoic field was largely reversed and includes the late Carboniferous-Permian Kiaman interval. The Mesozoic field was predominatly normal while in Tertiary times the balance of polarities is about even. It does not seem to have been generally noted that these changes in the bias of the magnetic field polarity correlate with the rapid episodes of Phanerozoic continental drift, while the intervals of broadly-constant bias correlate with the longer "quasi-static" intervals between (Briden 1967). In the Precambrian a smaller body of data span a much longer period of time and a stringent analysis such as that attempted by McElhinny is not possible. In Fig. 16 I have summed the proportion of normal and reversed magnetisations of the palaeopoles plotted in Figs. 2-9 to estimate the polarity bias of the Precambrian field and used the percen-

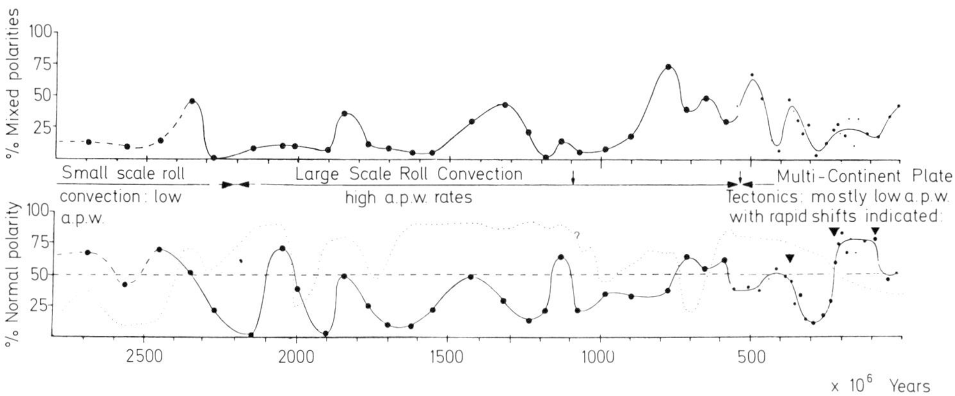

Fig. 16. The percentage of mixed polarities (representing the fractions of the palaeomagnetic studies plotted in Figs. 2 to 9 which record both normal and reversed directions) and the percentage of normal polarities in the same palaeomagnetic record plotted as a function of time before the present. The Precambrian analysis assumes that the a.p.w. curve plotted in Figs. 2 to 9 is a south a.p.w. path; this cannot yet be regarded as certain because there are significant gaps in the record at ca. 1500-1450 and 820-750 m.y. The Phanerozoic data are taken directly from McElhinny (1973). The dotted line in the lower figure shows the percentage of the continental crust in the southern hemisphere; this is uncertain after 1100 m.y. because it is not known when the Gondwanaland configuration was achieved (Sect. 3.6, it is calculated here on the assumption that it was attained by 1050 m.y.)

tage of studies exhibiting reversals to estimate the reversal frequency. The intervals chosen use segments of the a.p.w. path between well-constrained age dates and the number of poles used in the calculations range from 4 to 78.

In spite of the limitations of the data base, the resultant graphs (Fig. 16) bring out several features of the Proterozoic magnetic field: this field was predominantly of reversed polarity with a low reversal frequency between 2150 and 1000 m.y.; prior to this, polarities seem to have been about evenly balanced but the data points are too few to be sure of this. It is noteworthy that this long period of reversed polarity correlates with the bulk of the continental crust residing in the southern hemisphere (and the poles plotted in Figs. 2-9 are predominantly poles rather than antipoles). This enigmatic correlation is also suggested by the Phanerozoic record where the polarity bias changes from reversed to normal (Fig. 16) as the bulk of the continental crust moves from the southern hemisphere into the northern hemisphere.

Thus it is now possible to link some gross changes in the magnetic field characteristics to the dynamics of the crust and hence to changes in the driving mantle convection system. When a large scale mantle roll convection system was established by about 2200 m.y. (Sect. 5) the magnetic field polarity became strongly biased; it does not appear to have been upset by rapid a.p.w. movements directly because there is no clear correlation between reversal frequency and a.p.w. rate, but the polarity bias did change when a large part of the continental crust moved out of the southern hemisphere at about 2000 m.y. and 1800 m.y. The magnetic field experienced a marked decline in strength when the mantle convection system subdivided near the beginning of the Cambrian (Fig. 13) and it has only gradually achieved a magnitude comparable to the Proterozoic field (Fig. 15). Again the polarity bias of the field has been upset by dramatic changes in the mantle convection system which produced the rapid episodes of continental drift; the tendency for the magnetic field to change from predominantly reversed bias to predominantly normal bias as the crust moved into the northern hemisphere is a second order effect superimposed on these rapid changes in polarity bias.

Four inter-relating factors seem to be involved here: firstly, the concentration of the continents indicates a large scale prevailing mantle convection pattern; secondly, mantle convection determines the lateral temperature variation in the mantle at the core-mantle interface; thirdly, this temperature variation will assert some control on the liquid core motion, and fourthly, the core motion will decide the polarity bias.

5. Continental Consolidation and Lithosphere Development

5.1 The Evolution of Tectonic and Magmatic Environments

The consolidation of the continent crust and the history of tectonic and magmatic events within it document stages in heat release from the aesthenosphere and permit general conclusions to be drawn about lithosphere development. The Archaean ($\geq$2700 m.y.) tectonic régimes comprise dividions of high metamorphic grade granite-gneiss terrain separating greenstone belts (which essentially comprise thick successions of basic volcanics and volcanic-derived sediments metamorphosed to a much lower grade). Since these rocks developed as the proto-continental crust prior to the beginning of the palaeomagnetic record (Fig. 2), it is not possible to directly infer the form of the crust at these times. However, there are some suggestions that the crust had achieved the approximate form of Fig. 1 by about 2900 m.y. The oldest greenstone belts in the Kaapvaal and Zimbabwe cratons of southern Africa were formed between about 3550 and 2900 m.y. and have an essentially isotropic fabric on a large scale; they form broadly-

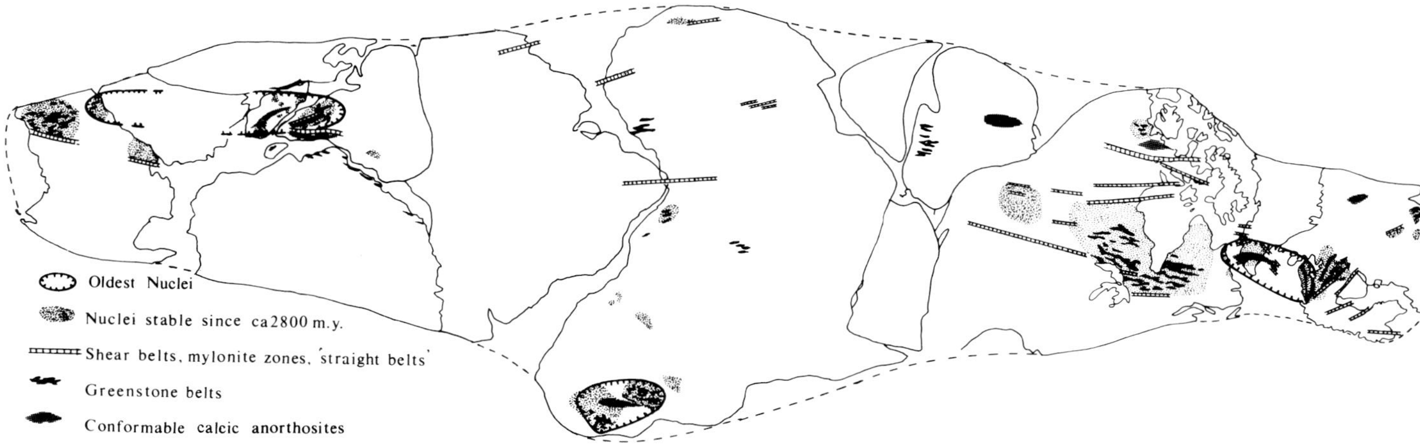

Fig. 17. The distribution of some Archaean and early Proterozoic tectonic and magmatic features on the Proterozoic Supercontinent

curved synclinal belts between high grade terrains with an outcrop pattern complicated by refolding and interaction with late- to post-tectonic intrusions (Windley 1977). The younger belts formed after 2900 m.y. in western Australia, south India, west and south Africa, in the Superior and Slave cratons of the Laurentian Shield, and in the Fennoscandian Shield are, however, notably anisotropic and have long axes oriented closely parallel to each other and to the long axis of the continental crust (Fig. 17). The greenstone belts have close analogies in rock types and chemistry with the present-day products of sea floor spreading, and may have developed between loosely-coupled or decoupled microcontinental blocks (Talbot 1973). The alignment evident in Fig. 17 indicates that aesthenosphere convection was taking place on at least two scales in late Archaean times, namely on a global scale required to concentrate the continental crust into a single unit, and on an aligned small scale represented by the permobile greenstone belt régime.

The next lineaments which accord closely with one another on the supercontinent reconstruction are the late Archaean to early Proterozoic straight belts (Fig. 17). These belts sometimes traverse distances of as much as 2000-3000 km across zones of already-stabilised crust with near isotropic fabric. Structures are deflected into these zones and reconstituted with a strong planar fabric (Watson 1973). They constitute major flaws running through the thickness of the continental crust which have controlled many later tectonic and thermal events including the subdivision of the crust into stabilised and reworked areas (Watson 1973, 1981); in early Proterozoic times they were important belts of volatile accession and relative weakness. The time interval represented by these belts is unknown but is reckoned to be several hundred million years near the Archaean-Proterozoic boundary; the shear belts associated with the Nagssugtoqidian in Greenland are dated 3000-2600 m.y. in age (Hickman 1979) and belts in the Canadian Shield appear to predate the Matachewan dykes at 2630 m.y.

These straight belts contrast with the later aspects of Proterozoic tectonism which were concentrated within "mobile belts". These latter are wide thermal-tectonic zones where older rocks have been reheated and/or tectonically reworked, and the oldest appear to be about 2100 m.y. in age. A prominent belt 2000-1800 m.y. in age ranges from the Fennoscandian-Ukrainian Shield (where it includes the Svecokarelian and Svecofennian terrains) through the Laurentian Shield (where it includes the Laxfordian, Nagssugtoqidian, Nain, Churchill and Bear terrains) and continues through the Siberian Shield (Sears and Price 1978); much of the West African Shield and adjacent parts of South America were affected by activity of similar age and more restricted belts occur in the China, India and Australian Shields. Together they form a broad zone of tectono-thermal activity (Fig. 18) and were contemporaneous with the first signs of subduction-related activity, mostly along the opposite margin of the continent. The younger mobile belts formed between 1350 and 950 m.y. are distinctly narrower in width but preserve the predominant trend parallel to the crustal axis.

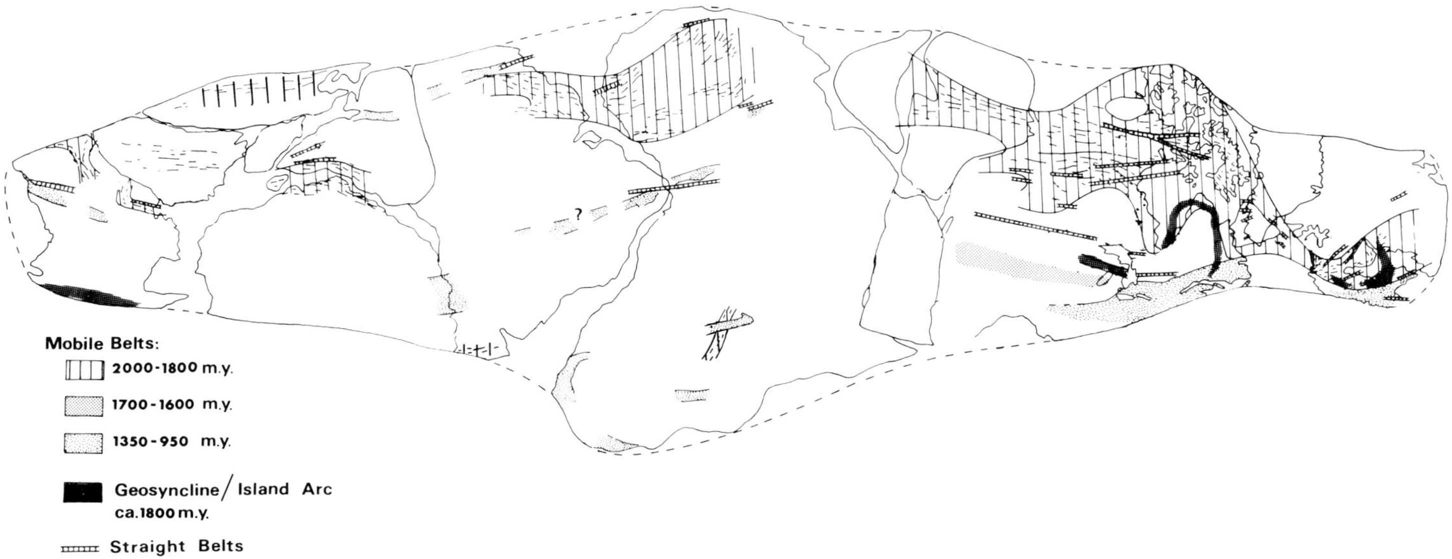

Fig. 18. The distribution of ensialic mobile belts formed between ca. 2200 and 950 m.y. on the Proterozoic Supercontinent

The Proterozoic era was also characterised by the intrusion of a distinctive set of igneous rocks. Early calcic anorthosites appear to have been intruded in late Archaean times and are mostly deformed but appear to exhibit some large scale trends that may have a regional significance (Windley and Bridgwater 1971, Sutton 1971 and Fig. 17). Younger anorthosites and rapakivi granites were intruded as large mushroom-shaped intrusions, and are remarkable not only because of their distinctive composition, but also because they were frequently intruded in an anorogenic environment (i.e., outside of the mobile belts) in regions with high local temperature gradients (Bridgwater and Windley 1973, Bridgwater et al. 1974). The zones affected by this kind of intrusion diminished gradually as the regions of high temperature gradients through the crust diminished, and the "chrontour" lines delimiting the termination of this activity define a gradual retreat towards one margin of the continental crust (Fig. 19). This margin thus appears to have remained relatively hotter than the opposite margin during Middle and Upper Proterozoic times. That this was the case is suggested by two further observations: firstly, rifting of the consolidated crust with the formation of deep aulacogens commenced firstly along the opposite margin of the crust at about 2000 m.y. in the Authapuscow aulacogen and spread extensively along the boundary between the Siberian and Laurentian Shields after about 1600 m.y. (Burke and Dewey 1973b). Presumably similar rifting affected the Australia-South America sector (although this is less well understood) and it was this colder margin of the Proterozoic crust that broke up to yield the smaller continental fragments in the present day scheme of plate tectonics. The second observation is that the hotter margin is the location of the first Proterozoic geosynclines and includes those belts of folding and sometimes calc-alkaline magmatism which can be linked to subduction of ocean crust beneath the continent (Fig. 19). These belts include the Mt. Isa (1800-1500 m.y.) and Adelaide (800-500 m.y.) geosynclines in Australia, the Transantarctic (Late Precambrian) belt, the Late Precambrian volcanic arcs along the Afro-Arabian margin (1000-600 m.y.), the Penokean (1800 m.y.), Ketilidian and Matatazai (1680-1600 m.y.) belts along the Laurentian margin and the Svecofennian arc (1700 m.y.) and Sveconorwegian margin (1000 m.y.) in the Fennoscandian Shield (Fig. 19). These zones were developing, in part, contemporaneously with ensialic mobile belts of comparable trend within the continental crust. Initial breakup of the continental crust (Sect. 3.6) appears to immediately postdate the formation of the Grenville-Sveconorwegian lineament, but the processes producing mobile belts were not entirely terminated by this event because most of the Pan African belts developed entirely as ensialic mobile belts between ca. 700 and 550 m.y. (Fig. 22, Piper et al. 1973, McWilliams and Kröner 1981).

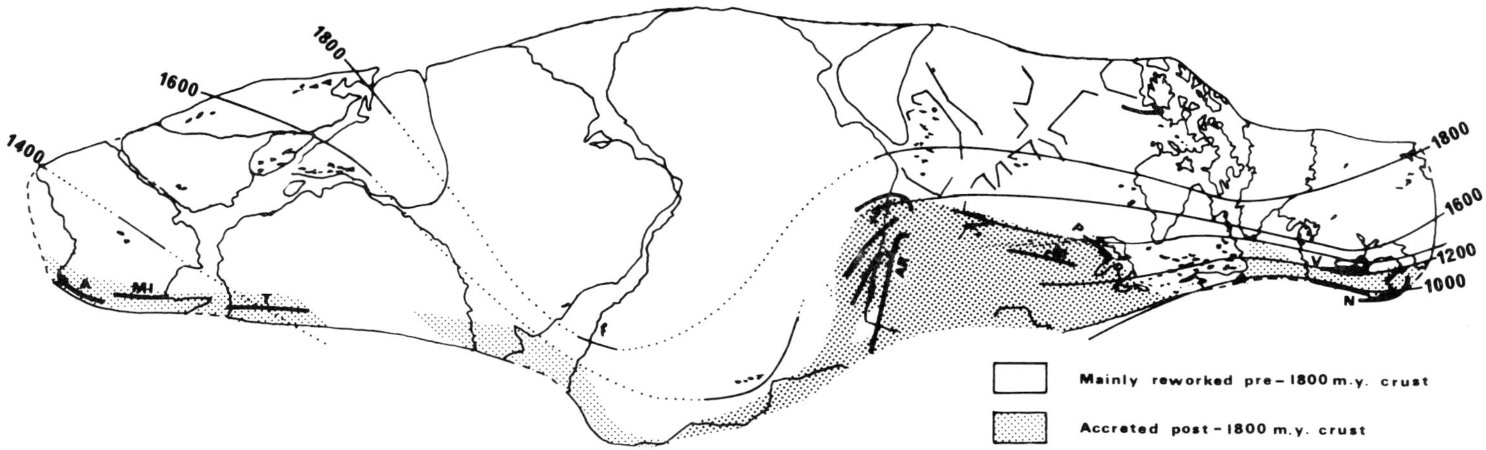

Fig. 19. The distribution of the Proterozoic massive anorthosites (ca. 2200-950 m.y.) with outcrops indicated in black; the contours define the approximate time of termination of this type of magmatism at exposed crustal levels in millions of years. The stippled areas are those zones of the continental crust which do not appear to contain reworked Archaean basement and may have been accreted since 1800 m.y. Features in these zones which can be related to the plate tectonic process of subduction of oceanic lithosphere beneath the continental crust are geosynclines indicated by the solid lines and lettered: A = Adelaide Geosyncline; MI = Mount Isa Geosyncline; T = Trans-Antartic; AR = Arabian Shield; M = Matatzai; P = Penokean; V = Svecofennian arc; N = Sveco-Norwegian arc

5.2 The Relationship to Lithosphere Development and Aesthenosphere Convection

The development of the continental crust into the Proterozoic supercontinent appears to have been controlled by two factors; the first was the progressive decline in radiogenic heat production from a value of 2-3 times the present value in Archaean times (Lambert 1976), and the second was the degassing of the mantle by partial melting which caused the solidus for mantle material (pyrolite) to migrate into higher temperature fields. When the pyrolite solidus and the mantle geotherm ceased to intersect in Archaean times, mantle-wide melting would have ceased and the fractionation of sialic material from it would presumably have been drastically reduced (Strong and Stevens 1974). At a somewhat later stage near the Archaean-Proterozoic boundary the geothermal gradient would have declined to the extent that it ceased to intersect the granulite stability field leading to the widespread dehydration of crustal and upper mantle amphibolites to granulites and the stabilisation of a thick granulitic crust (Tarling 1980). It was probably this event that led to the important transition in tectonic styles at ca. 2800 m.y., and with the stabilisation of large areas of sial only the oceanic lithosphere would have been an effective zone of convective heat transfer. This restriction of most of the mantle heat loss to the oceanic lithosphere presumably led to a temporary increase in temperature gradients here and a reduction in the strength of this part of the lithosphere. It is clear from the palaeomagnetic data that the continental crust was not again disrupted until late in Proterozoic times so it would appear that the greater heat loss facilitated by convection in the oceanic lithosphere combined with a continuing decline in the global heat production led to a faster increase in its strength relative to the continental lithosphere (Fig. 20). Only when the strengths of the oceanic and continental lithospheres were again comparable and the latter had cooled to an extent to permit large scale brittle fracture (Fig. 19) could changes in the mantle convection system lead to continental breakup (Tarling 1981).

The observed changes in the tectonic styles in the continental lithosphere can now be linked in a general way with the theory of mantle convection and the experimental observations on convecting fluids (Davies and Runcorn 1981). A declining radiogenic heat production within the mantle would have led to a progressive decrease in the mantle Rayleigh Number (Turcotte 1981). The actual value of this number will, of course, depend on the thickness of the convecting layer and the two models illustrated in Fig. 20 are calculated for whole-mantle single layer (Turcotte 1981) and multi-layer (McKenzie and Weiss 1975) convection. For either model the likely Rayleigh number exceeds the initial value for the onset of thermal convection. As the Rayleigh Number decreases the character of the convection is predicted to change (McKenzie and Weiss 1975, Davies and Runcorn 1981), from hexagonal to bimodal to roll convection (Fig. 20), i.e., the convection becomes increasingly anisotropic. The small-scale permobile

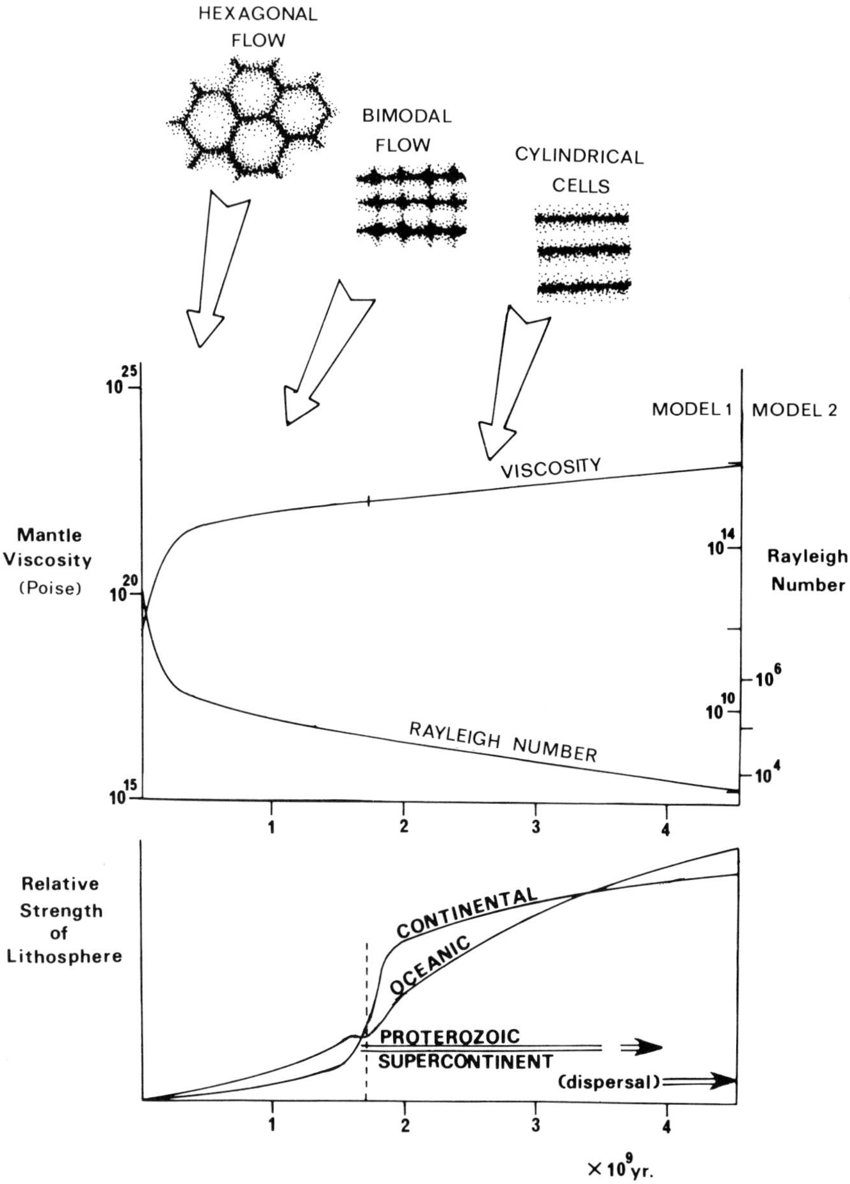

Fig. 20. The changing characteristics of thermal convection in the aesthenosphere from hexagonal to bimodal to cylindrical cells with the increase in mantle viscosity consequent on the decline in radiogenic heat production and the resultant decrease in mantle Rayleigh Number. The viscosity curve is calculated for whole-mantle convection with the upper boundary layer at T = 0°C (i.e., the surface plates play a dominant role in mantle convection as indicated by boundary layer theory). This is equivalent to the Model 1 for the increase in

(Continuation of Legend to Fig. 20 on p. 308)

(Legend to Fig. 20, cont.)
mantle viscosity (Turcotte 1981). Model 2 considers two layer convection in the mantle with the motion of the plate imposing a restraint and control on the driving convection pattern (McKenzie and Richter 1976). The large arrow shows the approximate assignments of convection patterns to different stages of lithosphere evolution, but it should be noted that the changes seem to have been transitional on different levels. Large scale roll convection seems to have concentrated the sialic crust (Fig. 1) in Archaean times prior to a transition from small scale bimodal to roll convection beneath the continental lithosphere. The lower graph is a purely diagrammatic presentation of the changing relative strengths of the continental and oceanic lithospheres (see Tarling 1981) with the consequences on breakup of the supercontinent and the establishment of multi-continent plate tectonics in Phanerozoic times

tectonic régime represented by the greenstone belts appears to illustrate this transition at about 2900 m.y. because the older belts are essentially isotropic while the younger belts are anisotropic but still seem to have been formed in an environment where the crustal blocks between were small and essentially decoupled. It can however, also be suggested that mantle convection existed on at least two levels at this time because a large scale roll convection system would presumably have been required in Archaean times to aggregate the globe-wide products of mantle fractionation into the primary lens-shaped body of crust (Fig. 17 and Fyfe 1974, 1976).

The straight belt régime of late Archaean-early Proterozoic times overlapped in time with the development of linear greenstone belts and the transition from bimodal to roll convection probably took place in different areas at different times. A subcontinental-crust lithosphere roll convection system appears to explain the remarkable linearity of these belts; the role of secondary convection in the zone between and above the downgoing limbs of the convection rolls is important in the model of McKenzie and Weiss (1975) and may explain the important restriction of volatile accession to these zones (Watson 1973). Since they represent the weaker belts of crust between larger blocks, the strike-slip motions might represent relative motions between the stabilised blocks because complete disruption does not appear to have been achieved. Little is known about the spacing between these belts but in West Greenland it appears to be about 100 km (Bak et al. 1975); this suggests the order of magnitude of sub-crustal roll convection in late Archaean times.

The ≤2200 m.y. Proterozoic mobile belts represent much wider zones of crustal reactivation and imply that the convection system operating beneath the crust had changed to a much larger scale by this time; since these belts continue to be axial and linear (Fig. 18) a roll system would appear to be still applicable and is anticipated from the continuing decline in radiogenic heat production. While changes in the thickness of the convecting layer below the lithosphere cannot be directly inferred, it seem likely that the major accretion events to the sial not only represent the active operation of this sub-continent convection, but bracket changes in the geometry of the convecting system. Moorbath (1975) recognises "accretion superevents" from isotopic data at 3800-3500 and 2800-2500 m.y., and the suggestion that later peaks in the frequency histogram of radiometric ages at 1900-1600 and 1200-900 m.y. (Dearnley 1966, Moorbath 1975) represent important accretions to the continental crust is supported by isotopic data (e.g., De Paolo 1981). The younger mobile belts (1350-950 m.y.) are both narrower and more closely spaced (Fig. 19) than the older belts and suggest that the roll cells had contracted by this time; they are unlikely to represent the identical systems as those responsible for the $\geq$1800 m.y. belts because there was only intermittent thermal/tectonic activity between 1800 and 1100 m.y. within the continental crust.

The complete replacement of this system of sub-crustal convection by a single global system is suggested by the effects of the ca. 1100 m.y. relative movements which had the resultant effect of shortening the length of the crustal body and producing a near-triangular shaped body with subduction-related miogeosynclines developing around the margins (Fig. 22). The next important change in aesthenosphere convection is represented by the subdivision of the covection system to form at least two separate cells in Lower Cambrian times (Fig. 13). With the continuing decline in radiogenic heat production and the thickening of the continental lithosphere, the relationship between this slow-moving multicell system and lithosphere tectonics has become increasingly obscure. Kanasewich et al. (1978) have analysed Phanerozoic plate distributions to suggest that the present convection system is symmetrical about the Earth's rotation axis and comprises three large cells, with the probability that only two cells were applicable to late Palaeozoic and Mesozoic times when the continental crust was aggregated together in the transient supercontinent of Pangaea. Kanasewich (1976) further shows that the present plate system is an organised one with two antipodal circular plates (centred on Africa and the Pacific) which are separated by a ring of elliptical plates; he infers that this organised system is only likely to exist if mantle convection currents now involve the entire mantle. Although theoretical and experimental considerations predict that the convection cells will continue to be cylindrical rolls until the Rayleigh Number becomes too small for aesthenosphere convection (Davies and Runcorn 1981), their form is now clearly less evident because the lithosphere is thicker and imposes a powerful restraint and control on the primary system.

6. Shelf Seas in Precambrian Times

6.1 Deduction from the Geological Record

The continental crust originated by fractionation from the mantle and seems to have been formed largely prior to 2500 m.y. with some important and discrete accretion events occuring later in Proterozoic times (see review by Windley 1977). Since the oceans also have a secondary origin and are a complementary product of the degassing of the mantle, their formation was presumably closely linked to that of the continental crust, and the continent does indeed appear to have presented a low freeboard to the oceans since the beginning of Archaean times (see Piper 1978 for discussion). The evidence for the extent of the shallow seas across the continental crust is however of a fragmentary nature and can only be surmised from the distribution of the sedimentary rocks.

Stromatolites are the fossil remains secreted by blue-green algae and bacteria and appear to form largely or wholly in the intertidal zone; they are only known from the Kaapvaal and Superior cratons in Archaean times but are much more common in the Proterozoic record (Windley 1977). The oldest ca. 3800 m.y. sedimentary rocks in West Greenland contain banded iron formations which are also common in Archaean greenstone belts. In the latter environment they occur in the higher levels of the volcanic-sedimentary successions, and a gradation from lower argillaceous deeper water environments to higher arenaceous and shallower water environments is frequently recognisable in these sequences (Anhaeusser 1971). The deposition of ironstone deposits was particularly prominent in Lower Proterozoic times between ca. 2600 and 1900 m.y. when their extent suggests the very wide distribution of shallow shelf seas during these times (Fig. 21). They are one product of deposition in shallow seas (and possibly alkaline lakes) during these times, and nearly always occur as part of a repeated sequence comprising dolomite, quartzite, red and black ferruginous shale, bedded iron ores and black shales; iron deposition as oxide, carbonate, silicate and sulphide represents a lateral facies transition from shallower to deeper water environments (Goodwin 1973a). The banded iron ores represent alternating siliceous and ferruginous layers, and although it is not certain that this banding is tidal, it does represent the regular episodic introduction of oxygen into the environment so that the dissolved ferrous iron (probably of volcanic origin) in solution could be deposited as ferric oxides or hydroxides (Trendall 1968). They appear to have been laid down in shallow water depths of less than 70 m of water (Dimroth 1968) but individual layers can be followed over distances of hundreds of kilometers, and suggest that the depositional basins had a very wide extent. Another remarkable feature of these deposits is the paucity of clastic deposits which implies that little land area was exposed and/or topographic gradients were too slight to permit much

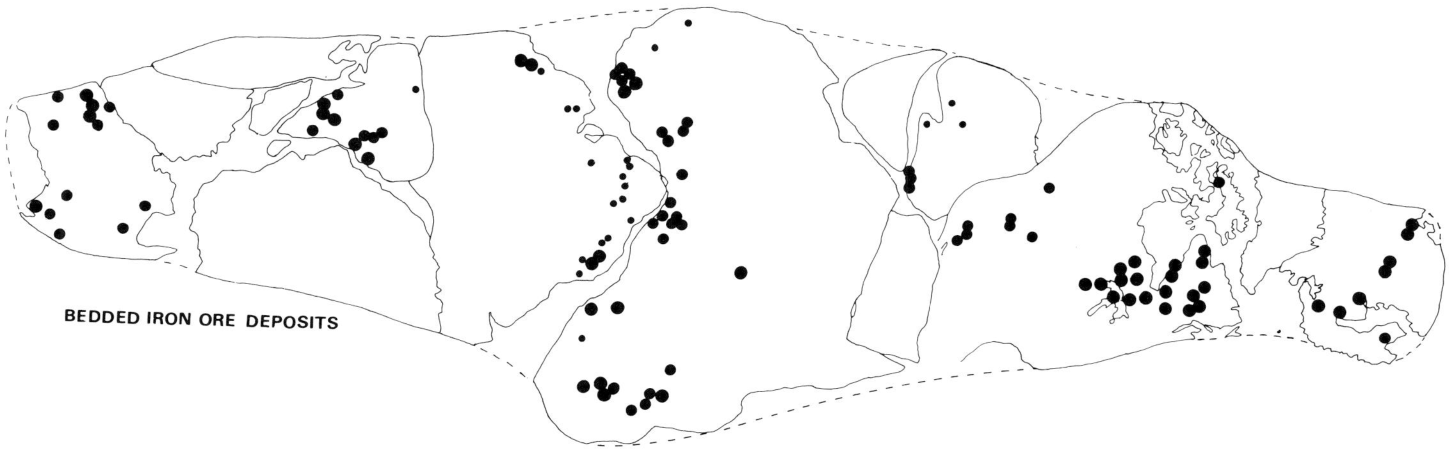

Fig. 21. The distribution of Precambrian bedded iron ore deposits within the Proterozoic Supercontinent. These are rocks of Lower Proterozoic age (>1900) deposited in shallow seas and possibly alkaline lakes. Large circles are deposits of more than 1000 million tons, intermediate circles are deposits of 100-1000 million tons and small circles are deposits of less than 100 million tons (After Goodwin 1973b)

degradation; in a regional context they seem to have formed in shallow seas surrounding Archaean nucleii (Goodwin 1973b and Fig. 21).

Following the crustal accretion event at ca. 1900-1800 m.y. there is a marked change in the characteristics of sedimentation. The disappearance of ironstones and the appearance of hematite-bearing red sediments seems to be related to the buildup of atmospheric oxygen and the consequent transition from mildly-reducing to mildly-oxidising conditions (Cloud 1973), but there is no doubt that the sedimentary record is much sparser between the ca. 1800 m.y. mobile episodes and Late Precambrian times. This sparse record suggests that seas covered very little of the continental crust because much of the ca. 1750-900 m.y. sedimentary record comprises deposits restricted to narrow and sometimes deep aulacogens (Stewart 1976) and is predominantly of a fluviatile (river-deposited) origin and formed in flood plain environments.

A change in environment is again observed in Late Precambrian times and its onset probably correlates with the first continental breakup at ca. 1100 m.y. (Sect. 3.6). It is characterised by the transgression of wide and shallow seas, and there is a long continuity of sedimentary deposits in circum-shield environments (Stewart 1976). Quartzites, limestones and dolomites typically grade laterally into miogeosynclinal sequences forming in developing geosynclines around the margin of the continent (Fig. 22). The next important environmental change was probably brought about by the continental breakup and dispersal in Lower Cambrian times. Long term sea level fluctuations are largely controlled by the length of the mid-ocean ridge system (Valentine and Moores 1970) and continental fragmentation has achieved major marine transgressions in Phanerozoic times. Whether a transgression or a regression results will depend on the ratio of the mid-ocean ridge volume to that of the complementary subduction zones, and a major regression is observed in most continents at the end of Precambrian times. The transgression of Lower-Middle Cambrian times seems to have begun when the new ocean basins had stabilised (Fig. 11), but this transgression is a complex phenomenon in detail probably because the breakup and ocean basin formation was a sequential rather than a single phenomenon (Fig. 9). This event seems to have increased the length of the marine shoreline by at least 50% (Fig. 11) and brought about a marked change in the chemistry of the oceans. An important indicator is the S^{34} isotope content of marine sediments which shows a dramatic increase at the Proterozoic-Phanerozoic transition (Holser 1977) and suggests that a rapid mixing as brines accumulated in enclosed basins isolated from the sea in late Precambrian times were released into the surface layer of the oceans as the continental crust was dismembered. Holser also postulates that later stages in this mixing cycle associated with sulphide-sulphate transfers restored nutrients in solution to surface waters and increased the deposition of organic carbon. The increase in calcium sulphate and phosphate in surface waters could accommodate the proliferation

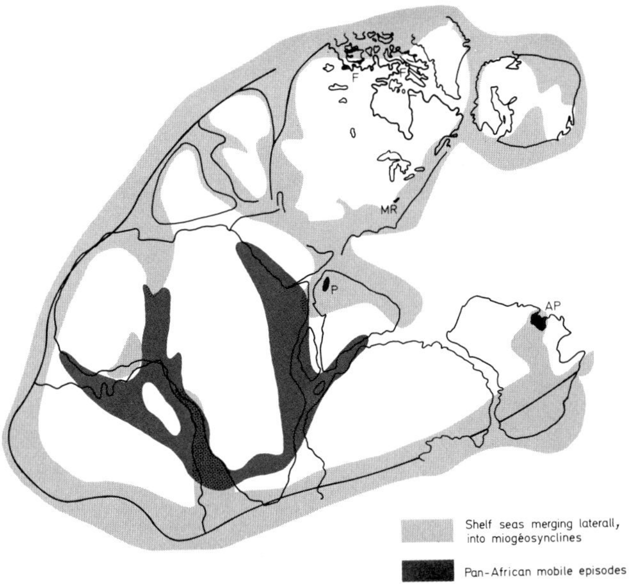

Fig. 22. Distribution of the continents in Late Precambrian times illustrating the approximate extent of the shelf seas before the Cambrian break up. This reconstruction is applicable to the Fenno-scandian-Ukrainian Shield after 1000 m.y. but can at present only be confirmed for the Australian and the India-Antarctica Shields after about 750 m.y. The dark shaded area illustrates the extent of the Pan-African mobile episodes at ca. 750-550 m.y., the last major mobile episode in the continental crust. The black areas are extensive basaltic magmatism of latest Precambrian age and are lettered: F = Franklin; MR = Mt Rogers; P = North Pakistan and A = Antrim Plateau; these episodes may have predated final breakup of this supercontinent by up to 50 m.y.

of apatite-secreting metazoa at this time: critical developments in these life forms were already underway some 60-100 m.y. prior to this transition (Lowenstam 1979) and it is the appearance of hard-bodied organisms that is specifically correlated with this event.

6.2 Implications to Tidal Friction

The present multi-continent plate tectonic system comprises a large number of dispersed continental fragments and a high total length of marine shoreline. As is shown in this paper, this situation is exceptional in the geological record: between 2800 and 570 m.y. the continental crust comprised a single unit with a possible short exception at ca. 1100 m.y., and a much more restricted marine shoreline. Furthermore the supercontinent was concentrated in high latitudes

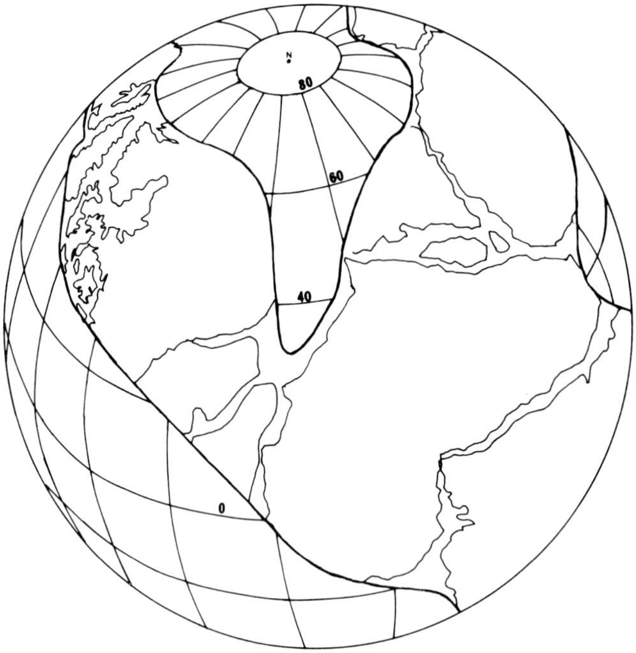

Fig. 23. The Proterozoic Supercontinent at ca. 700 m.y.

(Fig. 13) over much of this interval and broad equatorial shelf seas seem to have been unimportant. For these two reasons the tidal retarding couple was probably small in Proterozoic times and the lunar recession during this interval would thus have been limited; some rather sparse and contentious geological evidence (see Piper 1978) suggests that this was the case. Since transgressions and regressions are related to continental breakup and formation of new ocean ridge systems and the former, at least, did not occur prior to 1100 m.y., marine conditions were uniform for much longer periods of time than during the Phanerozoic; frequent changes of environment effected by transgressions or regressions do not appear to have characterised these times.

Calculations of the tidal retarding couple using the data of this paper could proceed, to a first approximation, on the assumption that 1. shallow shelf seas covered most of the Proterozoic supercontinent outside of the cratonic nuclei between 2700 and 1900 m.y. (Fig. 21), 2. shallow shelf seas covered only marginal zones of the crust between about 1900 and 1000 m.y. (geosynclines subduction-related belts in Figs. 18 and 19), 3. progressive transgression of shallow circumshield seas took place between about 900 and 570 m.y. (Fig. 22). Whilst these assumptions are highly generalised, they should help to quantify the major characteristics of lunar recession during the Proterozoic era.

Acknowledgements. I am grateful to Professor R.L. Wilson for valuable discussions about the data and interpretations presented here, and Professors Peter Brosche and Jürgen Sündermann for the opportunity to take part in the Tidal Friction Conference.

References

Anhaeusser CR (1971) Cyclic volcanicity and sedimentation in the evolutionary development of Archaean greenstone belts of shield areas. Geol. Soc Aust Spec Publ 3:57-70

Baer AJ (1976) The Grenville Province in Helikian times: a possible model of evolution. Philos Trans R Soc London Ser A 280:499-515

Baer AJ (1977) Speculations on the evolution of the lithosphere. Precambrian Res 5:249-260

Bailey DK (1974) Continental rifting and alkaline magmatism. In: Sorensen H (ed) The alkaline rocks. John Wiley, New York, pp 148-159

Bak J, Sorensen K, Grocott J, Kortsgaard JA, Nash D, Watterson J (1975) Tectonic implications of Precambrian shear belts in western Greenland. Nature (London) 254:566-569

Berger GW, York D, Dunlop DJ (1979) Calibration of Grenvillian palaeopoles by $^{40}Ar/^{39}Ar$ dating. Nature (London) 277:46-47

Bergh HW (1970) Palaeomagnetism of the Stillwater Complex, Montana. In: Runcorn SK (ed) Palaeogeophysics. Academic Press, London New York, pp 143-158

Bingham DK, Evans ME (1975) Precambrian geomagnetic field reversal. Nature (London) 253:332-333

Briden JC (1967) Recurrent continental drift of Gondwanaland. Nature (London) 215:1334-1339

Briden JC, Irving E (1964): Palaeolatitude spectra of sedimentary palaeoclimatic indicators. In: Nairn AEM (ed) Problems of palaeoclimatology. John Wiley Interscience, New York, pp 199-224

Bridgwater D, Windley BF (1973) Anorthosites, post-orogenic granites, acid volcanic rocks, and crustal development in the North Atlantic Shield during the Mid-Proterozoic. In: Lister LA (ed) Symposium on granite, gneisses and related rocks. Geol Soc S Afr Spec Publ 3:307-318

Bridgwater D, Sutton J, Watterson J (1974) Crustal downfolding associated with igneous activity. Tectonophysics 21:57-77

Buchan KL, Dunlop DJ (1976) Palaeomagnetism of the Haliburton intrusions: superimposed magnetisations, metamorphism and tectonics in the Late Precambrian. J Geophys Res 81:2951-2967

Bullard EC, Everett JE, Smith AG (1965) The fit of the continents around the Atlantic. Philos Trans R Soc London Ser A 285:45-51

Burke K, Dewey JF (1973a) An outline of Precambrian plate development. In: Tarling DH, Runcorn SK (eds) Implications of continental drift to the earth sciences, vol II. Academic Press, London New York, pp 1035-1045

Burke K, Dewey JF (1973b) Plume generated Triple Junctions: key indicators in applying plate tectonics to old rocks. J Geol 81:406-433

Burke K, Dewey JF, Kidd WSF (1976) Precambrian palaeomagnetic results compatible with contemporary operation of the Wilson Cycle. Tectonophysics 33:287-299

Carey SW (1976) The expanding Earth. Elsevier, Amsterdam

Carmichael CM (1967) An outline of the intensity of the palaeomagnetic field of the Earth. Earth Planet Sci Lett 3:351-354

Cloud PE (1973) Palaeontological significance of the Banded Iron Formation. Econ Geol 68:1135-1143

Cowie JW (1971) Lower Cambrian faunal provinces. In Middlemiss FA, Rawson PF, Newall G (eds) Faunal provinces in space and time. Seel House Press, Liverpool, pp 21-46

Davies PA, Runcorn SK (eds) (1981) Mechanisms of continental drift and plate tectonics. Academic Press, London New York, 362 pp

Dearnley P (1966) Orogenic fold belts and a hypothesis of Earth evolution. In: Ahrens LH (ed) Physics and chemistry of the Earth, vol VII. Pergamon Press, New York, pp 1-114

DePaolog DJ (1981) Neodymium isotopes in the Colorado Front Range and crust-mantle evolution in the Proterozoic. Nature (London) 291: 193-196

Dimroth E (1968) Sedimentary textures, diagenesis, and sedimentary environment of certain Precambrian ironstones. Neues Jahrb Geol Palaeontol 130:247-274

Doig R (1970) An alkaline rock province linking Europe and North America. Can J Earth Sci 7:22-28

DuBois PM (1962) Palaeomagnetism and correlation of Keweenawan rocks. Geol Surv Can Bull 71:1-75

Dunlop DJ (1981) Palaeomagnetic evidence for Proterozoic continental development. Philos Trans R Soc London Ser A 301:265-277

Elston DP, Grommé LS (1974) Palaeomagnetic polar wandering from Unkar Group and Nankoweap Formation eastern Grand Canyon, Arizona. In: Karlston TNV, Swann GA, Eastwood RL (eds) Geology of Northern Arizona with notes on archaeology and Palaeoclimate, part I. Regional studies. Geol Soc Am, Boulder, Colorado, pp 97-117

Embleton BJJ (1969) Evidence of rapid polar shift from the palaeomagnetic record of Eurasia. Nature (London) 222:1036-1037

Embleton BJJ, McElhinny MW (1975) The palaeoposition of Madagascar: palaeomagnetic evidence from the Isalo Group. Earth Planet Sci Lett 27:329-341

Embleton BJJ, Schmidt PW (1979) Recognition of common Precambrian polar wandering: a conflict with plate tectonics. Nature (London) 282:705-709

Engel AEJ, Kelm DL (1972) Pre-Permian global tectonics: a tectonic test. Bull Geol Soc Am 83:2325-2340

Engel AEJ, Itson SP, Engel CG, Stockney DH, Cray EJ, Jr (1974) Crustal evolution and global tectonics: a petrogenic view. Bull Geol Soc Am 85:843-858

Evans ME, Haye GS, Bingham DK (1980) The palaeomagnetism of the Great Slave Supergroup: the Akaitcho River Formation. Can J Earth Sci 17:1389-1395

Fyfe WS (1974) Archaean tectonics. Nature (London) 249:338

Fyfe WS (1976) Heat flow and magmatic activity in the Proterozoic. Philos Trans R Soc London Ser A 280:655-660

Games KP (1980) The magnitude of the archaeomagnetic field in Egypt between 3000 and 0 B.C. Geophys J R Astron Soc 63:45-56

Glikson AY (1979) The missing Precambrian crust. Geology 7:449-454

Glikson AY (1981) Uniformitarian assumptions, plate tectonics and the Precambrian Earth. In: Kröner A (ed) Precambrian plate tectonics. Elsevier, Amsterdam, pp 91-104

Glikson AY (1982) Geochemical, Isotopic and palaeomagnetic tests of Precambrian sial-sima patterns and early Earth radii. Bull geol Soc Amer (in press)

Goodwin AM (1973a) Archaean iron-formations and tectonic basins of the Canadian Shield. Econ Geol 68:915-933

Goodwin AM (1973b) Plate tectonics and evolution of Precambrian crust. In: Tarling DH, Runcorn SK (eds) Implications of continenatal drift to the earth sciences, vol II. Academic Press, London New York, pp 1047-1069

Greenwood WR, Hadley DG, Anderson RE, Fleck RJ, Schmidt DL (1976) Late Proterozoic cratonisation in south western Saudi Arabia. Philos Trans R Soc London Ser A 280:517-527

Harris NBW, Gass IG (1981) Significance of contrasting magmatism in north east Africa and Saudi Arabia. Nature (London) 289:394-396

Hickman MH (1979) A Rb-Sr age and isotope study of the Ikertôk, Nordre Strømfjord, and Evighedsfjord shear belts, West Greenland-outline and preliminary results. Grønl Geol Undersøgelse Rapp 89:125-128

Holland CH (ed) (1974) Cambrian of the British Isles, Norden and Spitsbergen. Wiley, New York, 300 pp

Holser WT (1977) Catastrophic chemical events in the history of the ocean. Nature (London) 267:403-408

Irving E, McGlynn JC (1976) Proterozoic magnetostratigraphy and the tectonic evolution of Laurentia. Philos Trans R Soc London Ser A 280:433-468

Irving E, Naldrett AJ (1977) Palaeomagnetism in Abitibi Greenstone Belt, and Abitibi and Matachewan diabase dikes: Evidence of the Archaean geomagnetic field. J Geol 85:157-175

Irving E, Park JK (1972) Hairpins and super-intervals. Can J Earth Sci 9:1318-1324

Kanasewich ER (1976) Plate Tectonics and planetary convection. Can J Earth Sci 13:331-340

Kanasewich ER, Havskiv J, Evans ME (1978) Plate Tectonics in the Phanerozoic. Can J Earth Sci 15:919-955

Kent PE (1973) East African evidence of the palaeoposition of Madagascar. In: Tarling DH, Runcorn SK (eds) Implications of continental drift to the earth sciences. Academic Press, London New York pp 873-875

Kirschvinck JL (1978a) The Precambrian-Cambrian boundary problem: magnetostratigraphy of the Amadeus Basin Central Australia. Geol Mag 115:139-150

Kirschvinck JL (1978b) The Precambrian-Cambrian boundary problem: palaeomagnetic directions from the Amadeus Basin, Central Australia. Earth Planet Sci Lett 40:91-100
Klootwyk CT (1980) Early Palaeozoic palaeomagnetism in Australia. Tectonophysics 64:249-332
Kobayashi K (1968) Palaeomagnetic determination for the intensity of the geomagnetic field in the Precambrian. Phys Earth Planet Inter 1:387-395
Kröner A (1976) Precambrian mobile belts of southern and eastern Africa - ancient sutures or sites of ensialic mobility? A case for crustal evolution towards Plate Tectonics. Tectonophysics 40:101-135
Lambert RStJ (1976) Archaean thermal régimes, crustal and upper mantle temperatures and a progressive evolutionary model for the Earth. In: Windley BF (ed) The early history of the Earth. John Wiley, New York, pp 363-387
Lowenstam HA (1979) What, if anything, happened at the transition from the Precambrian to the Phanerozoic. Precambrian Res 11:89-91
McElhinny MW (1971) Geomagnetic reversals during the Phanerozoic: Science 172:157-159
McElhinny MW (1973) Palaeomagnetism and Plate Tectonics. Cambridge Univ Press, Cambridge, 358 pp
McElhinny MW, Embleton BJJ (1976) Precambrian and early Palaeozoic palaeomagnetism in Australia. Philos Trans R Soc London Ser A 280:417-432
McElhinny MW, Evans ME (1968) An investigation of the strength of the geomagnetic field in the early Precambrian. Phys Earth Planet Inter 1:485-497
McElhinny MW, Senanayake WE (1980) Palaeomagnetic evidence for the existence of the geomagnetic field 3.5 Ga ago. J Geophys Res 85:3523-3528
McElhinny MW, Taylor SR, Stevenson DJ (1978) Limits to the expansion of Earth, Moon, Mars and Mercury and to changes in the gravitational constant. Nature (London) 271:316-321
McGlynn JC, Irving E (1978) Palaeomagnetism of the Pearson Formation, Great Slave Supergroup N.W.T. and the Coronation Loop. Can J Earth Sci 15:642-654
McKenzie DP, Richter F (1976) Convection Currents in the Earth's mantle. Sci Am 242:72-89
McKenzie DP, Weiss NO (1975) Speculations on the thermal and tectonic history of the Earth. Geophys J R Astron Soc 42:131-174
McWilliams MO, Dunlop DJ (1978) Grenville palaeomagnetism and Tectonics. Can J Earth Sci 15:687-695
McWilliams MO, Kröner A (1981) Palaeomagnetism and tectonic evolution of the Pan-African Damara Belt, Southern Africa. J Geophys Res 86:5147-5162
Moorbath S (1975) Evolution of Precambrian crust from strontium isotopic evidence. Nature (London) 254:395-398
Morgan GE (1976) Palaeomagnetism of a slowly-cooled plutonic terrain in Western Greenland. Nature (London) 259:283-385

Morgan GE, Briden JC (1981) Aspects of Precambrian palaeomagnetism with new data from the Limpopo Mobile Belt and Kaapvaal craton in Southern Africa. Phys Earth Planet Inter 24:142-168
Morris WA (1977) Palaeomagnetism of the Gowganda and Chibougamau Formations: evidence for 2200 m.y. folding and remagnetisation event of the Southern Province. Geology 5:137-140
Morris WA (1979) A positive contact test between Nipissing diabase and Gowganda argillites. Can J Earth Sci 16:607-611
Morris WA (1981) A positive fold test from Nipissing diabase. Can J Earth Sci 18/591-598
Morris WA, Roy JL (1977) Discovery of the Hadrynian Track and further study of the Grenville Problem: Nature (London) 266:689-692
Onstott TC, Hargraves RB (1981) Proterozoic transcurrent tectonics: palaeomagnetic evidence from Venezuela and Africa. Nature (London) 289:131-136
Palmer HC (1970) Palaeomagnetism and correlation of some Middle Keweenawan rocks, Lake Superior. Can J Earth Sci 7:1410-1436
Palmer HC, Halls HC, Pesonen LJ (1981) Remagnetisation in Keweenawan rocks Part I: Conglomerates. Can J Earth Sci 18:599-618
Park JK (1975) Palaeomagnetism of the Flin-Flon Snow Lake greenstone belt, Manitoba and Saskatchewan. Can J Earth Sci 12:1272-1291
Park JK (1977) A reconnaissance palaeomagnetic study of the Central Labrador Trough, Quebec. Can J Earth Sci 14:159-174
Patchett PJ, Bylund G (1977) Age of Grenville Belt magnetisations: Rb-Sr and palaeomagnetic evidence from Swedish dolerites. Earth Planet Sci Lett 35:92-104
Patchett PJ, Bylund G, Upton BGJ (1978) Palaeomagnetism and the Grenville orogeny: new Rb-Sr ages from dolerites in Canada and Greenland. Earth Planet Sci Lett 40:349-364
Pesonen LJ, Halls HC (1979) The palaeomagnetism of Keweenawan dykes from Baraga and Marquette Counties, northern Michigan. Can J Earth Sci 16:2136-2149
Pesonen LJ, Navanlinna H (1981) Late Precambrian Keweenawan asymmetric reversals. Nature (London) 294:436-439
Piper JDA (1975) The palaeomagnetism of Precambrian igneous and sedimentary rocks of the Orange River Belt in South Africa and South West Africa. Geophys J R Astron Soc 40:313-344
Piper JDA (1976a) Palaeomagnetic evidence for a Proterozoic Supercontinent. Philos Trans R Soc London Ser A 280:469-490
Piper JDA (1976b) Definition of pre-2000 m.y. apparent polar wander movements. Earth Planet Sci Lett 28:470-478
Piper JDA (1978) Geological and Geophysical evidence relating to continental growth and dynamics and the hydrosphere in Precambrian times: a review and analysis. In: Brosche P, Sündermann J (eds) Tidal friction and the Earth's rotation. Springer, Berlin Heidelberg New York, pp 197-241
Piper JDA (1980a) Palaeomagnetism of the Swedish rapakivi suite: Proterozoic tectonics of the Baltic Shield. Earth Planet Sci Lett 46:443-461
Piper JDA (1980b) Paleomagnetic study of Svecofennian basic rocks: Proterozoic configuration of the Fennoscandian, Laurentian and Siberian Shields. Phys Earth Planet Inter 23:165-187

Piper JDA (1981) The altitude dependence of magnetic remanence in the slowly-cooled Precambrian plutonic terrain of Central-West Greenland. Earth Planet Sci Lett 54:449-466

Piper JDA (1981b) Palaeomagnetism of the (Late Precambrian) West Greenland kimberlite-lamprophyre suite: definition of the Hadrynian Track. Phys Earth Planet Inter 27:164-186

Piper JDA (1982a) The Precambrian palaeomagnetic record: the case for the Proterozoic Supercontinent. Earth Planet Sci Lett 59:61-89

Piper JDA (1982b) Continental movements and breakup in late Precambrian-Cambrian times: prelude to Caledonian orogenesis. In: Gee D, Sturt BA (eds) The Scandinavian Caledonides and adjacent areas. Wiley, New York

Piper JDA, Briden JC, Lomax K (1973) Precambrian Africa and South America as a single continent. Nature (London) 245:244-248

Roy JL, Lapointe PL (1976) The palaeomagnetism of Huronian redbeds and Nipissing diabase; post-Huronian igneous events and apparent polar path for the interval -2300 to -1500 Ma for Laurentia. Can J Earth Sci 13:749-779

Roy JL, Robertson WA (1978) Palaeomagnetism of the Jcobsville Formation and the apparent polar path for the interval -1100 to -670 m.y. for North America. J Geophys Res 83:1289-1304

Schmidt PW (1980) Palaeomagnetism of igneous rocks from the Belcher Island, Northwest Territories, Canada. Can J Earth Sci 17:807-829

Schmidt PW, Embleton BJJ (1982) A geotectonic paradox: has the Earth expanded? J Geophys (in press)

Schults LD, Dunlop DJ (1981) Proterozoic magnetic overprinting of Archaean rocks in the Canadian Superior Province. Nature (London) 291:642-644

Schwarz EJ, Symons DTA (1969) Geomagnetic intensity between 100 million and 2500 million years ago. Phys Earth Planet Inter 2:11-18

Scotese CR, Bambach RK, Barton C, Van der Voo R, Ziegler AM (1979) Palaeozoic base maps. J Geol 87:217-277

Sears JW, Price RA (1978) The Siberian Connection: A case for Precambrian separation of the North American and Siberian cratons. Geology 6:267-270

Shackleton RM (1973) Correlation of structures across Precambrian orogenic belts in Africa. In: Tarling DH, Runcorn SK (eds) Implications of continental drift to the earth sciences. Academic Press, London New York, pp 1091-1095

Smith AG, Hallam A (1970) The fit of the southern continents. Nature (London) 225:139-144

Smith PJ (1967) The intensity of the ancient geomagnetic field: A review and analysis. Geophys J R Astron Soc 12:213-362

Smith RL, Stearn JEF, Piper JDA (1982) Palaeomagnetism of Upper Proterozoic sediments, NW Scotland. Scott J Geol (in press)

Stearn JEF, Piper JDA (1982) Palaeomagnetism of the Sveconorwegian mobile belt, SW Scandinavia. (in prep)

Stewart JH (1976) Late Precambrian evolution of North America: Plate Tectonic implications. Geology 4:11-15

Strong DF, Stevens RK (1974) Possible thermal explanation of contrasting Archaean and Proterozoic geological régimes. Nature (London) 248:37-39

Sutton J (1971) Some developments in the crust. Spec Publ Geol Soc Aust 3:1-10
Sutton J, Watson JV (1974) Tectonic evolution of continents in early Proterozoic times. Nature (London) 247:433-435
Symons DTA (1971) A palaeomagnetic study of the Nipissing diabase, Blind River, Elliot Lake area, Ontario. Geol Surv Can Pap 70-63: 18-30
Talbot CJ (1973) A plate tectonic model for the Archaean crust. Philos Trans R Soc London Ser A 237:413-428
Tarling DH (1980) Lithosphere evolution and changing tectonic régimes. J Geol Soc (London) 137:459-467
Tarling DH (1981) The tectonic evolution of the Earth's surface and changing lithospheric properties. In: Davis PL, Runcorn SK (eds) Mechanisms of continental drift and plate tectonics. Academic Press, London New York, pp 61-74
Trendall AF (1968) Three great basins of Precambrian banded iron formations; a systematic comparison. Bull Geol Soc Am 79:1527-1544
Turcotte DL (1981) Some major questions concerning mantle convection. In: Davis PL, Runcorn SK (eds) Mechanism of continental drift and plate tectonics. Academic Press, London New York, pp 173-182
Valentine JW, Moores EM (1970) Plate tectonic regulation of faunal diversity and sea level: a model. Nature (London) 228:657-659
Vidal G (1977) Late Precambrian microfossils. Geol Mag 114:292-294
Vidal G (1979) Acritarchs and the correlation of the Upper Proterozoic. Publ Inst Mineral Palaeontol Quat Geol, Univ Lund 219:1-22
Viezer J (1973) Sedimentation in geologic history: recycling versus evolution or recycling with evolution. Contrib Mineral Petrol 38:261-278
Viezer J (1976) ^{87}Sr:^{86}Sr evolution of seawater during geologic history and its significance as an index of crustal evolution. In: Windley BF (ed) The early history of the Earth. Wiley, New York, pp 569-578
Watson JV (1973) Effects of reworking on high-grade gneiss complexes. Philos Trans R Soc London Ser A 273:443-456
Watson J (1981) Flaws in the continental crust. Mercian Geol 8:1-6
Wilson RL (1970) Permanent aspects of the Earth's non-dipole magnetic field over Upper Tertiary time. Geophys J R Astron Soc 19:417-437
Wilson RL (1971) Dipole offset. The time average palaeomagnetic field over the past 25 million years. Geophys J R Astron Soc 22:491-504
Windley BF (1977) The evolving continents. John Wiley, New York, 385 pp
Windley BF, Bridgwater D (1971) The evolution of Archaean low- and high-grade terrains. Geol Soc Aust Spec Publ 3:33-46
Williams GE (1966) Palaeogeography of the Torridon Applecross Group. Nature (London) 209:1303-1306
Wynne-Edwards HR (1976) Proterozoic ensialic orogenesis: the millipede model for ductile plate tectonics. Am J Sci 268:289-308

Chemical and Isotopic Evidence for the Early History of the Earth-Moon System

H. Wänke and G. Dreibus[1]

1. Introduction

1.1 Primordial Abundances of Chemical Elements

The most precise information on the chemical composition of the primitive solar nebula comes from the analyses of primitive meteorites. The carbonaceous chondrites of the petrological type 1 (C1) are the best representatives of the primordial abundances of all except the most volatile elements as they represent the least fractionated group of meteorites (see Table 1).

We also have excellent agreement now between the abundances determined by analyses of C1 chondrites and the solar abundances derived from the study of spectral lines of the solar photosphere (Palme et al. 1981). Until a few years ago there was a disturbing disagreement between solar and C1 abundance value due to erroneously values of both kinds of data.

The most drastic differences excisted in the case of iron, for which the solar value was in error by as much as a factor of four until Garz and Koch (1969) redetermined the oscillator strength of iron which raised the solar iron abundance value by a large factor, so that it now matches the C1 abundance within 5%. On the other hand, the old value of fluorine in C1 chondrites exceeded the solar value by more than a factor of two. New and more reliable determinations of fluorine in C1 chondrites by Dreibus et al. (1979) reduced the fluorine abundance considerably, and today the solar and C1 abundances of these elements are within reasonable agreement. The most recent case concerns molybdenum, for which Palme and Rammensee (1981) reduced the C1 abundance value by a factor of 1.5, which now matches the solar value within 5%.

In the following chapters we shall frequently give elemental abundances h relative to C1, and Si. If there were no fractionation of chemical elements the abundance h of each element e in all objects p would be

---

1 Max-Planck-Institut für Chemie, Saarstraße 23, D-6500 Mainz, Fed. Rep. of Germany

equal to 1.00.

$$h = \frac{C\ (e,p)}{C\ (e,Cl)} \bigg/ \frac{C\ (Si,p)}{C\ (Si,Cl)} \quad C = \text{concentration} \quad .$$

Table 1. Meteorite classification

	I Undifferentiated meteorites						
Name	Carbonaceous chondrites			Ordinary chondrites			Enstatite chrondrites
Abbreviation	C			H, L and LL			E
Carbon content	up to 3%			~ 0.1%			~ 0.4%
Chemical status of:							
Iron	FeO, Fe_3O_4, FeS			FeO, FeS Fe^o(metal)			Fe^o, FeS, some Si as metal
Sulfur	S^- (FeS), SO_4^{++}			FeS			FeS but also CaS, MnS, CrS
	Elemental abundances relative to Si and Cl						
	C1	C2	C3	H	L	LL	E4
Refractories e.g., Al	1	1	1		0.7		0.5
Moderately volatiles e.g., Na	1	0.5	0.3		1		1
Volatiles e.g., In	1	0.5	0.25		0.01		1
Fe_{total}	1	1	1	1	0.8	0.7	1
Fe_{total} weight %	18	22	25	28	22	20	33
	II Differentiated meteorites (Meteorites differentiated by magmatic processes)						
a) Achondrites	Eucrites, diogenites, howardites. Aubrites. Shergottites, nakhlites, chassignites.						
b) Iron meteorites	Almost pure FeNi phase, but containing siderophile trace elements.						
c) Stony-iron meteorites	Pallasites, mesosiderites.						

1.2 Chemistry of Meteorites

As can be seen from Table 1 — which is thought to be a short introduction into meteoritics — even meteorites which were not subject to magmatic differentiations show fractionations of chemical elements. From the compositions of undifferentiated meteorites the following fractionations are evident to take place in the solar nebula prior to the formation of the parent bodies of these meteorites.

a) Fractionation of refractory elements as evident from their depletion in ordinary and enstatite chondrites.
b) Fractionation of moderately volatile, and volatile elements as evident from their depletion in C2, C3 and ordinary chondrites.
c) Metal-silicate fractionation as evident from the depletion of Fe, accompanied by a depletion of siderophile (iron-loving) elements in L (low iron) and LL (low-low iron) chondrites.

Differences in the bulk composition, i.e., fractionation of chemical elements prior to accretion are also to be expected in the case of differentiated planetary objects. For example the inner or terrestrial planets Mercury, Venus, Earth and Mars — and as a fifth object we may add the Earth's Moon — show considerable differences in their densities (Table 2). These differences are generally explained by a variation of the total Fe/Si ratio. However, if the high density of Mercury is to be explained by an extremely large FeNi core, then an increase of the solar Fe/Si ratio by a factor of five is required.

Table 2. Mean densities of terrestrial planets. (Ringwood 1979)

Planet	Mean density observed	Mean density at 10 kb	% Ni-Fe phase
Mercury	5.44	5.3	65
Venus	5.24	3.96	28.8
Earth	5.52	4.07	32.5
Moon	3.34	3.40	5
Mars	3.94	3.73	20

Reliable estimates of the bulk composition of differentiated bodies of our solar system exist only for those for which we have samples usable for laboratory analysis. Apart from Earth and Moon, it is the eucrite parent body (EPB) and the Shergotty parent body (SPB), of which we have samples in form of meteorites. Some of the iron and stony-iron meteorites are probably also related to the EPB.

Eucrites are basaltic rocks. Two other groups of achondrites — the diogenites and howardites — are closely related to them. The howardites, which represent in fact mechanical mixtures of eucrites and diogenites, were formed in the regolith of their parent asteroid (Dreibus and Wänke 1979). The reflectance spectra of the asteroid Vesta (diameter 500 km) is unique and matches exactly that of howardites (McCord et al. 1970). In fact, it might be that Vesta is indeed the parent body of the eucrite family (Hostetler and Drake 1978). With a Rb/Sr age of $4.60 \pm 0.07) \times 10^9$ years measured for the eucrite Juvinas (Allègre et al. 1975), the eucritic meteorites represent the oldest basalts of our solar system known so far.

Most differentiated meteorites have ages around 4.5 b.y. Exceptions are shergottites, nakhlites and chassignites — altogether nine individual meteorites — which are more than 2 b.y. younger. As these meteorites have in addition a number of other properties in common it became widely assumed that they were derived from a single parent body (SPB) (see Wood and Ashwal 1981). It has been suggested that planet Mars might be the parent body of these meteorites (Wasson and Wetherill 1979), and hence we might in fact already have samples from Mars in our meteorite collections.

1.3 Fractionation of Chemical Elements with Respect to Their Volatility

A number of authors (e.g., Wänke et al. 1973, Ganapathy and Anders 1974) tried to consider certain groups of elements related to each other by their condensation behaviour as building blocks of the planetary objects (Table 3). However, only for the refractory elements do we have well-founded evidence that they appear relative to each other in C1 abundance ratios in all objects of the solar system studied so far.

The condensation temperatures listed in Table 3 are only relevant for condensation in a solar nebula of primordial composition at a pressure of 10^{-3} atm, i.e., in accordance with the model of Cameron and Pine (1973). For a condensation of this type in the solar nebula, it is of course necessary that a peak temperature high enough for complete volatilization, i.e., 1800 K, was reached. Today there are severe doubts if such high temperature have ever been reached in the solar nebula, and Cameron (1978) himself has abandoned his earlier model.

Nevertheless, we have reasons to believe that large-scale fractionations governed by volatility have occurred prior to or during the accretion of planets. If not in the nebula, these fractionations may have taken place on early-formed km-sized planetesimals heated by the decay of the now extinct radioisotope ^{26}Al (half-life 700,000 years), as it was proposed by Wänke et al. (1981). In the model of

Table 3. Condensation sequence of elements. Valid for the conditions in a reducing solar nebula. Condensation as compounds or elements, e.g., Al in form of Al_2O_3; Ir as metal. (Grossman and Larimer 1974)

Temperature K		
> 1600	Refractories	Al, Ca, Ti, (some Mg and Si), Be Sc, V, Sr, Y, Zr, Nb, Mo, Ru, Rh, Ba, La, (and all REE), Hf, Ta, W, Re, Os, Ir, Pt, Th, and U
~ 1450	Metals	Fe, Ni, Co, Cu, Ge, Pd, Au, As, P
~ 1450	Mg-silicates	Mg, Fe(FeO), Si, Cr, Li, Mn
~ 1000	Moderately volatiles	F, Na, K, Rb, Cs, Zn
< 700	Volatiles	Ag, Sn, Sb, Te, S, Se, Cl, Br, I, Cd, In, Hg, Tl, Pb, Bi, C, N, H_2O

Ringwood (1970) these fractionations took place during planetary accretion. For some elements volatility is a strong function of the oxygen pressure, and we have therefore to expect changes in the condensation sequence. Heating experiments of Murchison C2 meteorite showed that under highly oxidizing conditions elements like S, As, Se, Mo, W, Re and Os can be evaporated at 1000°C, while at low oxygen fugacity In and Zn became highly volatile (Dreibus 1981, unpubl.). However, most of the refractory elements in Table 3 will almost certainly have high condensation points and elements like Na, K, Rb, etc. will have medium and Pb, In, Cd, etc. will have low condensation points. Also H_2O and H_2S will always be very volatile substances.

2. Basalts from Four Planets

As a first approach towards the understanding of the chemistry of Earth and Moon and its comparison to the chemistry of the EPB and SPB, we shall discuss the chemistry of basalts derived from these four objects. Naturally we cannot expect that a rock sample from planetary object which underwent magmatic fractionations can give us direct answer about the chemistry of its parent object, especially keeping in mind that for highly incompatible elements like La, U, Th, K, etc. fractionation considerably exceeding a factor of 100 can take place in magmatic processes. (Incompatible elements are elements which do not readily fit into major mineral phases. They have very low solid/liquid partition coefficients and hence will concentrate in the liquid phase of either partial melting or fractional crystallization processes. They are also easily extracted by other fluid phases like CO_2 or H_2O.)

Table 4. Composition of basalts from four planets

	SPB Shergotty	Earth ocean floor	Moon low-Ti	EPB Juvinas
Mg %	5.66	4.9	5.8	4.0
Al	3.87	8.2	4.9	6.92
Si	24.37	24.2	22.9	22.87
Ca	7.23	7.9	7.5	7.40
Ti	0.52	0.78	0.92	0.39
Fe	15.09	7.0	15.0	14.99
Cr ppm	1390	270	3880	2250
Mn	4030	1280	2050	4310
P	3460	520	270	400
S	1330	1100	650	1900
Na	9600	19300	2420	3140
K	1300	620	460	320
Rb	6.8	1.5	1.2	0.17
Ga	14.7	16.2	3.3	2.2
La	2.29	2.73	6.32	2.64
Yb	1.69	3.49	2.46	1.77
W	0.480	0.020	0.170	0.027
U	0.116	0.036	0.136	0.089

However, as we have selected samples with comparable concentrations of refractory incompatible elements (i.e., La), we can already draw qualitative conclusions on the chemistry of the four objects in Table 4 relative to each other. The refractory elements seem to have similar abundances on all four objects as both compatible (e.g., Ca) and incompatible refractory elements are present in all four basalts in concentrations which do not differ much from each other.

Moderately volatile elements like K are lowest in the eucrite Juvinas and highest in Shergotty; hence we can expect a similar sequence in the relevant objects. Manganese is considerably less abundant in terrestrial and lunar basalts as compared to Shergotty and Juvinas. Especially striking are the different concentrations of P and W for which elements differences about a factor of 10 are evident. We note that Shergotty is not only highest in P and W, but also in K. These three elements differ not only drastically in their geochemical character but also in their condensation behaviour.

3. Correlated Elements

Very helpful in respect to the attempt to estimate the bulk chemistry of a planetary object are the so-called correlated elements (Laul et al. 1972, Wänke et al. 1973). Correlated elements are pairs of elements which, independent of their absolute concentrations, always appear in constant abundance ratios in samples of a given planet.

One example of such an element correlation is the pair of FeO and MnO. As long as iron is in the 2+ state only and metals or sulfides are absent, theFeO/MnO remains constant because Mn^{2+} can easily replace Fe^{2+} in the two major mantle minerals, olivine and pyroxene, because of their similar ionic radii. Hence, the FeO/MnO ratio observed in rocks of a given planet should come close to the bulk ratio of these two elements in the silicate phase (mantle) of this planet.

Until recently, it was generally assumed that the different FeO/MnO ratios observed for different planets mainly reflect the depletion of Mn. This assumption was based on the behaviour of Mn in carbonaceous chondrites as a moderately volatile element. However, if was found that Mn is not depleted on the eucrite parent asteroid inspite of a high depletion of Na and K (see Sect. 4.3). As it can be seen from Table 4, eucrites have the highest Mn content of all basalts known so far. Hence, it seems more likely that the FeO/MnO ratio depends on the oxygen fugacity of the object in question, i.e., on its Fe_{metal}/FeO ratio. Under extremely reducing conditions, however, even Mn can be removed from the silicate phase in form of metal or sulfide (Keil 1968).

Other examples of correlated elements are the pairs of refractory incompatible elements Ba/La (Fig. 1), U/La (Fig. 2) and Th/U (see Sect. 2). Their appearance in C1 abundance ratios in all planetary objects strongly suggests that all refractory elements have been added to each object of our solar system in C1 abundance ratios.

Potassium is also a highly incompatible element and indeed the constancy of the K/U ratio in terrestrial rocks has long been known. Even more constant seems the K/La (Fig. 3) or K/Ba ratio. In all these pairs the moderately volatile element potassium is compared with a refractory element. As the depletion of volatile elements varies from planet to planet, the K/U or K/La ratios are also variable and in fact can be used to characterize the planets.

The W/La correlation has considerable bearing on the genesis of planetary objects (Rammensee and Wänke 1977). For these two refractory elements the correlation line for lunar samples lies a factor of 19 below the chondritic ratio. Tungsten is a refractory element like lanthanum. The reason for the depletion of W in the silicate (oxide) phase of planets is the siderophile character of this element. On

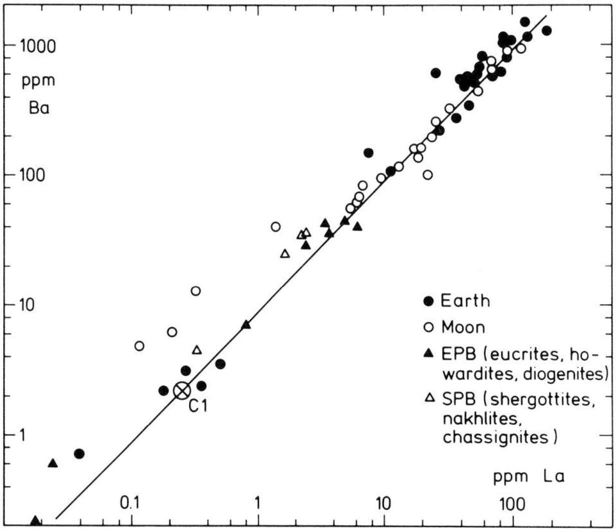

Fig. 1. Correlation of the refractory incompatible elements Ba and La in Earth, Moon, EPB (eucrite parent body) and SPB (shergotty parent body). Note the occurrence of Ba and La in strictly C1 abundance ratios on all four objects

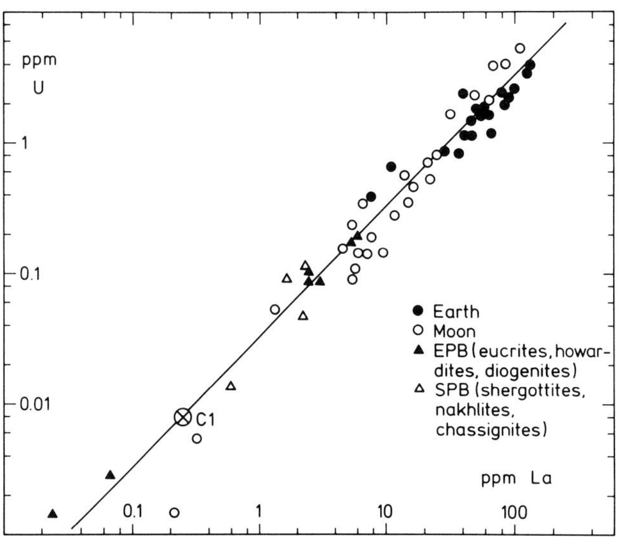

Fig. 2. Correlation of the refractory incompatible elements U and La in Earth, Moon, EPB and SPB. Note the occurrence of U and La in strictly C1 abundance ratios on all four objects

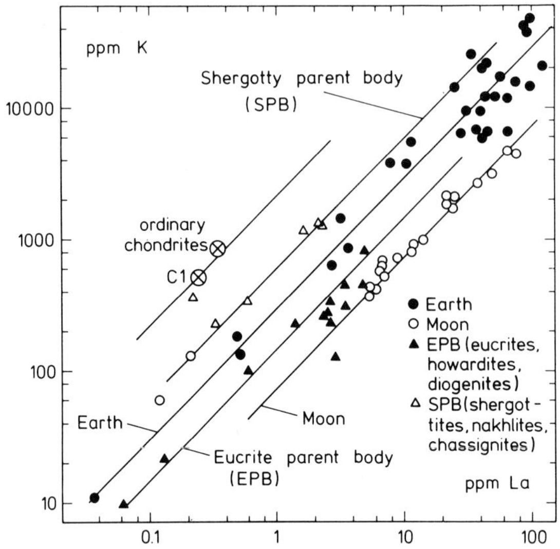

Fig. 3. Correlation of the incompatible moderately volatile element K with the incompatible refractory element La. Earth, Moon, EPB and SPB clearly have different K/La ratios all considerably below that in C1 and ordinary chondrites

the Earth and Moon about 95% W seems to have been separated together with a metal phase, only 5% went as oxide into the silicates and fractionated together with La in magmatic processes.

4. Chemistry and Formation of Earth and Moon

4.1 Chemistry of the Earth

The mass of the Earth's mantle amounts to about two-thirds of the total mass of our planet. Except for the most incompatible elements for which the contribution of the crust (0.4% of the Earth's mass) becomes important, the mantle determines the chemistry of the Earth for all lithophile elements. Furthermore, the concentrations of the siderophile elements in the mantle give us important clues as to the accretion sequence of the Earth. Today there is general agreement that the material of the crust is derived from the mantle mostly in form of partial melts.

The most common type of such partial melting products of the mantle are the basaltic magmas which frequently in violent volcanic eruptions carry with them more or less unaltered solid mantle material, the so-called ultramafic nodules. The nodules which are of most interest to us here are the garnet-lherzolites and the spinel-lherzolites. The former are characterized by the presence of garnet, which is only stable below a depth of 70 km. Hence, the garnet reflects the chemistry of this depth, while the spinel-lherzolites come from more shallow depth of 20 to 70 km. These lherzolite nodules very frequently became severely contaminated by various processes, before, during, or after their eruption. We have therefore to apply various criteria to distinguish between nodules which represent true mantle material, and those which contain alteration products. In fact, unaltered samples so far have been found only among the spinel-lherzolites.

From several hundred nodules studied Jagoutz et al. (1979) have succeeded in selecting five spinel-lherzolites which were free of contamination, except for small amounts of extremely incompatible elements. The eruption process of ultramafic nodules is somehow related to the genesis of the magmas which carry them to the surface, and therefore, it is no surprise that many of the nodules are depleted in those elements which are enriched in basalts relative to their mantle abundances.

Fortunately, we have clear cosmochemical constraints to distinguish those nodules which really represent primitive or at least choice to primitive mantle material. To distinguish primitive mantle material we only have to remember that all the refractory lithophile elements must have been present in the primitive mantle in their C1 abundance ratios (see Sect. 3).

Jagoutz et al. (1979) found five nodules that, according to these criteria, can be called primitive. Going to more and more incompatible refractory elements like Ti, Yb, Sm, Sr, or even La, Ba, U, etc., we can actually not expect to find relative C1 abundances as of the most incompatible elements the portions which reside in the crust may be in the order of 30% to 50%. From the five nodules of Jagoutz et al. (1979) only one (SC 1 from San Carlos, Arizona, USA) had almost C1 REE (Rare earth elements) abundance pattern, i.e., it was only slightly depleted in La.

There is, of course, the question whether these nodules are representative for the whole mantle. Drastic differences in the major and compatible trace element composition between, for example, the upper and lower mantle are, however, very unlikely. Any larger fractionation of these elements would lead to a considerable change in the ratios of compatible to less compatible elements or of the ratios of nonmafic to mafic elements (Ca/Sc, Al/Yb, etc.). But all these ratios in SC1 agree in fact within 10% with the C1 values.

The composition of the primitive Earth's mantle (mantle + crust) as derived from the work of Jagoutz et al. (1979) and Wänke (1982) is illustrated in Fig. 4 (see also Table 5). Normalized to Si and C1 the data show very similar enrichment factors between 1.28 and 1.53 for all refractory elements.

The depletion of Cr and Mn has been known for a long time but that of V was noticed only recently. Ringwood and Kesson (1977) favoured a depletion mechanism based on the higher volatility of Cr and Mn when compared with Si. Dreibus and Wänke (1979) presented evidence which

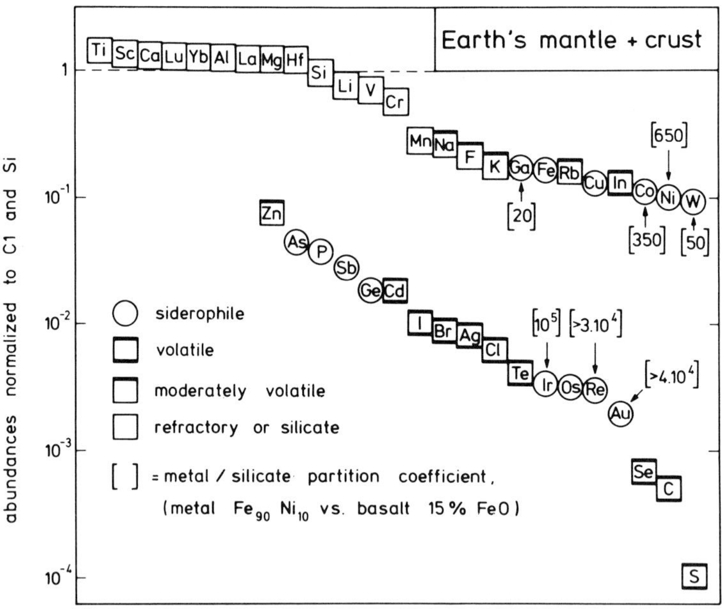

Fig. 4. Composition of the Earth's mantle (+ crust) normalized to C1 and Si derived from the analyses of primitive spinel-lherzolites. (Jagoutz et al. 1979, Wänke 1981 and unpublished data of the Mainz laboratory)

makes the depletion of Cr, Mn, and V by volatility highly improbable. They suggested that the amounts of Cr, Mn, and V, missing in the mantle went into the core in form of sulfides or oxides. Segregation into the core in the form of sulfides had already been suggested by Hutchison (1974), to explain the low concentration of Cr and Ti. However, undepleted nodules contain Ti in C1 abundance relative to the other refractory lithophile elements. We shall return to these three elements in Sect.4.4.

The similar abundances of the moderately siderophile elements Ga, Cu, W, Ni and Co relative to chondritic abundances are also very informative. The elements Ga, Cu, W, Co and Ni have highly different metal/silicate partition coefficients.

For a long time the high concentration of Ni in the Earth's mantle has been the concern of many scientists. Assuming a Ni concentration of 10% in the core with a metal/silicate partition coefficient of 650, we should expect 150 ppm Ni in the mantle under equilibrium condition. It has been argued that the partition coefficients were lowered due to high pressure or due to the presence of sulfur. Both these possibilities cannot be ruled out; however, they seem unlikely. Because

Table 5. Estimated compositions of Earth and Moon

	Earth mantle + crust		Moon metal and sulfur free			
	SCl Jagoutz et al. (1979), Wänke (1982)		Wänke et al. (1977, 1979)		This work	
	abs.	norm. to Cl + Si	abs.	norm.to Cl + Si	abs.	norm.to Cl + Si
Li ppm	2.17	0.75	-	-	-	-
F ppm	24	0.22	-	-	-	-
Na ppm	2700	0.27	450	0.045	421	0.043
Mg %	22.0	1.18	19.5	1.00	21.4	1.18
Al %	2.26	1.38	2.43	1.42	1.99	1.25
Si %	21.3	1.00	21.30	1.00	20.7	1.00
P ppm	74	0.037	-	-	-	-
S ppm	11	1.10^{-4}	-	-	-	-
K ppm	185	0.18	48	0.048	41	0.041
Ca %	2.53	1.41	2.71	1.44	2.25	1.29
Sc ppm	17.2	1.48	17.5	1.42	14.3	1.25
Ti ppm	1340	1.53	1200	1.42	1100	1.25
V ppm	78	0.71	82	0.71	74	0.69
Cr ppm	3100	0.59	2680	0.56	2564	0.50
Mn ppm	1010	0.28	1360	0.38	1240	0.35
Fe %	5.86	0.16	10.1	0.27	9.83	0.28
Co ppm	104	0.11	80	0.077	80	0.082
Ni ppm	2080	0.10	1630	0.073	1630	0.078
Cu ppm	28	0.13	-	-	-	-
Zn ppm	50	0.072	-	-	-	-
Ga ppm	3.2	0.18	-	-	-	-
Rb ppm	0.77	0.16	0.12	0.026	0.10	0.025
Sr ppm	22	1.30	26	1.42	21	1.25
In ppb	19.1	0.13	-	-	-	-
Ba ppm	5.7	1.30	7.1	1.42	5.1	1.25
La ppm	0.62	1.28	0.68	1.42	0.59	1.25
Eu ppm	0.15	1.30	0.15	1.42	0.14	1.25
Yb ppm	0.44	1.35	0.49	1.42	0.40	1.25
W ppb	16	0.090	14.5	0.078	13	0.073
Ir ppb	3.2	0.0034	-	-	-	-
Au ppb	0.54	0.0020	-	-	-	-
U ppb	21	1.30	27	1.42	20	1.25

of the very similar abundances of the elements Ga, Cu, W, and Ni we do not only require conditions to lower the partition coefficient for Ni, but conditions which make the partition coefficient of these five elements almost equal. Furthermore, a high abundance of sulfur is unlikely for the Earth which is highly depleted for a number of elements less volatile than sulfur (Ringwood 1979).

We should note also that the moderately volatile elements K, Na, Rb and F and the highly volatile In are similar in abundance to the moderately siderophile elements. The highly siderophile elements are again highly overabundant compared to the abundances expected from their partition coefficients. Although less abundant than the moderately siderophile elements by more than an order of magnitude, the highly siderophile elements Pd, Ir, Re, Os and Au appear again in almost chondritic abundances relative to each other. To summarize, we note the following trends in the elemental abundances in the mantle of the Earth (normalized to C1 and Si).

1. Refractory lithophiles (Al, Ca, Ti, Sr, Ba, REE, Th, U, etc.) *and Mg enriched*	× (1.3 ± 0.1)
2. V, Cr, Mn *depleted*	× (0.25 to 0.5)
3. Fe and moderately siderophiles (Ga, Cu, W, Co, Ni, etc.) depleted	× (0.1 to 0.15)
4. Moderately volatiles (Na, K, Rb, F, Zn) and the highly volatile element In *depleted*	× (0.1 to 0.15)
5. Highly siderophiles (Ir, Os, Re, Au, etc.) *strongly depleted*	× (0.002)
6. Highly volatiles (Cd, Ag, I, Br, Cl, Te, Se, C) *strongly depleted*	× (10^{-2} to 10^{-4})

The uniform enrichment of Mg and the refractory lithophile elements relative to Si could also be interpreted in terms of a Si deficiency (Ringwood 1958). It was suggested that the missing Si went into the core in metallic form, which would indicate gross chemical disequilibrium between mantle and core.

The mean enrichment factor of 1.3 for Mg and the refractory lithophile elements could correspond to an identical depletion factor for Si which leads to an absolute Si deficit of 6.3% in the mantle or a Si content of 12.5% in the core.

The calculated composition of the Earth based on the chemical composition of the Earth's mantle and crust and assuming strictly C1 abundances for the whole Earth is given in Table 6. The composition of the Earth's mantle obtained in this way agrees well with that derived from the spinel-lherzolite nodule SC1. The small differences especially in the MgO concentration can easily be explained by a slight difference in the MgO content between upper and lower mantle.

Table 6. Estimated composition of the Earth based on the observed chemical composition of the Earth's mantle and crust and assuming C1 abundances for the whole Earth. Moderately volatile and volatile elements are excluded. The mantle composition obtained from the spinel-lherzolite nodule SC1 (Jagoutz et al. 1979) are listed for comparison. All data in weight %

	Mantle:				Core:		
	Whole Earth	Mantle alone	Mantle from SC1		Whole Earth	Core alone	
SiO_2	29.78	44.78	46.38	Fe	27.04	80.69	Fe_{total}
MgO	26.32	39.58	37.13	Ni	1.69	5.04	= 31.0%
FeO	5.07	7.62	7.67	Co	0.09	0.27	
Al_2O_3	2.63	3.96	4.35	Si	4.19	12.50	
CaO	2.14	3.22	3.60	Mn	0.25	0.75	
Cr_2O_3	0.31	0.47	0.46	Cr	0.25	0.75	
NiO	0.16	0.24	0.28				
MnO	0.09	0.14	0.13				
	66.5	100.0	100.0		33.5	100.0	

Contrary to other models there is no need to arbitrarily raise the total Fe content beyond the C1 abundance. We find a core of 33.5 wt.%, containing 12.5% metallic Si, which will reduce the density of the core by about 10% as required according to current theories. The presence of Si in the core might result in the precipitation of Ni_2Si to form the solid inner core as suggested by Herndon (1979).

4.2 Accretion Sequence of the Earth and the Other Inner Planets

To explain the observed elemental abundance pattern of the Earth's mantle the following inhomogenous accretion model has been set up (Wänke 1981).

Accretion Sequence of the Earth. Accretion started with *highly reduced* material free of moderately volatile and volatile elements but containing all other elements in C1 abundance ratios (Fe and all siderophile elements as metals (W!), Si partly as metal, Cr, Mn and V as metals or sulfides). *Component A.*

After accretion of about 2/3 of the Earth and after core formation more and more *oxidized material* (Fe, Co, Ni, etc., Mn, Cr, V and Si, W! as oxides) was added containing all elements including moderately volatile and at least some volatile elements (In, etc.) in C1 abundances. *Component B.*

A model to build the inner planets from one reduced volatile-free and an oxidized volatiles containing component, but with different ratios of these two components, was previously proposed by Ringwood (1979). However, he assumed homogenous accretion in which both components are added to the growing planet simultaneously and equilibrated with each other. As outlined above, the observed chemical composition of the Earth's mantle — though not really in contradiction with homogenous accretion — seems to be explained in a more straightforward way by an accretion sequence in which the oxidized component is added only after the accretion of about 2/3 of the Earth's mass (from reduced material) and after segregation of the metal phase (i.e., after core formation).

The required change in oxygen fugacity of the accreting material fits into several possible models for the solar nebula. It is easy to visualize a scenario in which loss of H_2 from certain regions of the solar nebula is much more rapid than that of H_2O which could be held back in the form of ice grains.

A more severe constraint is the high depletion of all moderately volatile elements such as Na, K, etc. on all the inner planets as well as on the eucrite parent body (EPB) and, in fact, on all differentiated objects of our solar system. To fractionate these elements from the more refractory ones requires temperatures of the order of 1000 K. They will not be reached in most parts of the solar nebula, if its temperature indeed remains as low as claimed now, even by earlier proponents of a hot solar nebula (Cameron 1978). This is one of the reasons why, at present, there is no generally accepted theory of the formation of planets. While small objects could have lost moderately volatile elements during or after accretion, this seems to be impossible for large objects like the Earth.

As opposed to a hot solar nebula one could visualize a solar nebula, in which the only fractionation of importance is that of water, which would only be added in appreciable quantities to accreting planetesimals in regions where water occurred in the form of ice.

Fractionation of moderately volatile elements requires temperatures in the order of 1000 K. However, instead of having occurred in the solar nebula the high temperature stage could have occurred in primary objects of km-size, heated by the decay energy of ^{26}Al and/or chemical energy (Urey 1959, Clayton 1980). Extinct ^{26}Al was identified in certain meteorites in amounts sufficient to melt even objects in the size range of kilometers (Lee et al. 1976).

In interstellar matter a large fraction of carbon seems to be present in form of graphite (Mathis et al. 1977). Hence, in planetesimals which accreted from interstellar material unaltered by a hot solar nebula, the abundance of carbon my exceed that of silicon. Depending on the accretion temperature, i.e., depending on the location in the solar nebula, or depending on time of the cooling nebula, planetesimals will or will not contain H_2O (ice).

In the absence of water (inner part of the solar nebula) primary objects heated by the decay energy of ^{26}Al and by chemical energy (Clayton 1980) become strongly reduced by carbon. Hence, not only FeNi etc. but also large amounts of Si, Cr, Mn, V, etc. will be converted into metals or sulfides. Parallel to reduction, together with large quantities of CO and CO_2 volatile and moderately volatile elements will be lost from the primary objects. Hence, reduction and loss of volatiles are parts of one process. Component A could be generated in this way.

Planetesimals containing large amounts of H_2O will only reach considerably lower temperatures as the heat sources will mainly be used up for the evaporation of H_2O and the reaction $3C + 2\ H_2O \rightarrow 2\ CO + CH_4$. Highly oxidized planetesimals which retained most of their volatiles and being compositionally similar to Cl chondrites will result. Thus we might have a scenario which fulfils the conditions of the accretion history of the Earth and which will result in an increase in the bulk oxygen content from planet Mercury to Mars.

4.3 Chemistry of the Moon

Estimates of the mean chemical composition of the Moon have been published by various authors (see Table 7). As it can be seen, there is very good agreement for SiO_2, MgO and FeO. It is also generally taken as a fact that the volatile elements and also the moderately volatile elements like Na and K have low abundances on the Moon. Some controversy exists on the abundance of refractory elements like Al and Ca. For these elements the difference between the extrems amounts to a factor of 2.

The data from Wänke et al. (1977), which are used for comparison of the chemistry of Moon, Earth and differentiated meteorites were derived in the following way. The rocks returned from the lunar highlands — the oldest parts of the Moon — show a compositional trend which can be interpreted by the presence of a Mg-rich component with primitive characteristics. Wänke et al. (1974) called it "primary component", as it was thought to represent material from the last accretion stage of the Moon, which was mechanically mixed into the already mostly solidified surface rocks of feldspatic (anorthositic) composition. In the Mg-rich component a number of pairs of refractory elements like Al/Sc or Al/Ca were found to be present in their exact Cl ratios, but the Mg/Si ratio lay below the respective Cl value. The only mineral phase which leaves all refractory elements untouched, but strongly influences the Mg/Si ratio is olivine $(MgO, FeO)_2\ SiO_4$. Hence, it was concluded that part of the olivine originally present in the primary material was lost by segregation. Ringwood (1979)

Table 7. Estimates on the composition of the Moon (metal and sulfur free)

	Morgan et al. (1978)	Ringwood (1977a)	Wänke et al. (1977)	This work
SiO_2 [%]	43.3	44.6	45.6	44.2
TiO_2	0.4	0.3	0.2	0.18
Al_2O_3	7.6	3.7	4.6	3.76
Cr_2O_3	0.3	0.4	0.4	0.37
MgO	29.1	33.4	32.4	35.5
FeO	13.0	13.9	13.0	12.7
MnO	0.15	-	0.18	0.16
CaO	6.1	3.4	3.8	3.15
Na_2O	0.15	0.05	0.06	0.057
K [ppm]	78	100	60	41
U [ppb]	41	28	27	20

pointed out that during crystallization of an originally totally molten lunar magma ocean, after separation of olivine and anorthite the remaining liquid would for many elements be compositionally similar to the proposed primary component. As in their model calculations Wänke et al. (1977) treated anorthite as second independent component material from such residual liquids which can be described as primary matter minus olivine. From the observed Mg/Si ratio of 0.72 and a C1 value of 0.91 the amount of olivine lost was calculated to be 31%. According to this model the bulk composition of the Moon can be obtained by mixing 69% of the Mg-rich component and 31% olivine. Using for olivine a composition of a lunar dunite (sample 72417), the datas of the bulk composition of the Moon listed in Table 7 were obtained. Normalized to Si and C1 abundances we find the Moon to be somewhat enriched in refractory elements. The enrichment of refractory elements even exceeds their enrichment in the Earth's mantle slightly.

In the following section we will present strong evidence for the proposition that the Moon was formed from material of the Earth's mantle after core formation. As in most of the recent estimates on the composition of the Earth's mantle the Mg/Si ratio exceeds the C1 value considerably, one should assume Moon and Earth's mantle to have an identical Mg/Si ratio. The assumption of a Mg/Si ratio of the Earth's mantle as observed in the most primitive lherzolite nodule SC1 of Jagoutz et al. (1979), i.e., a Mg/Si ratio of 1.04, leads to a larger amount of olivine (48% instead of 31%), which is to be added to the Mg-rich component. This in turn leads to small changes in the data of the bulk chemistry of the Moon (Table 7). However, the abundances of the refractory elements in the Moon and the Earth's mantle now became equal within the uncertainty of these estimates. The well-known differences in the contents of volatile and siderophile elements of Earth and Moon can be explained straightforwardly. However, the difference in the FeO content (Earth's mantle 7.5% FeO, Moon 13% FeO) has remained mysterious. One possibility would be that the lower mantle contains considerably more FeO than the upper mantle. Another explanation for the difference in the FeO of Earth and Moon would be that the Earth's mantle had a FeO content equal to that found for the Moon (i.e., 13%) at the time when the material from which the Moon formed was separated. During geologic times the FeO concentration in the Earth decreased due to segregation of FeO from the mantle into the core which would lead to transfer of Si in form of SiO_2 from the core into the mantle according to the reaction $2FeO + Si \rightarrow 2Fe + SiO_2$. For details and consequences of such a process see Jagoutz and Wänke (1982).

4.4 Mn, Cr and V in the Earth's Mantle and the Moon

As pointed out by Dreibus and Wänke (1979), the large depletion of Mn, but also the smaller depletion of Cr and V on the moon are especially striking; even more because these three elements show similar depletions in the Earth's mantle. Previously it was generally assumed that the depletion of Mn is due to volatility. However, as shown by Dreibus and Wänke (1979), this explanation is completely inadequate, as the (EPB), which is similarly depleted in elements of medium volatility like Na and K, shows nearly no depletion of Mn. As lunar rocks and eucrites are also similar in respect to oxygen fugacity, volatility cannot be the reason for the depletion of Mn on the Moon and the Earth's mantle. In the case of Cr the volatility argument was never very convincing and fails completely in the case of V, which is in fact a refractory element and its abundance must be compared to Al or Ca.

The only explanation for the depletion of Mn, Cr and V one can think of is the one given by Dreibus and Wänke (1979), according to which large portions of these three elements were removed by the metal phase either in oxide form or as metals or sulfides. In the case of the Earth removal as oxides could be expected if the Earth's core contains a large amount of dissolved FeO as proposed by Ringwood (1977b), and if the Earth accreted homogeneously, i.e., metallic and oxidized iron were added simultaneously. In the inhomogeneous accretion model of Jagoutz et al. (1979) and Wänke (1981) outlined in Sect. 4.3 the accreting material was first highly reduced and volatile-free. This component A is thought to contain Mn, Cr and V mainly in reduced form. After the Earth has reached about 2/3 of its present mass and after core formation, more and more oxidized material was added (component B) containing Mn, Cr and V as oxides. In this case a large fraction of Mn, Cr and V of component A will enter the core in reduced form. Whichever of the two mechanisms was responsible for the depletion of Mn, Cr and V, both would only work in the case of the Earth, but not on the Moon.

We believe that the similar depletion of Mn, Cr and V on Earth and Moon to be the strongest argument for a close genetic relationship of Earth and Moon. The Moon must have formed somehow from the material of the Earth's mantle after core formation.

5. Origin of the Moon

Two mechanisms have been proposed for the removal of the protolunar material from the Earth's mantle.

a) Rotational instability.
b) Collision of the proto-Earth with another planetary sized body.

The direct fission model goes back to Darwin (1880), who developed a theory according to which Moon and Earth were originally combined in a single body which rotated with a period of about 4 h. Resonance between the free period of the proto-Earth and the solar tides leads to the development of a large tidal bulge, and finally the tidal bulge was ejected to form the Moon.

With variations the fission hypothesis by rational instability was brought up again by many authors since Darwin (Ringwood 1960, Wise 1963, Cameron 1963, Binder 1974). It was most thoroughly investigated by O'Keefe (1969, 1970, 1972) and O'Keefe and Sullivan (1978).

In most fission models involving rotational instability, rapid segregation of metallic FeNi to form the core leading to decrease of the moment of inertia is an essential step. However, the most severe objection to those models sets in at exactly this point, as core formation would have to be delayed until almost complete accretion of the proto-Earth. Such a delay seems very unlikely as it would require to keep the proto-Earth sufficiently cool until this moment.

A fission origin of the Moon triggered by collision of the proto-Earth by another planetary-sized object avoids most problems in the direct fission model and relies on processes which must inevitably have occurred during accretion of planets. According to Wetherill (1976) the mass of the second largest of the planetesimals from which the Earth accreted was in the mass range of 0.03 to 0.1 of the Earth's mass today. Objects of this size encountering the proto-Earth within a reasonable velocity range are also capable of ejecting material from the surface of the proto-Earth and placing it into orbit. A considerable portion of the mass leaving the Earth will be in volatilized state. Models for a lunar origin based on collision triggered fission were published by Hartmann and Davis (1975) and Cameron and Ward (1976). They can in fact be regarded as special but more plausible cases of Ringwood's model (1966, 1970), which invokes evaporation of the outer mantle by rapid accretion of the Earth to provide the parental circumterrestrial planetesimal ring (Ringwood 1979).

In this connection the almost identical age values of Earth and Moon are important. Based on single-stage lead isotope growth curves, Tera and Wasserburg (1974) obtained a model age of 4.42 b.y. for the Moon and Gancarz and Wasserburg (1977) found for the Earth a model age of 4.47 ± 0.05 b.y. This may be significantly less than the 4.56 b.y. found for the oldest meteorites (Tatsumoto et al. 1973). However, it is not yet understood if the Pb age of the Earth refers to core formation or to accretion. Nevertheless, a time difference in the order of 10^8 years would be consistent with an acceptable time scale for accretion of the Earth and formation of the Earth's core.

The observation that samples from Moon and Earth plot along the same oxygen isotope fractionation line was taken as strong indication of the origin of the Moon from the Earth's mantle (Clayton and Mayeda 1975, Clayton et al. 1976). This argument became somewhat weakened when subsequent investigations showed that aside terrestrial and lunar rocks, all samples from magmatically differentiated objects, plot very close to the same oxygen isotope fractionation line.

Acknowledgement. This work was supported by the Deutsche Forschungsgemeinschaft.

References

Allègre CJ, Birck JL, Fourcade S, Semet M (1975) Rubidium-87/Strontium-87 Age of Juvinas basaltic achondrite and early igneous activity in the solar system. Science 187:436

Binder AB (1974) On the origin of the Moon by rotational fission. Moon 11:53

Cameron AGW (1963) The origin of the atmospheres of Venus and Earth. Icarus 2:249

Cameron AGW (1978) Physics of the primitive solar accretion disk. Moon Planets 18:5

Cameron AGW, Pine MR (1973) Numerical models of the primitive solar nebula. Icarus 18:377

Cameron AGW, Ward WR (1976) Abstract. The origin of the moon. Lunar Sci 7:120

Clayton DD (1980) Chemical energy of cold-cloud aggregates: The origin of meteoritic chondrules. Astrophys J 239:L37

Clayton RN, Mayeda TK (1975) Genetic relations between the moon and meteorites. Proc 6th Lunar Sci Conf Geochim Cosmochim Acta Suppl 6:1761

Clayton RN, Onuma N, Mayeda TK (1976) A classification of meteorites based on oxygen isotopes. Earth Planet Sci Lett 30:10

Darwin GH (1880) On the secular changes in the orbit of a satellite revolving around a tidally disturbed planet. Philos Trans R Soc London 171:713

Dreibus G, Spettel B, Wänke H (1979) Halogenes in meteorites and their primordial abundances. In Ahrens LH (ed) Origin and distribution of the elements. Pergamon Press, Oxford, p 33

Dreibus G, Wänke H (1979) Abstract. On the chemical composition of the Moon and the eucrite parent body and a comparison with the composition of the Earth; the case of Mn, Cr, and V. Lunar Planet Sci X:315

Ganapathy R, Anders E (1974) Bulk compositions of the Moon and Earth, estimated from meteorites. Proc 5th Lunar Sci Conf Geochim Cosmochim Acta Suppl 5:1181

Gancarz AJ, Wasserburg GJ (1977) Initial Pb of the Amitsoq gneiss, West Greenland and implications for the age of the Earth. Geochim Cosmochim Acta 41:1283

Garz T, Koch M (1969) Astron Astrophys 2:274

Grossman L, Larimer JW (1974) Early chemical history of the solar system. Rev Geophys Space Phys 12:71

Hartmann WK, Davis D (1975) Satellite-sized planetesimals and lunar origin. Icarus 24:504

Herndon JM (1979) The nickel silicide inner core of the Earth. Proc R Soc London Ser A 368:495

Hostetler CJ, Drake MJ (1978) Quench temperatures of Moore County and other eucrites: residence time on eucrite parent body. Geochim Cosmochim Acta 42:517

Jagoutz E, Wänke H (1982) Has the Earth's core grown over geologic times? Abstr Lunar Planet Sci XIII:358

Jagoutz E, Palme H, Baddenhausen H, Blum K, Cendales M, Dreibus G, Spettel B, Lorentz V, Wänke H (1979) The abundances of major, minor and trace elements in the Earth's mantle as derived from primitive ultramafic nodules. Proc 10th Lunar Sci Conf Geochim Cosmochim Acta Suppl 11:2031

Keil K (1968) Mineralogical and chemical relationships among enstatite chondrites. J Geophys Res 73:6945

Laul J, Wakita H, Showalter DL, Boynton WV, Schmitt RA (1972) Bulk, rare earth, and other trace elements in Apollo 14 and 15 and Luna 16 samples. Proc 3rd Lunar Sci Conf Geochim Cosmochim Acta Suppl 3:1181

Lee T, Papanastassiou DA, Wasserburg GJ (1976) Demonstration of ^{26}Mg excess in Allende and evidence for ^{26}Al. Geophys Res Lett 3:41

McCord TB, Adams JB, Johnson TV (1970) Asteroid Vesta: Spectral reflectivity and compositional implications. Science 168:1445

Mathis JS, Rumpl W, Nordsieck KN (1977) The size distribution of interstellar grains. Astrophys J 217:425

Morgan JW, Hertogen J, Anders E (1978) The Moon: Composition determined by nebular processes. Moon Planets 18:465
O'Keefe JA (1969) Origin of the Moon. J Geophys Res 74:2758
O'Keefe JA (1970) The origin of the Moon. J Geophys Res 75:6565
O'Keefe JA (1972) The origin of the Moon: Theories involving joint formation with the Earth. Astrophys Space Sci 16:201
O'Keefe JA, Sullivan E (1978) Fission origin of the moon: Cause and timing. Icarus 35:272
Palme H, Rammensee W (1981) The cosmic abundance of molybdenum. Earth Planet Sci Lett 55:356
Palme H, Suess HE, Zeh D (1981) Abundances of the elements in the solar system, Section 3.4. Landoldt-Börnstein New Ser VI/2a. Springer, Berlin Heidelberg New York, p 257
Rammensee W, Wänke H (1977) On the partition coefficient of tungsten between metal and silicate and its bearing on the origin of the Moon. Proc 8th Lunar Sci Conf. Geochim Cosmochim Acta Suppl 8:399
Ringwood AE (1958) The constitution of the mantle - III; Consequences of the olivine-spinel transition. Geochim Cosmochim Acta 15:195
Ringwood AE (1960) Some aspects of the thermal evolution of the Earth. Geochim Cosmochim Acta 20:241
Ringwood AE (1966) Chemical evolution of the terrestrial planets. Geochim Cosmochim Acta 30:41
Ringwood AE (1970) Origin of the Moon: The precipitation hypothesis. Earth Planet Sci Lett 8:131
Ringwood AE (1977a) Basaltic magmatism and the composition of the Moon I. Major and heat-producing elements. Moon 16:389
Ringwood AE (1977b) Composition of the core and implications for origin of the Earth. Geochem J 11:111
Ringwood AE (1979) Origin of the Earth and Moon. Springer, Berlin Heidelberg New York
Ringwood AE, Kesson SE (1977) Basaltic magmatism and the bulk composition of the Moon II. Siderophile and volatile elements in Moon, Earth and chondrites. Implication for lunar origin. Moon 16:425
Tatsumoto M, Knight RJ, Allègre CJ (1973) Time differences in the formation of meteorites as determined from the ratio of lead-207 to lead-206. Science 180:1279
Tera F, Wasserburg GJ (1974) U-Th-Pb systematics on lunar rocks and inferences about lunar evolution and the age of the moon. Proc 5th Lunar Sci Conf. Geochim Cosmochim Acta Suppl 5:1571
Urey HC (1959) Primary and secondary objects. J Geophys Res 64:1721
Wänke H (1981) Constitution of terrestrial planets. Philos Trans R Soc London A303:287
Wänke H, Baddenhausen H, Dreibus G, Jagoutz E, Kruse H, Palme H, Spettel B, Teschke F (1973) Multielement analyses of Apollo 15, 16, and 17 samples and the bulk composition of the Moon. Proc 4th Lunar Sci Conf. Geochim Cosmochim Acta Suppl 4:1461
Wänke H, Palme H, Baddenhausen H, Dreibus G, Jagoutz E, Kruse H, Spettel B, Teschke F, Thacker R (1974) Chemistry of Apollo 16 and 17 samples: Bulk composition, late stage accumulation and early differentiation of the moon. Proc 5th Lunar Sci Conf. Geochim Cosmochim Acta Suppl 5:1307

Wänke H, Baddenhausen H, Blum K, Cendales M, Dreibus G, Hofmeister H, Kruse H, Jagoutz E, Palme C, Spettel B, Thacker R, Vilcsek E (1977) On the chemistry of lunar samples and achondrites. Primary matter in the lunar highlands: A re-evaluation. Proc 8th Lunar Sci Conf. Geochim Cosmochim Acta Suppl 8:2191

Wänke H, Dreibus G, Palme H (1979) Non meteroritic siderophile elements in lunar highland rocks: Evidence from pristine rocks. Proc 10th Lunar Planet Sci Conf. Geochim Cosmochim Acta Suppl 11:611

Wänke H, Dreibus G, Jagoutz E, Palme H, Rammensee W (1981) Chemistry of the Earth and the significance of planets and meteorite parent bodies. Abstr 12th Lunar Planet Sci Conf Lunar Planetary Inst, Houston, p 1139

Wasson JT, Wetherill GW (1979) Dynamical, chemical and isotopic evidence: regarding the formation locations of asteroids and meteorites. In: Gehrebs T (ed) Asteroids. Univ Arizona, Tucson, p. 926

Wetherill GW (1976) The role of large bodies in the formation of the Earth and Moon. Proc 7th Lunar Sci Conf. Geochim Cosmochim Acta Suppl 7:3245

Wise D (1963) An origin of the moon from the earth: Some new mechanisms and comparisons. J Geophys Res 74:6034

Wood AC, Ashwal LD (1981) SNC Meteorites: Igneous rocks from Mars? Proc 12th Lunar Planet Sci Conf Geochim Cosmochim Acta Suppl 16:1359

Concluding Remarks

Our meeting revealed considerable quantitative improvements in the classical and non-classical methods of measuring the angular-momentum exchange within the earth-moon-system. The value of these measurements is enhanced by progress in the determination of meteorological influences on the earth's rotation and of concomitant effects from celestial mechanics. The general improvement does not apply to the most desirable paleontological information, but in this field the performance of a variety of studies seems feasable, which - as was demonstrated - should give more insight into the growth mechanism. The theoretical interpretation of the tidal friction process has again confirmed the dominating role of the oceans in comparison with the solid earth. Computer models of the oceanic tides are able to give realistic estimates of the torques between the moon and the oceans whereas the treatment of the interaction between the oceans and the solid earth is problematic due to the implied small scales. The necessary maps of the old oceans can now be provided for a much longer time in the past than was thought before and with increasing accuracy in the more recent geological epochs. On the other hand, the importance of the resonance frequencies of the oceans for our topic has become clearer, and as a consequence, the models for old oceans must be computed for a large number of consecutive geological epochs in narrow time steps. Fed with such an input, the long time integration of the lunar orbit could perhaps bring forth more stronger restrictions for the early state of the earth-moon-system. The remarkable (but not complete) chemical similarity of the earth's mantle and the moon and the dynamical information are together the main boundary conditions for all theories of the origin. It should be stressed, however, that our topic is not identical with this 'final' question and that our contributions should not only be evaluated with regard to that question. The connections between the earths magnetic field and rotation and the paleotides themselves are fascinating subjects with great implications for the understanding of our present earth.

P. Brosche

Intense Atmospheric Vortices

Proceedings of the Joint Symposium (IUTAM/IUGG) held at Reading (United Kingdom) July 14–17, 1981

Editors: **L. Bengtsson, Sir J. Lighthill**

1982. 195 figures. XII, 326 pages
(Topics in Atmospheric and Oceanographic Sciences)
ISBN 3-540-11657-5

With contributions by F. K. Browand, M. Challa, D. R. Davies, R. P. Davies-Jones, Du Xing-yuan, M. P. Escudier, H. P. Evans, A. E. Gill, W. M. Gray, R. Hide, G. J. Holland, E. J. Hopfinger, J. B. Klemp, Y. Kurihara, L. M. Leslie, D. K. Lilly, B. A. Lugovtsov, T. Maxworthy, A. D. McEwan, K. V. Ooyama, R. P. Pearce, R. L. Pfeffer, R. Rotunno, R. C. Sheets, J. Simpson, R. K. Smith, J. T. Snow, R. E. Tuleya, Xiao Wen-jun, Xie An, Zhou Zi-Dong

The concept of vorticity is of central importance in fluid mechanics, and the change and variability of atmospheric flow is dominated by transient vortices of different time and space scales. Of particular importance are the most intense vortices, such as hurricanes, typhoons and tornadoes, which are associated with extreme and hazardous weather events of great concern to society.

This book examines the different mechanisms for vorticity intensification that operate in two different kinds of meteorological phenomena of great importance, namely the tropical cyclone and the tornado. The understanding of these phenomena has grown in recent years due to increased and improved surveillance by satellites and aircraft, as well as by numerical modelling and simulation, theoretical studies and laboratory experiments. The book summarizes these recent works with contributions from observation studies (from radio sonde data, aircraft, satellites and radar) and from stiedies concerning the physical mechanism of these vortices by means of theoretical, numerical or laboratory models. The book contains articles by the leading world experts on the meteorological processes and on the fundamental fluid dynamics mechanism for vorticity intensification.